Mining North America

Mining North America

An Environmental History since 1522

EDITED BY

J. R. McNeill
and George Vrtis

UNIVERSITY OF CALIFORNIA PRESS

University of California Press, one of the most
distinguished university presses in the United States,
enriches lives around the world by advancing scholarship
in the humanities, social sciences, and natural sciences. Its
activities are supported by the UC Press Foundation and
by philanthropic contributions from individuals and
institutions. For more information, visit www.ucpress.edu.

University of California Press
Oakland, California

Library of Congress Cataloging-in-Publication Data

Names: McNeill, John Robert, editor. | Vrtis, George,
    editor.
Title: Mining North America : an environmental history
    since 1522 / edited by J. R. McNeill and George Vrtis.
Description: Oakland, California : University of
    California Press, [2017] | Includes bibliographical
    references and index.
Identifiers: LCCN 2016055852 (print) | LCCN 2016057573
    (ebook) | ISBN 9780520279162 (cloth : alk. paper) |
    ISBN 9780520279179 (pbk : alk. paper) | ISBN
    9780520966536 (ebook)
Subjects: LCSH: Mineral industries—Environmental
    aspects—North America—History.
Classification: LCC HD9506.A2 M5453 2017 (print) |
    LCC HD9506.A2 (ebook) | DDC 338.2097—dc23
LC record available at https://lccn.loc.gov/2016055852

Manufactured in the United States of America

26 25 24 23 22 21 20 19 18 17
10 9 8 7 6 5 4 3 2 1

Contents

Illustrations

TABLE

Acknowledgments

The idea for this book emerged during a workshop titled "History Underground: Environmental Perspectives on Mining," held at the Rachel Carson Center for Environment and Society in Munich, Germany, during the summer of 2011. The goal of that workshop was to take stock of the global environmental history of mining, and so for three days participants presented papers that took us to four continents and deep into the past. Five of the essays in this volume were first presented at that workshop, and so we wish to thank the Rachel Carson Center and its directors, Christof Mauch and Helmuth Tricschler, for their generous support of the workshop, as well as Frank Uekötter, Donald Worster, and Bernd Grewe, all of whom served as session chairs and contributed valuable insights throughout the workshop.

We are also grateful to several other organizations for supporting this book. Our home institutions—Georgetown University and Carleton College—contributed funds for maps and the index and provided congenial homes in which to work. The University of California Press offered enthusiasm for the book that never wavered, despite its gestation taking longer than any of us wished. We are especially grateful to Niels Hooper, our editor at the press, for his encouragement and patience. Thanks also to Bradley Depew, who guided the book through all phases of assembling the manuscript, reviewing the artwork, designing the cover, and moving the book into production. In addition, we wish to express our gratitude to Steven Baker, who did a superb job copyediting the book and enhancing the clarity of our prose.

We also wish to thank our authors for joining this effort to unearth the environmental history of mining in North America. Each has made a unique contribution to this book, and the volume would be the weaker without any one of them. We particularly wish to single out Steven Hoffman, our wonderful and perceptive colleague from the University of St. Thomas in St. Paul, Minnesota, who contributed the essay on the mining of Alberta's tar sands. Sadly, Steven passed away in 2015 before the book was completed. We are grateful to his sister Mary Lammert for giving us permission to publish his essay.

Lastly, to our families—Julie, Will, Patrick, Jesse, and Kat; and Anne, Meadow, and Henry—our heartfelt thanks for supporting us during the long journey this book required.

Introduction

Of Mines, Minerals, and North American
Environmental History

GEORGE VRTIS AND J. R. MCNEILL

Modern North America is a world built of minerals. Buildings, transportation networks, communication systems, waterworks, machinery, and appliances of every kind—all of these depend on minerals clawed from beneath the earth's surface. A modern American automobile now contains at least thirty-nine different minerals, everything from aluminum and bauxite to tungsten and zinc. Computers and smartphones can contain even more, including precious metals such as gold, silver, and platinum, as well as rare earth elements such as europium and yttrium that provide color for their liquid-crystal displays.[1] Electrical lines depend on copper or aluminum; pipes are formed from several different metals or plastic compounds; and anything made of concrete relies on gravel, sand, and varied mixtures of calcium, silicon, and the sulfate mineral gypsum. Even North America's entire industrial food system leans heavily on minerals. The macronutrient fertilizers nitrogen, phosphorous, and potassium are key elements supporting plant growth and yield. Without them, conservative estimates suggest, crop yields would plummet by as much as half.[2]

Modern North America also depends on its vast array of minerals in staggering quantities. Today, a common American automobile contains more than a ton of iron ore hardened into steel, 240 pounds of aluminum, 42 pounds of copper, and 41 pounds of silicon.[3] In 2013, the United States, Mexico, and Canada together produced more than 7.1 million automobiles and 9.3 million commercial vehicles. In terms of copper alone, these production figures translate into something on the order of 344,000 tons of copper. Computers, smartphones, and many other everyday consumer electronic goods contain far less material per unit, but they are now produced and consumed in the hundreds of millions.[4] In the United States, the number of electronic devices sold annually doubled between 1997 and 2009, when it reached 438 million units.[5] Although the figures are difficult to compare due to reporting variances, the overall domestic production and raw material processing of nonpetroleum minerals (with the exception of tar sands) in the United States, Canada, and Mexico in 2013 is estimated to have contributed 4 percent, 3.4 percent, and 4.9 percent of total GDP, respectively. If the value added by industrial processing were included in this calculus, the total contribution of minerals to the U.S. economy in 2013 would leap to 14.5 percent of GDP.[6] Comparable figures are likely for the Canadian and Mexican experiences.

These are dizzying figures, but together they begin to illustrate the importance of minerals—and thus *mining*—in forging modern life in North America. Although indigenous peoples have dug into the earth for useful and treasured materials across the continent for thousands of years, the advent of the industrial age marked a new relationship among North American societies, minerals, and mines that eventually remade those societies and the natural world in fundamental ways. As minerals moved from the periphery to the core of economic production, they reordered whole economies, reshaped deep-seated cultural and social patterns, redirected scientific and technological initiatives, redistributed military and economic might, refocused political power and international relations, and recast the health and fortunes of local communities and entire regions. These same dynamics also reverberated through the environment, influencing North American ecological communities and physical processes on a scale that rivaled anything in the continent's thirteen or fourteen millennia of human experience. From Canada to Mexico, mining has honeycombed vast portions of the continent, leveled old mountains and created new ones, fashioned massive open-pit wastelands and buried entire valleys in waste rock. It has also denuded forests, accelerated soil erosion, imperiled wildlife populations, polluted water and air with all sorts of chemical wastes, and created some of North America's most troubling and enduring environmental problems. Indeed, as even a cursory examination of the U.S. Environmental Protection Agency's Superfund National Priorities List reveals, mining sites litter the listing, and the two sites with the highest hazard ranking scores are both tied to mining.[7]

And yet, despite the central importance of mining in shaping modern North American societies and environments, it has received relatively little attention from environmental historians when compared to other major pillars of the North American economy, such as agriculture or even forestry. Considered as a whole, two major trends are evident in our historiography thus far. First, the environmental history of mining in North America has unfolded around three main groups of minerals— precious metals such as gold and silver, industrial metals such as copper and iron ore, and solid-fuel minerals such as coal and uranium—and most of the scholarship has focused on just a single mineral within each of these categories. Notable studies such as Andrew Isenberg's *Mining California,* Thomas Andrews's *Killing for Coal,* and Timothy LeCain's *Mass Destruction* have all followed this pattern.[8] They have also, as Paul Sutter recently observed, "breathed new life into these seemingly dead spaces."[9] These studies and others have begun to help historians imagine and rethink how mining, and the far-reaching environmental changes it precipitates, is woven into our understanding of urban environments, socioeconomic conditions, labor relations, science and technology, industrial disease, consumerism, and environmental politics, among other important areas of inquiry.

The second major trend evident in the historiography is its focus on national contexts, with scarce attention given to larger continental, comparative, or global perspectives. Though not surprising given the nature and organization of the historical profession, most of the work published thus far has unfolded within the confines of a single nation-state. In terms of North America, innovative exceptions include John Wirth's *Smelter Smoke in North America,* Kathryn Morse's *The Nature of Gold,* and Samuel Truett's *Fugitive Landscapes.* Each of these studies has skillfully navigated North American borderlands to reveal aspects of the international character of mining, mineral processing, and environmental change, and the way such factors as pollution, labor migration, capital flows, and varied cultural, social, and legal traditions have all contributed to the strands of change.[10]

At still wider scales that reach beyond North America in comparative, continental, or global frameworks, the environmental history of mining has made a fruitful, if limited, start. It is treated rather briefly in J. R. McNeill's *Something New Under the Sun* and John Richards's *The Unending Frontier,* while being more richly developed in Anthony Penna's *The Human Footprint* and Kate Brown's *Plutopia.* At the same time, mining did not garner a "turning point" in Frank Uekoetter's *The*

Turning Points of Environmental History, nor did it receive its own chapter in J.R. McNeill and Erin Stewart Mauldin's *Companion to Global Environmental History* or Andrew Isenberg's *Oxford Handbook of Environmental History.* Taken all together, efforts to cross North American and other geopolitical borders to develop broader, transnational perspectives on the environmental history of mining have only just begun to illuminate the larger patterns, forces, and issues at work in shaping our national, continental, and global histories.[11]

. . .

This book, then, aims to contribute to the emerging field of mining environmental history by focusing on North America and exploring relationships among North American societies, mining, and environmental change since the Spanish first began developing mines in the region north of Mexico City in 1522. In the fourteen original essays that follow, scholars from Canada, Mexico, and the United States tackle a wide array of issues and developments in the environmental history of North American mining. While authors were given the freedom to explore the questions and issues that interested them most, they were also asked to consider several common focal points. Among these were physical environmental change, human health issues, regional comparisons, and the various ways that people thought about mining, the natural world, and the environmental changes that intertwined both. By encouraging attention to these common focal points, our goal was to illuminate important aspects of our subject, lend coherence to the volume as a whole, and most important, begin to open windows on how and why North America's differing cultural, social, political, and physical contexts influenced the various ways that mineral development, environmental change, and environmental concern have historically influenced one another.

This last goal, we believe, is a particularly interesting and important one. At a time when North American societies are becoming ever more dependent upon minerals and producing ever more radical environmental changes in their pursuit and use of them, they are also simultaneously forging and deepening social and cultural patterns based on those developments. In a world of infinite resources and infinite pollution sinks, this might never become a problem. But ours is not that world, and concerns are mounting on many fronts. To better understand those contemporary concerns and the layers of history wrapped up in them, we hope that the continental and comparative perspectives deployed here prove illuminating for environmental historians, historians of mining, and everyone concerned about today's ongoing struggles to secure maximum quantities of minerals with minimal scarring of nature.

The book is organized into three thematic sections that unfold in a roughly chronological sequence. Those thematic sections, in order, are titled "Capitalist Transformations," "Industrial Catalysts," and "Health and Environmental Justice." As even a quick perusal of these themes suggests, the essays in this book address several core features of North American environmental history. Capitalism, industrialization, environmental justice—these themes run right through the heart of environmental history, animating some of the field's most profound insights and historiographical interventions.[12] While many of these essays speak to more than one of these themes (and to other areas of inquiry as well), we have positioned the essays in the book based on the major thrust of their argument. In working through the balance of this introduction and the essays that follow, then, we hope that readers will recognize the unmistakable centrality of mining to these larger historiographical conversations and to the North American experience. Indeed, we hope that after reading through this book, they will find it hard to imagine North American environmental history—and nearly every other field of North American history, too—in quite the same way.

. . .

To support this bold claim, the book begins by considering the relationship among capitalism, mining, and environmental change. Nowhere does that relationship have deeper roots in North America than in Mexico's precious-metals mining belt, which is the focus of Daviken Studnicki-Gizbert's essay. Since the Spanish first began to search for gold and silver nearly half a millennium ago, these mountains and valleys have witnessed a succession of mining regimes that cycled through phases of production, decline, abandonment, and reanimation that quickened and intensified as the centuries wore on. Instead of a simple calculus of mineral exhaustion and the extension of the mineral frontier into new territory, capitalist impulses drove episodic increases in capital investment, energy inputs, and scientific and technological advances that allowed miners to navigate declining ore concentrations by intensifying production. The results have been counterintuitive. Rather than production declining over time, each successive mining regime has actually proven more productive than the last. The current regime is dominated by Canadian transnational corporations, which are now extracting gold in concentrations measured in

Yellowknife,
Northwest Territories
Gold

C A N A D A

Peace & Athabasca Rivers/Cold Lake
Alberta, Saskatchewan
Tar sands oil

Mesabi Range,
Minnesota
Iron ore

Elliot Lake,
Ontario
Uranium

Asbestos,
Quebec
Asbestos

Deer Lodge Valley,
Montana
Copper

Penokee-Gogebic
Wisconsin, Michigan
Iron ore

Comstock Lode,
Nevada
Gold and silver

Front Range, Colorado
Gold and silver

UNITED STATES

Colorado Plateau:
Utah, Colorado,
New Mexico, Arizona
Uranium

Las Animas County: Colorado
Coal

Grants, New Mexico
Uranium

Gulf of
Mexico

Real de Cartorce
San Luis Potosí
Silver

Cerro de San Pedro
San Luis Potosí
Gold and silver

PACIFIC
OCEAN

MEXICO

0   600
MILES

FIGURE I.I. Locations of the mining sites and minerals examined in this book. Map by
Jerome Cookson.

parts per tens of millions, a sign that this long cycle of mineral production may at last be nearing its end, though one may well have thought the same a century ago.

As this series of mining regimes took shape across northern Mexico, they also forged an entirely new ecological order, or what Studnicki-Gizbert calls "a mining ecology." Imagined as a series of time-lapsed images, the centuries-long pattern would show mountains being slowly turned inside out or leveled altogether, towering waste-rock piles rising alongside mines, forests being leveled and then reappearing, soil erosion quickening, stream flows becoming increasingly erratic, marshes drying out, aquifer water levels dropping, wildlife disappearing, gaping holes opening in the earth's surface where open-pit mines were being sunk, and new toxic concentrations building up in and around mining districts and slowly spreading their way outward via watercourses, prevailing wind patterns, and other ecological and cultural transport systems. Many of these environmental changes are developed in even greater detail in Antonio Avalos-Lozano and Miguel Aguilar-Robledo's essay, which focuses on a single silver-mining district within the larger world of northern Mexico's precious-metals mining belt. Drawing on methodologies and source materials from paleoecology, environmental chemistry, and climatology, among other fields, the authors underscore the sheer scale of environmental change that shook the region. In terms of tailings, for instance, they calculate that more than half a million metric tons of pollution-riddled waste rock—more than 1.1 billion U.S. pounds—was dumped into local ecosystems between 1773 and 1827 and remain there today. Taken together, these first two essays underscore the profound environmental transformations unleashed by mining and capitalist developments, as well as some of the less obvious ways that the logic of capitalism pushed that process forward over the past five centuries.

The environmental changes that gold and silver mining precipitated in northern Mexico were also, of course, driven by industrial developments, which are the focus of part two of this book. Although industrialization is a sprawling concept that gathers up a wide array of developments, it is also a difficult one to imagine in North America without mining, minerals, and environmental change animating its core. In many ways, what the industrial age did, from an environmental perspective, was to take all the ecologically transforming impulses of capitalism and couple them to new and ever more powerful tools, energy sources, and organizational forms. The results, as the essays in this section show, encompassed not only the sheer ecological tumult that ensued but, perhaps even more provocatively, the ways in which a host of environmental, economic, and cultural developments forged complicated new eco-social dialectics that have tended to persist through time.

These dynamics are clearly illustrated in one of the largest and most valuable precious-metals mining regions in nineteenth-century North America, the high-country mining districts of the Colorado Front Range. From the mid-1860s to the end of the century, as George Vrtis argues in his essay, the interlocking features of the industrial world came together in the Front Range to intensify mining and environmental change on a scale never before seen in the region or in much of North America at the time. As these processes unfolded, they also gave rise to an emerging eco-cultural consciousness and to new understandings about mineral deposits that reverberated across the region. Environmental concerns, anxieties, and, ultimately, legislative and other sociopolitical developments all took shape in the Front Range as the ecological toll mounted. While some of these concerns—such as deforestation, pollution, and waste—have long been recognized as veritable pillars of the American conservation movement, Vrtis suggests that mining also raised other, more far-reaching worries about the superabundance of the earth, scarcity, and the future than has been traditionally recognized.

The central role played by environmental change and natural forces in the Front Range also helped structure the very nature of mining operations elsewhere in North America. As Robert N. Chester III reveals in his essay on the largest silver strike in American history—the great Comstock Lode in northwestern Nevada—the region's volatile geological and hydrological features played pivotal roles in shaping industrial mining processes and organizational development. In response to geological peculiarities such as porous quartz matrices and malleable clays that were prone to cave-ins, or seemingly random seams of boiling-hot groundwater that scalded miners and flooded mines, Comstock investors and managers poured in massive amounts of capital and technology to enable mining, facilitate expansion, and mitigate some of the worst environmental hazards. As these environmental and economic developments converged on the Comstock, mining became increasingly dominated by large technological systems and highly capitalized corporations that not only contributed to the growth of corporate power and the vast environmental changes that ensued, but also shaped the daily lives and safety of miners, well into the twentieth century.

Just as the collision of environmental and cultural factors helped shape the construction of power from above in the industrial age, so too did it help fashion forms of power from below. In southern Colorado, as Thomas G. Andrews observes in his essay, a series of three deadly 1910 coal mine explosions laid bare the perilous subterranean environmental conditions, political tensions, and social solidarity that had long been taking shape in Colorado's coalfields. From the beginning of industrial coal mining in Colorado in the 1870s, the state's collieries had powered the evolution of mineral-intensive industrialization and political power across the Rocky Mountain region, while simultaneously filling cemetery plots with miners and fueling the emergence of labor solidarity among its highly diverse workforce. In the wake of the 1910 mine disasters and the public outcry that followed, these opposing forces bubbled to the surface, gathered attention at state and national levels, and helped the United Mine Workers of America regain a foothold in southern Colorado and push for new mine safety regulations. Each of these developments, and the long histories behind them, also helped seed the tumultuous Colorado coalfield war of 1913–14—the deadliest labor strike in U.S. history—and revealed the deep-rooted connections between mining, energy, and social power in industrializing North America.

Those connections eventually lead back to the very minerals and other forms of matter that lie at the center of these histories. As Timothy James LeCain argues in his essay on Montana's Deer Lodge Valley Smoke War in the late nineteenth century, the ultimate source of human social power in this clash between miners and ranchers was a multifaceted kind of material power that inhered in the region's copper, Longhorn cattle, sulfur, arsenic, and other types of nonhuman matter. Over time, as the Anaconda Copper Mining Company's giant Washoe smelter pumped out massive volumes of smoke filled with sulfur and arsenic, the material power of copper began to undercut the material power of Longhorn cattle, as well as the social power of the ranchers who depended on those cattle. In 1909, the courts ruled on behalf of Anaconda, arguing in part that the copper produced by the smelter was now economically essential to the region, the state, and the nation. Rather than some sort of crude environmental determinism, LeCain's analysis opens a far more complicated and thought-provoking dialogue on the ways that matter helped form and define the very basis of human social power in fundamental ways.

The interactions between miners and ranchers is a reminder that industrial mining districts were not only places of work but also places that people called home. The desire to forge a human community atop a foundation of finite natural resources has always challenged mining settlements, and as Jeffrey T. Manuel reveals in his essay, the Lake Superior iron ore mining districts in northern Minnesota were no exception. As workers flocked to Iron Range mines in the early twentieth century and the population boomed, civic boosters invested heavily in lavish public facilities and promoted future economic development even as the very basis of the economy—the high-grade iron ore that first attracted miners and townsfolk—showed signs of depletion. These seemingly inchoate developments, Manuel argues, fit into the era's prevailing belief in science and technology to overcome natural limits and resolve pollution problems, as well as a deep-seated cultural yearning for permanence and long-term economic growth. In response to this mix of cultural and environmental signals, the Iron Range transitioned to mining lower-grade ores and generating mountains of tailings riddled with worrisome asbestos-like fibers that threatened the long-term stability of the very homes and communities that miners, town boosters, and residents sought to secure in the first place.

The book's third and final part picks up on these themes and ties them into one of the leading edges of environmental history and the contemporary environmental movement: the quest for health and environmental justice. As in thousands of other places trampled beneath the boots of North American industrial centers, the harshest environmental risks and burdens of mining often fell most directly upon those nearest the bottom of the economic order. Similar to industrial laborers elsewhere, miners, their families, and neighbors were often subjected to the worst environmental effects of mining. But the essays in this section also complicate this well-known narrative. The power to create and mold mining district environments reached well beyond the logic of industrial capitalism, extending into the murky and deeply intertwined social, cultural, scientific, and political relations that animated mining sites across North America.

Each of these dynamics is evident in one of the most insidious and dangerous forms of mining and mineral processing that North America has ever known: uranium mining. As Eric Mogren argues in his essay on uranium mining in the American West, once American policymakers defined uranium as necessary for national security during World War II, a host of social, cultural, and political considerations locked in and effectively downplayed or dismissed the environmental and health risks of uranium mining. And those risks were considerable. In addition to the common hazards that came with hard-rock mining, uranium miners and nearby communities were also exposed to radiation that damaged cells and triggered malignancies. In terms of lung cancer alone, one 1960 epidemiological study revealed that uranium miners had a 450 percent greater chance of contracting the disease than other miners. Those risks fell particularly hard on Native Americans in the Colorado Plateau, where many of the mines were located. In the peak period of uranium production, from 1961 to 1962, for instance, three-quarters of the uranium mines were on Native or federal lands and many Native peoples, especially Navajos, worked in the mines. Not until 1978 and the passage of the Uranium Mill Tailings Radiation Control Act would the federal government officially end its policy of placing uranium production above the environmental and health concerns of laborers and local communities.

The Cold War environmental history of uranium mining in Canada reveals a similar pattern of environmental inequality and regulatory neglect. As the nuclear arms race between the United States and the Soviet Union heated up after World War II, U.S. and Canadian policymakers created similar federal regulatory agencies to develop and enforce oversight policies for uranium mining and milling. Despite these similarities, as Robynne Mellor reveals in her essay comparing the American and Canadian experiences with uranium mining, the environmental and health effects from radiation exposure (and other harmful forms of pollution) were arguably worse in the United States. The reasons for this, Mellor argues, are not entirely clear, but she does point to two aspects of this historical puzzle. First, the relatively more benign Canadian experience was not the result of better regulatory oversight. In fact, she suggests, Canada had a worse regulatory record than the United States. And second, in both countries, the pernicious and destructive effects of uranium mining and milling fell hardest on indigenous peoples such as the Anishinaabe in Ontario and the Navajo in New Mexico.

The permissive and inequitable environmental regulatory regime that characterized both American and Canadian uranium mining for much of the twentieth century resonates with other mining districts and other types of mining across North America. In Canada's Northwest Territories, as John Sandlos and Arn Keeling reveal in their essay, a similar process of ineffective pollution control and local contamination emerged around the territorial capital of Yellowknife. From 1938 to 2004, three gold mines produced an estimated 15 million ounces of gold and also mobilized hundreds of thousands of tons of the toxic element arsenic. While the arsenic posed a threat to the entire Yellowknife population, it fell disproportionately on the adjacent Yellowknives Dene First Nation community, which relied on the local environment for foraging, grazing, and drinking water derived from winter snowmelt. The existing records show that some Yellowknives and their domesticated animals were sickened and even killed by acute arsenic poisoning. But the toxic effects of gold mining at Yellowknife have not ended there. As Sandlos and Keeling suggest, they have led to major disruptions in the way Yellowknives view and interact with the land, and have thereby served as agents of a sort of industrial neocolonialism that is once again threatening the Yellowknives with fear, cultural loss, and dispossession. With 237,000 tons of arsenic trioxide dust buried in the abandoned mine shafts beneath just one of the region's mines, the neocolonial legacy of gold mining and environmental inequality at Yellowknife is likely to endure well into the future.

Lurking within the toxic history of mining at Yellowknife and other places in North America lie complicated and contested social and political dynamics that helped shape those outcomes. These dynamics are clearly evident in Nancy Langston's essay on indigenous communities and iron-ore mining in the Lake Superior basin. Disputed scientific and technological understandings about pollution and health risks, divergent cultural and social views about the meaning and significance of the natural world, differing legal and racial experiences—each of these disparities and the ways they were rationalized, Langston argues, shaped the toxic outcomes that have long befallen indigenous peoples in the Lake Superior basin, such as the Bad River Band of the Lake Superior Chippewas. Those contested interpretations are anything but confined to the past. They also bleed into the present, promoting visions of resource use, environmental quality, regulatory behavior, and cultural identity that continue to favor some individuals and groups of people at the expense of others.

The contested views and interpretations that shaped environmental inequality in the Lake Superior basin and other North American mining districts were also influenced by the interplay of global and local forces. When rich mineral deposits were discovered in North America, they routinely became subject to the shaping power of markets, economies, politics, and other forces far beyond their districts. Perhaps no other mineral in world history has embodied that power more than petroleum, and as Steven Hoffman reveals in his essay, it continues to do so in the current exploitation of Alberta's vast tar sands deposits. As Canada seeks to increase the production of tar sands oil and move from being a regional player in the global petroleum market to rivaling Saudi Arabia and Russia, it is wedding itself to international capital, global markets, foreign-owned corporations, and other nations' politics, each with its own internal logic. While the benefits remain uncertain, some of the costs, Hoffman argues, are becoming clear. The mining of tar sands deposits is routinely referred to as one of the most environmentally destructive industrial projects ever undertaken, generating land, water, and air pollution that rivals even the most egregious forms of environmental degradation involved in producing conventional crude. Similar to the impacts of uranium, gold, and iron ore examined in this book, the brunt of the harmful environmental consequences are falling on Alberta's numerous First Nation communities that live in the areas most affected by tar sands development. Their voices, and others tied to Canada's environmental and human rights communities, are currently being shoved aside in the rush to develop Alberta's tar sands.

The interaction of global and local forces in shaping North American mining communities and environmental inequality have also taken more subtle and surprising shapes. As Jessica van Horssen argues in her study of the largest chrysotile asbestos mine in world history—the Jeffrey mine in the aptly named town of Asbestos, Quebec—local townspeople influenced mining and social developments in important and rather unexpected ways. Nowhere is this more evident than in the pivotal 1970s, when the dangerous health risks of asbestos became too well known to ignore any longer. In the early 1970s, several medical studies were undertaken of Jeffrey mine laborers and Asbestos residents that showed a clear link between asbestos and the risk of developing certain types of cancer. In one study, for instance, the odds of developing the asbestos-related cancer mesothelioma was three orders of magnitude greater (1 in 10 versus 1 in 10,000) for those working in the asbestos industry than for the general population. Despite these shocking risks, and despite class-action worker compensation lawsuits against asbestos manufacturers elsewhere, and despite the global collapse of the asbestos industry that followed in the wake of these studies, Jeffrey mine workers and the people of Asbestos continued to support the industry, ignore its risks and dangers, and cling to the community that asbestos built.

. . .

In the end, the story of the Jeffrey mine and the many other stories that make up this book reveal the complex environmental and cultural dynamics that entangled once obscure geological deposits and human history into a single web of interrelationships. Those interrelationships extend in all directions. They reach from the depths of the earth to the upper reaches of the atmosphere, from the realms of science and technology into the spheres of culture, economics, and politics and beyond. Indeed, it is difficult to imagine an aspect of North American life and environmental change over the past half millennium that has not some-how been touched by mining and the minerals it produces.

That deep and pervasive influence is the central message of this book. Mining and minerals stand at the center of modern North American life, shaping the way such key historical developments as capitalism and indus-trialization have contributed to environmental change. Equally important, uncovering the role that mining and minerals have played in our environ-mental history draws attention to the centrality of the environment in the endless struggle for social power. Although the environmental justice movement is a relatively recent development, these essays make clear that environmental inequality is anything but new and that it has served as a battleground for social conflict in mining districts—far from the city cent-ers where we are so accustomed to seeing it—across North America.

Perhaps the most important lesson this book teaches, then, is this: mining has always been hard on the environment and on people, and if we want to protect both, we need to think hard about what we are doing, why we are doing it, and how we might do it better. Historians can help with each of these inquiries, as we hope this book shows. But research opportunities abound. We still need to better understand the ways we perceive mines and the landscape transformations they entail. We also need to better understand the historical roots of our mineral-based societies. And we need to further unearth the stunningly laissez-faire environmental history of mining regulation. Historians may not be able to change the way we mine minerals from the earth or the social ramifications that flow from mining, but we can offer historical narra-tives that make clear the linked environmental and social costs of each. If we are to mine smarter in the future, we must change practices that prevail in the present, but are rooted, sometimes deeply, in the past.

NOTES

1. On the mineral content of a modern American automobile and consumer electronic goods, including computers and cell phones, see Committee on Criti-cal Mineral Impacts et al., National Research Council of the National Acade-mies, *Minerals, Critical Minerals, and the U.S. Economy* (Washington, DC: National Academies Press, 2008), 50–57. On the significance of minerals to modern life generally, see T. E. Graedel, E. M. Harper, N. T. Nassar, and Barbara K. Reck, "On the material basis of modern society," *Proceedings of the National Academy of Sciences* 112:20 (May 19, 2015): 6295–300.

2. W. M. Stewart, D. W. Dibb, A. E. Johnston, and T. J. Smith, "The Contri-bution of Commercial Fertilizer Nutrients to Food Production," *Agronomy Journal* 97:1 (January 2005).

3. Committee on Critical Mineral Impacts et al., National Research Council of the National Academies, *Minerals, Critical Minerals, and the U.S. Economy,* 51.

4. On North American motor vehicle production by nation-state, see OICA (International Organization of Motor Vehicle Manufacturers), "2013 Production Statistics," oica.net/category/production-statistics/2013-statistics/ (accessed November 2015).

5. U.S. Environmental Protection Agency, *Electronics Waste Management in the United States through 2009* (Washington, DC: U.S. Environmental Protection Agency, 2009), epa.gov/waste/conserve/materials/ecycling/manage.htm#report (accessed November 2015).

6. For the United States, see *Mineral Commodity Summaries 2014* (Reston, VA: U.S. Geological Survey, 2014), 5. For Canada, see *Facts and Figures of the Canadian Mining Industry, 2014* (Ottawa, ON: Mining Association of Canada, 2014), 6, 11. For Mexico, see *Anuario estadístico de la minería Mexicana, 2013,* Servicio Geológico Mexicano Publicación No. 43 (Hidalgo, CP: Servicio Geológico Mexicano, 2014), 5.

7. U.S. Environmental Protection Agency, *National Priorities List,* epa.gov/superfund/superfund-national-priorities-list-npl (accessed December 2016). Based on the EPA's hazard ranking system, the top two sites are Big River Mine Tailings/St. Joe Minerals Corporation (site score 84.91) in Deslodge, Missouri, and the Murray Smelter (site score 86.60) in Murray City, Utah. See also Environmental Defense's Scorecard: The Pollution Information Site, "Rank Superfund Sites," scorecard.goodguide.com/env-releases/land/rank-sites.tcl (accessed November 2015).

8. Andrew C. Isenberg, *Mining California: An Ecological History* (New York: Hill and Wang, 2005); Thomas G. Andrews, *Killing for Coal: America's Deadliest Labor War* (Cambridge, MA: Harvard University Press, 2008); and Timothy J. LeCain, *Mass Destruction: The Men and Giant Mines That Wired America and Scarred the Planet* (New Brunswick, NJ: Rutgers University Press, 2009). Other noteworthy studies that examine the environmental history of mining in North America include the following, organized by nation-state. For the United States, see Mimi Sheller, *Aluminum Dreams: The Making of Light Modernity* (Cambridge, MA: MIT Press, 2014); Brad Tyer, *Opportunity, Montana: Big Copper, Bad Water, and the Burial of an American Landscape* (Boston: Beacon Press, 2013); Kent A. Curtis, *Gambling on Ore: The Nature of Metal Mining in the United States, 1860–1910* (Boulder: University Press of Colorado, 2013); Andrew Scott Johnston, *Mercury and the Making of California: Mining, Landscape, and Race, 1840–1890* (Boulder: University Press of Colorado, 2013); Katherine G. Morrissey, "Rich Crevices of Inquiry: Mining and Environmental History," in *A Companion to American Environmental History,* ed. Douglas Cazaux Sackman (Malden, MA: Wiley-Blackwell, 2010), 394–409; and Peter Goin and C. Elizabeth Raymond, *Changing Mines in America* (Santa Fe, NM: Center for American Places, 2004). For Canada, see Arn Keeling and John Sandlos, eds., *Mining and Communities in Northern Canada: History, Politics, and Memory* (Calgary, Alberta: University of Calgary Press, 2015); Laurel Sefton MacDowell, *An Environmental History of Canada* (Vancouver: University of British Columbia Press, 2012); Liza Piper, *The Industrial*

Transformation of Subarctic Canada (Vancouver: University of British Columbia Press, 2009); and Graeme Wynn, *Canada and Artic North America: An Environmental History* (Santa Barbara, CA: ABC-Clio, 2006). For Mexico, see Daviken Studnicki-Gizbert and David Schecter, "The Environmental Dynamics of a Colonial Fuel-Rush: Silver Mining and Deforestation in New Spain, 1522 to 1810," *Environmental History* 15:1 (January 2010): 94–119; and Chantal Cramaussel, "Sociedad colonial y depredación ecológica: Parral en el siglo XVII," in *Estudios sobre historia y ambiente en América I: Argentina, Bolivia, México, Paraguay,* ed. Bernardo García Martínez and Alba Gonzalez Jacome (México, DF: Colegio de México, Instituto Panamericano de Geografía y Historia, 1999), 93–107.

    9. Paul S. Sutter, "The World with Us: The State of American Environmental History," *Journal of American History* 100:1 (June 2013): 116.

    10. John D. Wirth, *Smelter Smoke in North America: The Politics of Transborder Pollution* (Lawrence: University Press of Kansas, 2000); Kathryn Morse, *The Nature of Gold: An Environmental History of the Klondike Gold Rush* (Seattle: University of Washington Press, 2003); and Samuel Truett, *Fugitive Landscapes: The Forgotten History of the U.S.-Mexico Borderlands* (New Haven: Yale University Press, 2006).

    11. J. R. McNeill, *Something New Under the Sun: An Environmental History of the Twentieth-Century World* (New York: W. W. Norton, 2000); John F. Richards, *The Unending Frontier: An Environmental History of the Early Modern World* (Berkeley: University of California Press, 2003); Anthony N. Penna, *The Human Footprint: A Global Environmental History* (Malden, MA: Wiley-Blackwell, 2010); Kate Brown, *Plutopia: Nuclear Families, Atomic Cities, and the Great Soviet and American Plutonium Disasters* (New York: Oxford University Press, 2013); Frank Uekoetter, ed., *The Turning Points of Environmental History* (Pittsburgh: University of Pittsburgh Press, 2010); J. R. McNeill and Erin Stewart Mauldin, eds., *A Companion to Global Environmental History* (Malden, MA: Wiley-Blackwell, 2012); and Andrew C. Isenberg, ed., *The Oxford Companion to Environmental History* (New York: Oxford University Press, 2014). Other important works that examine the environmental history of mining beyond North America in comparative, continental, or global frameworks include Kendall W. Brown, *A History of Mining in Latin America: From the Colonial Era to the Present* (Albuquerque: University of New Mexico Press, 2012); Frank Uekoetter, ed., "Mining in Central Europe: Perspectives from Environmental History," *RCC Perspectives* 10 (2012); and Nicholas A. Robins, *Mercury, Mining, and Empire: The Human and Ecological Cost of Colonial Silver Mining in the Andes* (Bloomington: Indiana University Press, 2011).

    12. For instance, William Cronon's *Changes in the Land: Indians, Colonists, and the Ecology of New England* (New York: Hill and Wang, 1983) and Donald Worster's *Dust Bowl: The Southern Plains in the 1930s* (New York: Oxford University Press, 1977) catalyzed a major, ongoing focus in environmental history on the often ecologically degrading effects of capitalism.

PART ONE

Capitalist Transformations

CHAPTER 1

Exhausting the Sierra Madre

Mining Ecologies in Mexico
over the Longue Durée

DAVIKEN STUDNICKI-GIZBERT

The Cerro de San Pedro is the name of what used to be a small Mexican mountain. It is also the eponym of a small mining town perched in a highland valley overlooking the city of San Luis Potosí, Mexico. Today it is the object of a large-scale open-pit gold and silver mining project, one profoundly reconfiguring local topographies, hydrological systems, and the district's geochemical composition. Every day over the past six years the New Gold mining corporation has detonated massive charges of ANFO, Geldyne, Powerfrac, and Pentex. The mountain of San Pedro is no longer. It has now been reduced to neat set of benches that contour around an ever deepening and enlarging pit. The excavated material is trucked out to leaching piles, where it is sprayed with a cyanide-water solution to filch out microscopic particles of gold, or it is dumped in piles of waste that fill the valleys and arroyos of the surrounding watershed. The impacts on local waters are tremendous. Aside from the acidification and heavy-metal release that are contaminating the water, the pit is creating a massive well effect that is drawing in the region's subterranean water flows. By plugging up drainages, the mine is obstructing the movement of surface water. The liners underlying the leaching pads are sealing off one of the regional aquifer's most important recharge zones, and yet simultaneously the project draws in enough water from it to provision an estimated 50,000 people in the neighboring city of San Luis Potosí.

This has long been the way with mining. Since the arrival of Spanish miners in 1592, the extraction of metals from the subsurface has destroyed and re-created landscapes. Indeed, the layered traces of these past transformations remain visible today, even in the midst of the massive upturning of the land. The hillsides that surround the current operations are covered with a mix of exposed sheets of host-rock, patches of soil, and hardscrabble clusters of mesquite, scrub oak, long and spiky ocotillo plants, and a variety of cactuses. This kind of vegetative cover was historically produced during the Spanish period (1592–1821) as the demands for fuel drove deforestation across the region. Nineteenth- and early-twentieth-century mining left other kinds of traces. Much of the surviving built environment dates from that period as does an important network of dams, reservoirs, and millraces for washing ores or for high-pressure hydraulics. Mining also left large piles of tailings, mine waste, and *scoria* (pebble-sized twists of furnace discards), all heavily mineralized, open to the elements, and blooming into yellows, ochres, and a near blue-green.

The story of Cerro de San Pedro is but a thread in a much larger story that has come to define the landscapes of the Mexican mining belt, a territory composed of hundreds of mines scattered along the Sierra Madre Occidental and Oriental, the Central Mesa, and the highlands to the west and east of the Valley of Mexico. Mining's hold on this territory has proven remarkably persistent. The mining and processing of metals—copper and gold especially—began in the pre-Columbian period, but with Spanish conquest and colonization mining rapidly expanded to become one of the principal pillars of the economy. The core areas of colonial mining—Parral, southern Sonora, central Jalisco, Zacatecas, San Luis Potosí, Guanajuato, Taxco, Pachuca, the highlands of Guerrero and Oaxaca—were all brought into activity within a matter of decades between the 1520s and the 1590s. Its limits—the north of Sonora and the southeast of Chiapas—were set by the mid-seventeenth century and then stayed put. During the Bourbon revival of the eighteenth century, a royal commission inventoried some 453 active mining districts in the viceroyalty. Almost all of these had been established in the sixteenth and early seventeenth centuries.[1] At the turn of the twentieth century, at the height of the U.S. period in Mexican mining, a survey cosponsored by the American Institute of Mining Engineers and the Mexican Republic's Ministerio de Fomento counted 401 mines in exactly the same areas.[2] In 2011, Mexico's geological survey (SGM) listed 718 mines, the majority run by Canadian-based transnational corporations.[3] They are all situated in the same districts mined for the past five hundred years.

In short, Mexico hasn't seen green-field or first-strike mining since the colonial period. Instead it has been marked by a succession of mining regimes—colonial, industrial, and the *mega-minería* of today—that have occupied the same territories. This pattern is not peculiar to Mexico. Many of the core mining territories of the Americas passed through this same succession, over roughly similar time frames. We are familiar with the notion of mines booming and busting, but what the *longue durée* history of Mexico shows—and this is what deserves closer attention—is that this cycle repeated itself, and then repeated again. It is not what we might expect. Like other extractive industries, mining is based on the removal of a nonrenewable resource. Mineral stocks diminish from the very first day they are exploited to the moment of their exhaustion. This fact is reflected in the decline of ore grades over time. But mineral production in Mexico did not follow this decline in any neat and linear way. Quite the contrary: it moved through a series of cycles, each defined by the expansion, maturity, and decline of a particular mining regime.

This historical pattern raises the two key issues. The first concerns the characteristics and drivers of mining's cyclical history. Although ore grades have indeed declined across the centuries, each successive mining regime has proven to be more productive than its predecessor, producing more metal in any given year and removing a greater amount from the deposit overall. Thus, instead of exhaustion we see periodic phases of reanimation, acceleration, and amplification. This was the outcome, I argue here, of historical capitalism's reworking of resource extraction in this part of the world. Instead of extending a commodity frontier into untapped regions, capitalist forms of mining in Mexico (and most of Latin America) avoided the pinch of increasing metal scarcity through intensification.[4] At different junctures—during the Bourbon reforms of the eighteenth century, the first liberal period of the nineteenth century, and again in the 1990s—we see capitalist mining pushing past the limits of exhaustion thanks to new and more powerful assemblages of laws, technology, and energy flows.

The second issue concerns the environmental consequences of this parade of mining regimes over half a millennium. As each regime established itself within the Mexican mining belt, it created a distinct and coherent set of relationships with local waters, soil, and life. These relationships formed a mining ecology: the matrix of interacting components—socioeconomic relations, physical and biotic systems, as well as energy and material flows—that constituted the landscape wrought by mining. Seen in this light, mining ecologies are comparable to agro-ecologies in that they are part of the larger set of "second natures" created by human activity, but distinct from these in that they are created around the extraction and use of nonrenewable materials. Myrna Santiago's social and environmental history of the Mexican oil patch shows another extractivist ecology, an "ecology of oil," an assemblage of property regimes, laws, race and labor relations, and political economy that together conditioned the environmental consequences of resource extraction.[5]

This chapter focuses on the play of time, and follows the historical progression of Mexico's different mining ecologies over nearly five centuries: the colonial regime (1522–1821); the ecology formed around industrial mining during the late nineteenth and twentieth centuries (1883–1960s); and the open-pit mining ecology of our neoliberal present (1990s to today). To provide detail and continuity, it centers its account on the story of the Cerro de San Pedro, folding this narrative within a broader discussion of regional trends and variations. These show a centuries-long pattern of stepwise increases in capital investments, energy inputs, and material flows coupled with a corresponding intensification of mining's environmental footprint. These are the results of mining's centuries-long struggle against exhaustion.

A COLONIAL MINING ECOLOGY, 1522 TO 1821

Spanish colonization of what would become the Viceroyalty of New Spain was famously motivated by the quest for precious metals. At first, gold and silver were obtained as part of the spoils of conquest, but plunder quickly gave way to a more systematic search for, and exploitation of, precious-metals deposits. Only months after the final battles for Tenochtitlán (in August 1521), Hernán Cortés was dispatching expeditions to the gold-bearing placers of the Río Balsas and the Río Papaloapán in Oaxaca, as well as parties of miners and foundry men, to investigate reports of metal deposits in Súltepec-Taxco in the mountains 80 kilometers to the southwest of Mexico.[6] These areas became New Spain's first colonial mining districts, with production beginning in 1522 and 1524, respectively. Colonial silver mining was a paradigmatic example of an early modern commodity frontier. It moved north to Pachuca–Real del Monte, Zacatecas, Guanajuato, and Sombrete in the 1540s and 1550s, then to San Luis Potosí, Durango, and Parral (1590s to 1630s), and finally to Sonora—the northern edge of New Spain—by 1633.[7] Within a hundred years, the northern (Sonora) and southern (Oaxaca) limits of

New Spain's mining belt had been set, creating a mining region some two thousand kilometers long.

Spanish miners and officials developed a regime of high and long-lasting productivity. For close to three hundred years, colonial Mexican mining increased its production of silver, gold, and other metals (especially copper, mercury, and lead). There was a long lull in the seventeenth century, but it was overcome by the efforts of the Bourbon state in the mid- to late eighteenth century. By Independence in 1821, Mexico had produced close to 49,000 metric tons of silver, with annual production peaking at 611 tons in 1804.[8]

Colonial mining was one of the important forms of proto-industrial production in the early modern Atlantic world. It combined the European arts of mine building and metallurgy, private capital investment of colonial and Atlantic merchants, the work of the state to ensure the supply and regulation of labor (peasant-miners, corvée workers, or African slaves), and the intensive use of energy from hydraulics and biofuel consumption. The product was distributed globally through networks that reached as far as China as early as the late sixteenth century.[9]

Each individual mining operation was relatively small. Exploitation of the mine's one to three tunnels depended on the work of between eight and a few dozen men supported by mules and oxen and, where conditions permitted, a water mill. Milling, amalgamation, and smelting took place in the *hacienda de beneficio* (metallurgical works). Although there were important exceptions, these facilities were usually modest in size: a mill, a patio, one to three small smelting furnaces, and a dozen or so workers. The average mill processed slightly less than half a ton of ore per day—a good indication of the local rhythms of extraction.[10] Colonial mining scaled up by joining such operations together. The larger mining districts tapping larger deposits (e.g., Guanajuato, Zacatecas, and Pachuca–Real del Monte), combined dozens of such mining operations mobilizing tens of thousands of *operarios*. Medium-sized districts such as Parral, Cerro de San Pedro, Zimapán, and Taxco counted between five and ten thousand workers.

Cerro de San Pedro, 1591–1821

Colonial mining set in motion a series of ecological changes whose trajectories were complex and contingent on local variables of topography, climate, soil composition, and vegetation cover. This process is best viewed locally. The mines of Cerro de San Pedro, a middling-size mining district in Mexico's near north, were established in a valley perched among the rounded peaks of a subrange of the Sierra Madre Oriental. When the Spanish miners arrived in 1592, the sierras were covered in a mix of mesquite-oak and pine-oak forests with scrub, cacti, and denser stands of mesquite and willow along the watercourses of the larger valley. Lying beneath the plants were rendzina soils: a relatively thin layer of dark-red, fertile, and humus-rich soil spread out over the calcareous bedrock.[11] Sources mention surface water flows of intermittent arroyos, perennial streams and rivers, and extensive marshes where the watershed flattened out into the plain of San Luis Potosí.[12] Until the twentieth century, surviving wetlands continued to be one of the main recharge areas for the region's aquifer.[13] In precolonial times the variety of biomes had made the area a favored ground for local Guachichil and Pamé hunters and gatherers.[14]

Mining began in 1592, and within a year more than twenty mines were working their way into the mountain of San Pedro. Mills and foundries were established to process the ore, either in the neighboring valley of San Luis or wherever there was enough water flow to wash the ores and power the crushing mills. In 1630 there were more than fifty such mills. An estimated five thousand people labored at Cerro de San Pedro itself: mine workers, *carboneros* (charcoal makers), mill workers, muleteers, and artisans.

Colonial mining affected the regional environment in two main ways. The first was fuelwood consumption. An average of 126 square kilometers of forest were cut every year in and around Cerro de San Pedro for fuelwood.[15] The surrounding highlands were cleared by *carboneros* in a matter of years. Observers described the local landscape as completely denuded of any tree or sizable shrub save "a few surviving yuccas upon the bald hills."[16] Rapid and thorough deforestation had important repercussions on the local landscape that have endured to this day. From a human ecology perspective, the elimination of forests destroyed the subsistence base of local Guachichil and Pamé peoples, facilitating their incorporation into the Spanish colonial sphere as slave laborers and their eventual demise as an autonomous, living culture. From a landscape ecology perspective, the most important effect of deforestation was the massive erosion of the thin cap of rendzina soils. Precipitation in this area mainly comes in the form of intense summer rainstorms that can drop between 40 and 90 millimeters of rain at a time.[17] Without the protective cover of the leaf canopy or the anchoring function of root mats, the soils washed out at prodigious rates, leaving behind slabs of local blue limestones. Soil-building processes in the region's semiarid climate are extremely slow and were further impeded by the subsequent introduction of sheep and goat herding. Today, the soils of the Cerro de San Pedro range are limited to small pockets of very sandy and mineralized lithosols in the crevices of the bedrock and around the root mats of some tenacious shrubs and cacti.

With the forest cover gone and the soil cover going, local watersheds were less capable of modulating the pulses of seasonal and episodic rainfall. Severe flooding episodes powerful enough to wash out buildings and structures in the valley city of San Luis were recorded throughout the colonial period and continue to the present day.[18] Between the floods, the overall trend was toward a general desiccation of the landscape. The lack of forest and soil cover resulted in local watersheds draining faster and drying up sooner than they had before. Marshlands described in the early seventeenth century no longer appeared on maps drawn in the eighteenth and nineteenth centuries. Steady, year-round water flow in local rivers diminished and eventually disappeared. In the early eighteenth century an official report on the state of agriculture in the area in and around San Luis Potosí identified as central problems the chronic flooding, the constant droughts, and the generally thin and "sterile" soils.[19]

A second, more chronic kind of environmental change came from the production of mining waste. This material came in various forms: discarded rocks piled up at or near mine entrances; finer tailings generated after processing; and the scoria raked out from the smelters, all of which were laced with lead, zinc, manganese, and mercury. Its environmental effects were a function of quantity and concentration. In this first cycle of mining at Cerro de San Pedro, both of these measures were low. At an average ore grade for the period just shy of 7 kilograms per ore-ton, the total waste produced over two hundred years of mining would have amounted to some 675,000 ore-tons.[20] This waste accumulation made available for dispersion about 1,000 tons of lead and close to 5,000 tons of zinc.[21] Over 9,000 tons of mercury were brought into the mines of San Pedro as well.[22] These wastes were released across the fifty or so *haciendas de beneficio* and the dozens of mine-mouths in the district. Some were 20 to 50 kilometers away from one another. This made for a notably dispersed set of pollution points. The effects of occupational exposure to mercury and lead were much more intense since mill workers operated in direct contact with high concentrations of these metals, either in solution, in the slurry of the patio process, or as lead- and mercury-bearing vapors during smelting and refining.[23]

The patterns of environmental change described for the mines of Cerro de San Pedro obtained across Mexico's mining belt. Mass fuel-wood consumption cleared an estimated 315,642 square kilometers of forests during the colonial period, a territory equivalent to the state of New Mexico or Poland. For the period 1570 to 1820, 64,470 tons of mercury were imported into the mining districts of Mexico.[24] Julio Camargo estimates that Mexican mining added an additional 13 percent to the globe's mercury emissions during this period.[25] The areas cleared of forests and the amounts of contaminants released into local environments were tremendous and in their respective ways transformed local ecologies.

The question of exhaustion was already posed in the early sixteenth century when the first viceroy, Antonio de Mendoza, warned in the 1520s that wood stocks were in danger of disappearing and would shut down the mines of Taxco if they did.[26] The German polymath Alexander von Humboldt, visiting at the turn of the nineteenth century, wondered how mining survived in a country "which wants combustibles, and where the mines are on table lands destitute of forests."[27] The first part of the answer is that Mexican smelters tapped wood reserves up to a hundred kilometers away, a distance unimaginable to Europeans.[28] The second is to be found on the world market for silver. Silver prices never dropped during this period, despite the constant increase in production. Mexican charcoal makers were simply paid enough to make the four- or five-day trip by ox train for fuelwood worthwhile. Tens of thousands of tons of mercury were shipped thousands of kilometers from Spain and Peru to allow the continued exploitation of declining grades of ore. Below the surface, renewed investment in the eighteenth century (and the Crown's military disciplining of mine workers in the 1760s) increased the ranks of the miners working veins ever deeper in the earth.[29]

The crisis of colonial mining came during the Independence era (1810–1824), when uprisings severely disrupted the flows of capital and trade. By 1825 the break from Spain severed the transatlantic networks of Basque and other Peninsular (Spanish-born) merchants. British and other Europeans took their place but failed miserably and never recouped their investments.[30] Without capital, Mexican mine operators could not pay workers or repair equipment and infrastructure damaged during the insurrection.[31] From 1810 to the early 1880s, many mines were left to the *gambusinos*, small parties—often kin and neighbors—of hard-scrabble miners who picked through what had been left by the colonial-era operations.[32] Other mines were abandoned entirely, especially those of the Sierra Madre that faced a revived indigenous resistance from the Yaquis, Raramuri, Apache, and Comanche peoples.[33] Either way, silver production dropped steadily. The nadir was reached in 1870, when the entire country's silver production amounted to less than 28 tons.[34] At their height in the eighteenth-century boom, the mines of San Luis Potosí alone produced four times as much in a year.

QUICKENING: THE INDUSTRIAL MINING REGIME, 1880S TO 1970S

The reanimation of the Mexican mining sector came in the late nineteenth century through industrialization and foreign capital. These twinned forces transformed the organization of the extractive complex toward the mass-processing of lower-grade ores.[35] New sources of energy—principally coal, gas, and hydroelectricity—provided the mechanical power and heat. New techniques and infrastructure provided the instruments, and the formation of a new and expanded class of industrial mine workers provided the labor. Such stepwise increases in energy, material, and human scales were financed by mass investment from abroad, mainly the burgeoning American capital markets of the Gilded Age. The expansion also resulted in new engagements with ecologies and human populations, both adjacent to and far beyond the mines.

The catalyst for the transformation of Mexican mining came from the regime of Porfirio Díaz (1876–1911). Díaz normalized relations with international capital markets and ushered in a series of reforms aimed at opening the mining sector to foreign investment.[36] Carlos Pacheco, minister of industrial development and colonization, was convinced that mining would prime the pump of the Republic's economic development, settle its national debt, and modernize Mexico's agricultural and manufacturing sectors.[37] Pacheco played a key role in realizing the new political economy of Mexican mining. He commissioned a nationwide survey of Mexico's mining districts and fed the results into Antonio García Cubas's landmark geographical survey of the country.[38] He also organized and steered political debate over the legislative reform of mining law, which reached its intended conclusion in 1883.[39] The mining code was then reformed twice more, in 1886 and 1892. The guiding aim of the 1892 code was "Facility to acquire, liberty to exploit, and security to retain." The new codes opened Mexican mining to foreign ownership for the first time in the country's history. Foreigners could stake as many claims as they wished. This allowed a single company to consolidate control over an entire deposit and establish an operation of corresponding size. The claims themselves measured hundreds of acres in surface area and included expropriation rights.[40]

American mining companies quickly capitalized on the opening provided by the Mexican state. Financed by U.S. investors, these companies began acquiring Mexican mines and then bringing them back into operation. The volume of U.S. investments in Mexican mining grew twentyfold (from $2 million to $55 million) in the six years following the first mining code reform of 1886. It then doubled every decade until the Revolution began in 1911.[41] At first, American mine operators concentrated on the waste produced by Spanish-era mining operations. It was a considerable amount, and it was basically free for the taking. The Pachuca River Concentrating Company, for instance, acquired rights to tailings piles generated by over three centuries of mining at Pachuca–Real del Monte.[42] Then the companies began buying up thousands of acres in claims, thus acquiring entire mining districts. At its height, the American Smelting and Refining Company (ASARCO), the company that made most of the Guggenheims' fortune, controlled a nationwide network of dozens of mining districts and smelters spread from Sonora to Oaxaca.[43]

The central effect of the American mining companies' entry was to scale up the extractive system. Extracting and processing far greater volumes of ore were the means to overcome the limits imposed by low ore grades. This was done through the industrialization of mining operations using new and more powerful forms of extraction and refining. Pneumatic drills, air ventilators, water pumps, power hoists, rail carting, and aerial tramways—all of these worked together to extract an average of 1,350 tonnes (metric tons) of ore per day (a colonial mine head did well if it took out at least half a ton). This ore was then submitted to new forms of refining such as cyanide flotation and mass smelting. Driving the increasing scale of operations was the arrival of coal and, especially, hydroelectric power.[44] By the late 1920s (that is, before the advent of oil-fueled electricity generation), hydroelectricity accounted for 83 percent of Mexico's industrial power. Mining companies consumed over a third of this total.[45]

The scale of these operations required unprecedented amounts of capital—measured in the millions to tens of millions of dollars—that had even American investors balking.[46] It was certainly unavailable to Mexican miners and entrepreneurs. A contemporary observer, V. M. Braschi, warned that the Mexican mining industry was becoming "a series of large mines controlled by the smelting interests and a few independent large companies."[47] He was quite right. By the 1910s, foreign companies controlled between 97 and 98 percent of a steadily growing industry.[48]

Cerro de San Pedro, 1870–1948

Just as industrialization changed Mexican mining, it also reconfigured its environmental impacts. Like many other mines in nineteenth-century Mexico, Cerro de San Pedro was largely abandoned after Independence. For decades it was mainly worked by *gambusinos*. The return of large-scale mining at Cerro de San Pedro began in earnest in 1890 with the arrival of Robert S. Towne's Compañía Metalúrgica Mexicana. Towne's operations in the San Luis Potosí area also included the coal-burning smelter built on the western edge of the city, as well as the Mexican Northern Railroad and the Monterrey Mineral and Terminal Railway, which allowed him to process ores from mines across the region. A sizable timber concession in the nearby Sierra de Alvarez supplied timber and fuelwood for the company's operations. Inside the smelter new, coal-fired crushers milled the ore that was then fed into eleven separate furnaces. Every day, the smelter-mill processed just shy of 1,200 tons of ore. At this volume of production it took Towne's operations only a year and a half to work as much ore as the colonial mines of Cerro de San Pedro did in more than two hundred years. Eighty percent of the ore came from the mines of Cerro de San Pedro. Towne built a branch rail line to allow an ore train to run daily between the mines and smelters. Shafts were widened to allow the operation of new power-hoists. Pumping stations and new adits resolved the flooding problem. Pneumatics, trolleys, and bogie carts increased the rate at which miners could break up and extract material from the mine-face. And, not least, all of this work, while supported by the tools and power sources of the industrial age, was driven by a large force of miners, hoistmen, carpenters, mechanics, furnace-men, and laborers. Reduced to a small village in 1890, a decade later Cerro de San Pedro was a bustling town of close to eight thousand people. Another thirteen hundred workers worked down at the smelter in the city of San Luis Potosí.

The return of mining, this time on an industrial scale, ushered in new transformations of the landscape and ecology of Cerro de San Pedro. Again, the key issues were energy and mining wastes. The industrialization of the extraction process was undergirded by a shift in energy sources from local fuelwoods to coal and *chapopote* (a mineralized tar)

FIGURE 1.1. ASARCO operations at Cerro de San Pedro, 1920s.
Courtesy of Armando Mendoza, Cerro de San Pedro.

hauled in by train from hundreds of kilometers away: the coal from the newly opened coalfields of Coahuila and the *chapopote* from Veracruz. Bringing energy from afar allowed local forests to continue a recovery that had begun with the early-nineteenth-century mining collapse. That Towne should have invested in timber concessions in the neighboring Sierra de Alvarez is revealing in this regard. Cleared by *carboneros* during the colonial period, the Sierra's forests had returned during the nineteenth century to the point that structural-grade timbers could be cut for use in the mines. This demand for lumber was important, and photographs of the period show large piles of timber stacked beside the railyards and warehouses of the mines (see figure 1.1), but it did not scour the forests to anywhere near the same extent as colonial charcoal making had.

The Mexican mining industry, however, was simply trading old sins for new. The move away from fuelwood consumption was made possible by the development of coal and hydroelectricity. These had their own impacts on the environment. The environmental costs of energy were externalized, displaced to the sites of hydroelectric projects, generators, and coalfields located elsewhere in the Republic and beyond.

The industrial scales of late-nineteenth-century mining greatly increased the generation of mining waste. Towne's smelter in San Luis Potosí produced close to a thousand tons of waste every day: tailings and slag left over from the refining process. Piles were also deposited around the various mine heads in Cerro de San Pedro proper. These consisted of extracted rock judged to be of too low a grade to be hauled down to the smelter. The quantities of slag and tailings sufficed to create embankments of between ten and thirty meters high at the southern outlet of the Cerro de San Pedro valley. Both tailings and waste rock piles stocked important quantities of lead, zinc, cadmium, copper, and arsenic that spread into the environment as waterborne sediments or wind-borne dust.

Another source of heavy-metal contamination came from emissions. The precise quantity and composition of heavy-metal emissions for the San Luis Potosí smelter remains to be determined but it is reasonable to think of quantities of over 1,000 kilograms of lead, zinc, and arsenic.[49] Contemporary environmental chemistry studies (2006 and 2008) show us the toxic legacies of the Towne-ASARCO smelter. They measure lead, cadmium, arsenic, and copper accumulated in local soils, rooftops, and air. These metals exist at high concentrations down to ten centimeters beneath the surface in local soils, which suggests long-term deposition.[50] The studies also demonstrate a clear spatial pattern of distribution, with the highest concentration values recorded in the city's poorer southeastern neighborhoods.[51] These neighborhoods are also at the outlet of the watershed that drains the Cerro de San Pedro valley and might therefore be receiving episodic pulses of waterborne contamination from the waste piles. Decades of heavy-metal contamination are recorded in the very bodies of those who live in these neighborhoods as well as in the immediate vicinity of the smelter complex. Their hair and blood contain levels of lead and arsenic well in excess of national and World Health Organization norms.[52]

Studies of other sites of industrial mining in Mexico paint a similar picture of heavy-metal contamination of waters and soils. The average mine produced roughly 400,000 tonnes of waste a year.[53] Tailings were rarely contained in impoundments, or if they were, were not reliable over the long term. At the mines of Zimapán, Hidalgo, for instance, arsenic from the tailings has been seeping into the water table since the 1930s. Enough accumulated to contaminate the aquifer of an entire valley.[54] Similar forms of water contamination by arsenic, lead, and other heavy metals from early-twentieth-century mining have been measured in other mining districts in Mexico, such as the Mineral de Pozos (Guanajuato), Matehuala (San Luis Potosí), and Santa Barbara (Chihuahua).[55] As for soils, heavy metals have been recorded at every site that has been studied: Baja California Sur, Torréon, Monterrey, Matehuala, La Paz (SLP), and Zimapán all register concentrations of heavy metals well above national or international norms.[56] Finally, since much of the tailings waste bears sulfides, industrial mining also set in motion processes of acidification and heavy-metal leaching through acid rock drainage (ARD). In places such as Zimapán or Taxco values of 2.6 pH and 2.8 pH were obtained in local soils.[57] This is a degree of acidity somewhere between that of vinegar and lemon juice, a level well past what most plants can tolerate.

Industrial mining in Mexico survived the shocks of Revolution, World War I, and the progressive reassertion of national control over the sector that followed the Great Depression. ASARCO, for instance, was able to negotiate the continuation of its operations with successive revolutionary leaders. The American period began to draw to a close only in the 1940s as the mines were "Mexicanized," taken over either by mineworker cooperatives or by emergent conglomerates of Mexican capitalists.[58] Whether under Mexican corporate or cooperative ownership, the form of production—and thus the environmental impacts—remained the same: electric and coal-powered extraction, cyanide flotation systems, and high-energy smelting. However, the 1940s marked the beginning of a decades-long decline in Mexican mining. Deposits were worked out, ore grades dropped, and access to large volumes of investment capital was harder and harder to obtain. By the early 1970s, Mexican gold production was back down to less than ten tons a year.[59] As during the previous slack period in Mexican mining (1821–83), the shortage of capital meant that miners had gone about as far as they could go with the technology available. In the second slack period (1940–1993) profitable ores grew scarcer. Getting what metal remained would require new technology, more energy, and new injections of capital.

THE INVISIBLE GOLD RUSH, 1993 TO THE PRESENT

The latest Mexican mining boom began in the early 1990s when the sector was reopened to foreign corporations, but its deeper roots can be found at the Carlin Trend in Nevada. It was there in the 1960s that a small group of geologists and metallurgists working for the Newmount Mining Corporation had tested the theory that gold could be found in large disseminated porphyry deposits. It existed in tiny flecks of metal less than a micron in size (most bacteria measure one to ten microns across)—invisible gold. The means of obtaining it was simple and awesome. Open-pit mining techniques, borrowed from the copper industry, were combined with mass heap leaching with cyanide solution. The scales were massive, but the production costs were much lower than industrial tunnel mining and the total amount obtained from the deposit—in a shorter amount of time—was much higher than what could be drawn out under colonial or industrial regimes. Lower costs, higher amounts: it was an irresistible combination for a growing number of mining companies.

Mexican mining was opened to these new techniques and new corporations during the neoliberal reforms of the 1990s. A year before NAFTA, a new mining code was put in place, the Ley Minera of 1993. This was part of a broader trend at the time that saw the systematic liberalization of mining codes around the world.[60] The Ley Minera of 1993 was essentially a repeat of the Porfiriato's 1892 code, with its emphasis on freedom of access, liberty of operation, and stability of tenure. But it introduced a number of key innovations tailored to the new open-pit complex. For instance, claims on underground resources granted preeminent rights to the surface—a first in nearly five centuries of mining in Mexico. Mining claims were suddenly accorded precedence over forms of tenure such as the common-hold *ejidos* established after the Revolution, conservation and protected areas, and, in case of expropriation, private property. The new code also guaranteed that title and concession could not be legally revoked before term (companies found to be in breach of the law could only be fined), and it lengthened the duration of title. These were also critical measures for open-pit projects, given the massive amounts of financing they require and the need to amortize such investments over the long term.[61]

The principal players within this latest cycle in Mexican mining are corporations based in Canada. Within four years beginning in 1991, the number of properties owned by Canadian companies in Mexico almost quintupled (52 to 244).[62] In 2005, Canadian corporations controlled close to two-thirds of the Mexican mining industry, developing some 420 projects across the Republic. The latest figures (2010) show that 78.5 percent of the 718 mining projects then under development were under their control.[63] The predominance of transnational mining companies was in part attributable to the high capital costs associated with operating open-pit mines. These are significantly higher than for industrial tunnel mining—$500 million to $1.5 billion is the general range of what it costs to start an open-pit mine—and expenditures need to be paid out for a number of years before any gold is produced.[64] The technology involved is common to the industry as a whole, so what advantages Canadian corporations enjoyed were due to their access to the Toronto Stock Exchange (TSX). Thanks to a more favorable fiscal regime and an existing depth of technical and financial expertise in the sector, the TSX positioned itself as the world's principal hub for the capitalization of mining exploration and development.[65] In 2005 alone, the TSX provided close to $4 billion in financing to mining companies, by far the largest amount raised on the world's various exchanges.[66] The payoff so far has been excellent. The average cost of producing an ounce in an open-pit mine has always remained well below the metal's market price. In 2012, it varied between $225 and $275 an ounce when the price of gold was above the $1,600 mark.

The financial scale of open-pit mining is directly related to the physical and energy scales at which it operates, which are in turn made necessary by the physical characteristics of the porphyry deposits that it targets. Porphyries are large disseminated ore deposits that can stock enormous quantities of gold and silver. Take for example Goldcorp's Peñasquito project at Mazapil (Zacatecas), which began operations in 2010. It contains an estimated 17.8 million ounces of gold and 1,277 million ounces of silver, as well as important amounts of marketable zinc and lead.[67] The Peñasquito mine is located in the same zone that had previously been worked by ASARCO in the early twentieth century and by Basque miners back in the 1570s. This is the third pass over the same district, and the extremely low ore grades worked by Goldcorp reflect this.

There exists a rule in geological economics known as Lasky's Law, which holds that "the tonnage of ore increases geometrically as grade decreases arithmetically."[68] Given gold concentrations now as low as 1.8 parts per billion, Goldcorp will excavate and process truly astronomical quantities of the Sierra Madre in order to extract the $33 billion of gold and silver remaining in the deposit. Something on the order of 130,000 tonnes of ore will be processed on a *daily* basis. This does not include the large amounts of waste rock and overburden moved in the process. At the end of the project's remarkably short twenty-one-year life, Goldcorp will have moved close to a trillion tonnes of ore, created two large open pits (the larger measuring 4.5 kilometers in diameter and close to 1.5 kilometers in depth), and disturbed large swaths of the Mazapil valley for tailings impoundments, waste rock piles, a cyanide heap leaching pad, processing plants, and the other remaining infrastructure for the mine. For Mexico it is a form of mass terra-forming of unprecedented proportions.

The movement of so much physical mass requires enormous amounts of mechanical energy. In the case of Peñasquito, 25,769,628 megajoules per day will be used to run the mine's drills, haul trucks, crushers, mills, and grinders.[69] To put this energy requirement in historical perspective, a single Komatsu haul truck consumes as much energy during a single shift as an entire colonial mining operation consumed over fourteen years. Heat energy continues to be required for smelting, but in the open-pit complex this accounts for a small fraction of the total energy budget. Most is spent on moving matter.

All of this intensive mining activity carries important environmental costs, both locally and globally. Given that energy for open-pit mining principally comes in the form of diesel, and given the quantities involved, large gold-mining corporations have gone so far as to purchase their own oil companies to reduce and stabilize energy costs.[70] The extraction and processing of the fuel consumed by Mexican open-pit mines generate environmental impacts borne in oil fields and refineries across the world. Mining has now become one of the most important emitters of greenhouse gases (GHGs). In Canada, mining and oil extraction account for 21 percent of the country's GHG emissions.[71] Hydroelectricity continues to supply important energy inputs to open-pit mining, and here too environmental consequences are distanced from the mine itself. In Brazil, Chile, Panama, and Ecuador, hydroelectric development is being pushed by *mega-minería*'s high demands for energy.

In the immediate vicinity of the mine, the generation of waste rock and tailings creates the potential for acid rock drainage on a massive scale. Thousands of tons of heavy metals such as arsenic and lead are brought up to the surface, where they can move into local ecosystems and waterways. At the same time, the operation of open-pit mines requires large volumes of water—measured in the tens of millions of liters—on a daily basis to run solutions through the leach pads or for other extractive processes. The location of many open-pit mines in Mexico in semiarid highland areas creates important pressures on aquifers with low recharge rates. The excavation of deep pits also creates large well effects that further disturb local hydrological patterns. Finally, the compounds used to extract metal from the ore—especially cyanide used in heap leaching—are fatal if they accidentally come into contact with people or animals. Environmental toxicologists are only recently becoming aware of the negative effects to plant and water life exposed to more persistent forms of "weak" cyanide complexes.[72]

Cerro de San Pedro, 1996 to the Present

In the early 1990s, Cerro de San Pedro was inhabited by about 250 souls. They lived off the sparse resources that the land could still offer: goat herding, beekeeping, gathering, and family-scaled *gambusino* mining. Then Jorge Barnet arrived. He was a local *"corredor de minas,"* or mine-broker, who proceeded to buy out all the existing mining titles in the valley. Given the general opinion that the mines of San Pedro had given all they had, he obtained the titles for a pittance. He then sold his package to a consortium of Canadian companies that included Glamis, Cambior, and Metallica Resources and that in 2008 morphed into New Gold.

The project of reanimating gold mining at Cerro de San Pedro was stalled for almost ten years because of a series of lawsuits won by a determined local opposition. But in 2006, these decisions were overridden or ignored by state and federal officials, and the project began in earnest. The scale of the operation is awesome and so is the speed. It has taken New Gold six years to level the Cerro de San Pedro. It is now beginning the pit, which is projected to reach 600 meters in depth over the next six or seven years. When production ends, the company will have displaced a mountain and a half and recovered a cube of gold measuring less than 1.5 meters a side.

In the course of its operating life, the mine at Cerro de San Pedro will use 32 million liters of water per day. This is obtained from the regional aquifer, which has been dropping steadily over the past decades because of the demands of the city of San Luis Potosí (with a current population over 1 million). The leach pads cover close to seven square kilometers with impermeable sheeting that is then loaded with ore and cyanide solutions. They sit directly over an area of permeable rock that has traditionally been one of the aquifer's principal recharge areas. The pit itself will, when abandoned, put further stresses on local water because the water welling up from the pit floor will be both contaminated and available for high rates of evaporation.

The waste piles, an inescapable feature of each mining regime, have now become large enough to in-fill entire sections of the local drainage. When rain falls through this body of sulfide-bearing rock, it allows ARD generation on a massive scale. It is true that the rate of ARD and heavy-metal leaching are slowed by the modest overall precipitation levels of this area's semiarid climate. These effects might take decades or even centuries to be perceived. But since there is no adequate remediation or treatment plan for these wastes, these processes will take place sooner or later and will last for millennia.[73]

THE LONG-TERM VIEW

Looking at the history of Cerro de San Pedro over the long term allows one to follow the formation and reformation of mining ecologies over four hundred years. The account given here could be provided for mining districts across Latin America and other parts of the world where mining has been present for centuries. It shows how mining was able to continue the extraction of a nonrenewable metal deposit at an exponentially increasing rate. Whereas in the colonial period a mine expended a bit more than 1,000 MJ of mechanical energy per day processing half a ton of ore, a contemporary open-pit mine burns through 26,000 times that amount in moving 130,000 tons of material daily. Each regime succeeded in extracting metal at faster and faster rates. In the case of Cerro de San Pedro, average annual production figures rose from 23.4 tons of silver per year during the colonial period to 86 today.

This took place in a context of ever-decreasing ore grades, the standard measure of a deposit's exhaustion. Exhaustion and abandonment did indeed occur, once in the 1820s and again in the 1950s. The current mine at Cerro de San Pedro is slated to shut down in 2018. Each time, what brought mining back into operation was the application of more powerful energy sources and more efficient extractive systems. At each turn, these efforts were enabled by new political economies that geared themselves to extractivism, whether under the Spanish Habsburgs and Bourbons in the sixteenth and eighteenth centuries, the Porfiriato of the late nineteenth century, or the neoliberal government of today. Each was underwritten by ever-growing amounts of financial capital gathered on increasingly globalized networks.

What are the consequences of this history for landscape ecologies? The effects are layered. One can still see and feel the same rocky scrubland "created" after years of forest stripping in the colonial period. The same goes for the Towne-ASARCO mine's tailings, its waterworks, its buildings and mine heads, and the lead- and zinc-laced dust in the valley's neighborhoods and villages. Today's project will also leave its mark— now on a colossal scale. It is remaking the topography itself: erasing mountain drainages, flattening mountains, plugging up an entire region's aquifer. And then there's all that isn't seen. If in the colonial period, the mines introduced about 8 kilograms of heavy metals into the environment per day, today's mine brings to the surface 91 tonnes of heavy metals every twenty-four hours. This isn't good news for the 2 million citizens of the neighboring city of San Luis Potosí, whose aquifer is drying up fast and whose air, soils, and waters are already ranked among Mexico's most polluted.

Each mining regime cycled through the same phases of (re)activation, production, decline, and abandonment. But each boom was faster than the last. Over the long term one observes a kind of compression in time even as all the other scales (energy, material, capital, environmental effects) increased exponentially. Throughout this history, networks of capital and capitalist development were the main drivers of this process. This brings us to an important point made by Karl Marx in relation to how capital steals time from social and natural bodies. He was mainly thinking about human beings, showing how capital generates further profits by developing production systems ever more efficient at tapping into the vitality of labor, pushing it past the limits of health and social reproduction. Interestingly, he equated this process to the exhaustion of natural vitality: "It [capital] [. . .] attains this end by shortening the extent of the laborer's life, as a greedy farmer snatches increased produce from the soil by robbing it of its fertility."[74] If we apply Marx's insight to the natural bodies of mining districts, we see that extractive systems have become ever more powerful and intensive as greater and greater energy and matter are mobilized over shorter and shorter periods of time. The drafts on the natural capital of local and regional landscape ecologies have likewise grown and, in so doing, have sped up the exhaustion of the land. In the case of Cerro de San Pedro, mining has pushed the regional ecology past the thresholds of natural renewal and resilience. In certain mining districts of Mexico, it is true, the marks of past mining regimes are fading from the land. But in Cerro de San Pedro they are all still present and clear, and, given the rates of the geophysical and biochemical processes involved, they will likely last for millennia.

NOTES

1. See Alvaro Sánchez-Crispín, "The Territorial Organization of Metallic Mining in New Spain," in *In Quest of Mineral Wealth: Aboriginal and Colonial Mining and Metallurgy in Spanish America*, ed. Alan K. Craig and Robert C. West (Baton Rouge: Geosciences Publications, Dept. of Geography and Anthropology, Louisiana State University, 1994), 163.

2. "Special Map Supplement: The Mines and Railways of Mexico," *Transactions of the American Institute of Mining Engineers* 32 (1903).

3. Servicio Geológico Mexicano, *Anuario estadístico de la minería mexicana 2010* (México, DF: Secretaía de Economia—Coordinación General de Minería, 2010), 13–14.

4. The historical geographer Jason W. Moore lays out a clear definition of historical commodity frontiers and their place in the development of capitalism in "Sugar and the Expansion of the Early Modern World-Economy: Commodity Frontiers, Ecological Transformation, and Industrialization," *Review (Fernand Braudel Center)* 23 (2000): 409–33.

5. Myrna Santiago, *The Ecology of Oil: Environment, Labor, and the Mexican Revolution, 1900–1938* (New York: Cambridge University Press, 2006), 4. Other recent work on the environmental history of mineral extraction, not all of it framed in the same ecological terms, includes Kathryn Morse, *The Nature of Gold. An Environmental History of the Klondike Gold Rush* (Seattle: University of Washington Press, 2003); Timothy LeCain, *Mass Destruction: The Men and Giant Mines That Wired America and Scarred the Planet* (New Brunswick, NJ: Rutgers University Press, 2009); Peter Van Wyck, *The Highway of the Atom* (Montreal: McGill-Queens University Press, 2010); David Stiller, *Wounding the West: Montana, Mining, and the Environment* (Lincoln: University of Nebraska Press, 2000); Richard V. Francaviglia, *Hard Places: Reading the Landscape of America's Historic Mining Districts* (Iowa City: University of Iowa Press, 1991); Stephen Rippon, Peter Chlaughton, and Chris Smart, *Mining in a Medieval Landscape. The Royal Silver Mines of the Tamar Valley* (Exeter, UK: University of Exeter Press, 2009); Nicholas A. Robins, *Mercury, Mining, and Empire: The Human and Ecological Cost of Colonial Silver Mining in the Andes* (Bloomington: Indiana University Press, 2011).

6. Hernán Cortés, "De los muchos descubrimientos de minas y las expediciones que envio Hernan Cortes" in Javier Moctezuma Barragan y Sergio Pelaez Parell, eds., *Antologia minera de Mexico. Primera estación, siglo XVI* (México, DF: Sec. de Energia, Minas e Industria Paraestatal, 1994), 35–37.

7. Sánchez-Crispín, "Territorial Organization," 157–58; Peter Gerhard, *A Guide to the Historical Geography of New Spain* (New York: Cambridge University Press, 1972); Robert C. West, *Sonora: Its Geographical Personality.* (Austin: University of Texas Press, 1993), 45. On commodity frontiers, see Jason W. Moore, "Silver, Ecology, and the Origins of the Modern World, 1540–1640," in *Rethinking Environmental History: World System History and Global Environmental Change,* ed. Alf Hornborg, J.R. McNeill, and Joan Martinez-Alier (Lanham, MD: AltaMira Press, 2007), 123–42.

8. See the silver production series developed by Richard Garner, based on data compiled from the Royal Treasury Houses of New Spain, at http://www.insidemydesk.com/hdd.html.

9. On the early modern arts of mining, see Antonio Barrera-Osorio, *Experiencing Nature: The Spanish American Empire and the Early Scientific Revolution* (Austin: University of Texas Press, 2006), 65–70; Modesto Bargallo, *La amalgamación de los minerales de plata en Hispánoamerica colonial* (México, DF: Compania Fundidora de Fierro y Acero de Monterrey, 1961), and his *La minería y metalúrgia en la América Española durante la epoca colonial* (México, DF: Fondo de Cultura Económica, 1955), 107–60, 173–79, 196–203. For capital investments, see Juan Carlos Sola Carbacho "El papel de la organización familiar en la dinámica del sector mercantil madrileño a finales del siglo XVIII," *Historia Social* [Madrid] 32 (1998); Edith Boorstein Couturier, *The Silver King: The Remarkable Life of the Count of Regla in Colonial Mexico* (Albuquerque: University of New Mexico Press, 2003); Frédérique Langue, *Mines, terres et société à Zacatecas (Mexique) de la fin du XVIIe siècle à l'indépendance* (Paris: Publications de la Sorbonne, 1992); Peter Bakewell, *Silver Mining and Society in Zacatecas, 1546–1700* (New York: Cambridge University Press, 1970), esp.

ch. 2; David Brading, *Miners and Merchants in Bourbon Mexico, 1763–1810* (New York: Cambridge University Press, 1971); and John Tutino, *Making a New World: Founding Capitalism in the Bajío and Spanish North America* (Durham, NC: Duke University Press, 2011). On labor, see Silvio Zavala and Maria Costel, *Fuentes para la historia del trabajo en Nueva España* (México, DF: Centro de estudios históricos del movimiento obrero mexicano, 1980); Bakewell, *Silver Mining and Society.* On the provisioning of slaves to the mines, see Carlos Sempat Assadourian, *El tráfico de esclavos en Cordoba. De Angola a Potosi. Siglos XVI–XVII* (Cordoba: Universidad Nacional de Cordoba, 1966). On the Crown's authority over and regulation of the mines, see María del Refugio Gonzáles Domínguez, "Notas para el estudio de las Ordenanzas de Minería en México durante el siglo XVIII," *Revista de la Facultad de Derecho de México* 26 (January–June 1976): 157–67; and *Ordenanzas de la minería de la Nueva España formadas y propuestas por su Real Tribunal* (México, DF: UNAM, 1996). On hydraulics, see Diana Birrichaga Gardida, "El dominio de las 'aguas ocultas y descubiertas': Hidráulica colonial en el centro de México, siglos XVI–XVII," in *Mestizajes tecnológicos y cambios culturales en Mexico,* ed. Enrique Florescano and Virginia García Acosta (México, DF: CIESAS, 2004), 120; Robert C. West, *The Mining Community in Northern New Spain: The Parral Mining District* (Berkeley: University of California Press, 1949), 41–42; Sebastian de la Torre y León, "Informe sobre las minas de Bolaños, 1774," in *Las minas de Nueva España en 1774,* ed. Alvaro López Miramontes and Cristina Urrutia de Stebelski (México, DF: Instituto Nacional de Antropología e Historia, 1980), 44–45. On fuel wood, see West, *Parral Mining District*; Chantal Cramaussel, "Sociedad colonial y depredación ecológica: Parral en el siglo XVII," in *Estudios sobre historia y ambiente en América I: Argentina, Bolivia, Mexico, Paraguay,* ed. Bernardo García Martínez and Alba González Jácome (México, DF: Colegio de México, Instituto Panamericano de Geografia e Historia, 1999). On smelting, see Barba, *Arte de los metales: En que se enseña el verdadero beneficio de los de oro y plata por açogue, el modo de fundirlos todos, y como se han de refinar y apartar unos de otros* (Madrid: Imprenta del Rey, 1640), 100, 106–7, 131.

10. For San Luis Potosí, see Guadelupe Salazar González, *Las haciendas en el siglo XVII en la región de San Luis Potosí. Su espacio, forma, función, significado y la estructuración regional* (San Luis Potosí: Universidad Autónoma de San Luis Potosí, 2000). For Parral, Chihuahua, see West, *Parral Mining District.*

11. Author's field observations, Sierra de Alvarez, San Luis Potosí, February 2007 and July 2008; Instituto Nacional de Estadística y Geografía, *Carta de Suelos, San Luis Potosí, 250:000* (México, DF: Instituto Nacional de Estadística y Geografía, 1993).

12. Pedro de Arismendi Gogorron, *Denuncia de asientos de agua.* 27.04.1613, Archivo Historico del Estado de San Luis Potosí—Alcaldía Mayor (hereafter AMSLP), 1613 (2), exp. 40, CONAGUA, Servicio Meteorológico Nacional, 2011.

13. Instituto Nacional de Estadística y Geografía, *Carta estatal, hidrología subterránea, San Luis Potosí, 1:700,000* (México, DF: Instituto Nacional de Estadística y Geografía, 1993).

14. According to archaeological evidence and historical accounts, the most common animals were deer (both *Odocoileus hemionus* and *O. Virginianus*), hares, rabbits, squirrels, and porcupine. Leonardo Lopez Lujan, *Nomadas y sedentarios. El pasado prehispanico de Zacatecas* (México, DF: INAH, 1989), 21; Francois Rodriguez Loubet, *Les Chichimeques. Archéologie et ethnohistoire des chasseurs-collecteurs de San Luis Potosi, Méxique* (México, DF: Centre d'etudes mexicaines et centraméricains, 1985), 159, 162.

15. Daviken Studnicki-Gizbert and David Schecter, "The Environmental Dynamics of a Colonial Fuel-Rush: Silver Mining and Deforestation in New Spain, 1522 to 1810," *Environmental History* 15 (January 2010): 94–119.

16. Alejandro Montoya, "Población y sociedad en un Real de Minas de la Frontera Norte Novohispana. San Luis Potosí de finales del siglo XVI a 1810" (Ph.D. dissertation, Université de Montréal, 2003), 84; the quote is from *Ciudades, Villas y Lugares, Reales de Minas y Congregaciones de Españoles de el Obispado de Mechoacan, 1649*, MS 1106 A, f. 47r, Ayer Collection, Newberry Library.

17. Normales climatológicas, CONAGUA, Servicio Meteorológico Nacional, http://smn1.conagua.gob.mx/index.php?option=com_content&view=article&id=172&tmpl=component (accessed January 2, 2017).

18. *Informes y licencia de los pobladores de Tlaxcalilla sobre las lluvias,* AMSLP, 1626 (3), exp. 21, 21.08.1626; *Petición para reparar gran parte de la iglesia que se cayó a raíz de las lluvias,* AMSLP, 1626 (3), exp. 23, 21.08.1626; *Pleito sobre la edificación de una casa en un solar en la plazuela de Mascorro que sirve como desague de la lluvia,* AMSLP, 1673 (3), 29.08.1673; Montoya, *Población y sociedad,* 238.

19. *El alcalde mayor procede hacer informacion sumaria de oficio sobre la abundancia o escasez de mantenimientos que en esta ciudad hay [. . .],* AMSLP, 1739 (1), exp. 23, 20.10.1739.

20. Recorded silver production for the period 1628 to 1820 totaled 4,727,280 kilograms; see silver production data for Caja de San Luis Potosí, compiled by Garner, http://www.insidemydesk.com/cajas/S/slpotosi.txt (accessed May 13, 2011).

21. Richard J. Lambert et al., *Technical Report on the Cerro San Pedro Mine, San Luis Potosí, Mexico* (Toronto: Scott, Wilson, Roscoe, Postle Associates Inc., 2010), 17–5.

22. Estimates of colonial Hg:Ag production ratios of range from 2:1 (Bakewell) to 2.3:1 (Chaunu). Given the presence of galena and other lead-silver compounds in the Cerro de San Pedro deposit, mercury consumption was probably lower than the norm. See Peter Bakewell, "Registered Silver Production in the Potosí District, 1550–1735," *Jahrbuch für Geschichte Lateinamerikas* 12 (1975): 85; Pierre Chaunu, *Seville et l'Atlantique (1504–1650). La conjoncture. Tome VIII 2:2* (Paris: SEVPEN, 1959), 1976–77.

23. Kenneth Brown "Worker's Health and Colonial Mercury Mining at Huancavelica, Peru," *The Americas* 57:2 (April 2001): 467–96; Robins, *Mercury, Mining, and Empire.*

24. J.O. Nriagu, "Mercury Pollution from the Past Mining of Gold and Silver in the Americas," *Science of the Total Environment* 149 (1994): 167–81;

Julio A. Camargo, "Contribution of Spanish-American Silver Mines (1570–1820) to the Present High Mercury Concentrations in the Global Environment: A Review," *Chemosphere* 48 (2002): 53–54. See also N. Pirrone et al., "Historical Atmospheric Mercury Emissions and Depositions in North America Compared to Mercury Accumulations in Sedimentary Records," *Atmospheric Environment* 32 (1998): 929–40.

25. Camargo, "Contribution of Spanish-American Silver Mines," 53.

26. *Relación de Antonio de Mendoza a Luis de Velasco* (1550–1551), in *Los Virreyes Españoles en America durante el Gobierno de la Casa de Austria. Mexico 1*, ed. Lewis Hanke, Biblioteca de Autores Espanoles 273 (Madrid: Atlas, 1976), 40; Cramaussel, "Sociedad colonial y depredación ecológica," 99.

27. He was referring specifically to the districts of Zacatecas, Guanajuato, and Pachuca. Alexander von Humboldt, *Political Essay on the Kingdom of New Spain*, Vol. 3, trans. John Black (London: Longman, Hurst, Rees, Orme and Brown, 1814), 280.

28. Studnicki-Gizbert and Schecter, "Colonial Fuel-Rush," 103. For Chihuahua, see Cramaussel, "Sociedad colonial y depredación ecológica," 100. For comparison, see G. Hammersley's work on early modern English iron smelting, "The Charcoal Iron Industry and Its Fuel, 1540–1750," *Economic History Review* 26 (1973): 605.

29. Felipe Castro Gutiérrez, "El liderazgo en los movimientos poplares de 1766–1767," in *Organización y Liderazgo en los movimientos populares Novohispanos,* ed. Felipe Castro Gutiérrez et al. (México, DF: UNAM, 1991), 203–18; Noblet B. Danks, "The Labor Revolt of 1766 in the Mining Community of Real del Monte," *The Americas* 44:2 (October 1987): 143–65.

30. Horace Marucci, "American Smelting and Refining Company in Mexico, 1900–1925" (Ph.D. diss., Rutgers University, 1995), 41.

31. Stephen Haber, *The Politics of Property Rights: Political Instability, Credible Commitments, and Economic Growth in Mexico, 1876–1929* (New York: Cambridge University Pres, 2003), 36 ff.

32. J. M. G. del Campo, *Reseña del Mineral del Cerro de San Pedro Octubre 31, 1881, San Luis Potosi,* folder 1289, p. 17, Collection of Latin American Manuscripts, Yale University; Walter Harvey Weed, "Notes on Certain Mines in the States of Chihuahua, Sinaloa and Sonora, Mexico," *Transactions of the American Institute of Mining Engineers* 32 (1903): 406.

33. Marucci, "American Smelting and Refining Company in Mexico," 35.

34. This amount was produced by 255 operating mines. See ibid., 28.

35. At Cerro de San Pedro the grades, or metal-to-ore ratios, were as follows: average 2,300 grams of silver per metric ton (Ag/t) of ore; high 14,394 grams Ag/t; low: 120 grams Ag/t. For gold, the low-end cut-off grade was 1.5 grams Au/t.

36. William E. French provides a good overview in his "Mining and the State in Twentieth-Century Mexico," *Journal of the West* (1988).

37. Dennis Kortheuer, "Santa Rosalia and Compagnie du Boleo: The Making of a Town and Company in the Porfirian Frontier, 1885–1900" (Ph.D. diss., University of California, Irvine, 2001), 71, 81.

38. Antonio García Cubas, *Cuadro geográfico, estadístico, descriptivo é histórico de los Estados Unidos Mexicanos* (México, DF: Oficina Tipografica de la Secretaría de Fomento, 1884), 181–232.

39. Santiago Ramírez, *Apuntes para un proyecto de codigo de mineria: presentados al señor Ministro de Formento General D. Carlos Pacheco* (México, DF: Oficina Tipografica de la Secretaría de Fomento, 1884).

40. Richard Chism, "A Synopsis of the Mining Laws of Mexico," *Transactions of the American Institute of Mining engineers 32* (1902): 5, 8, 40. Compilations of total areas of claims for the twentieth century are in Haber, *Politics of Property Rights,* 267, 277. The total area under concession was 446,000 hectares.

41. The breakdown of these investments is as follows: pre-1880, $1 million; 1886, $2 million; 1888, $3 million; 1892, $55 million; 1901, $101 million; 1911, $249 million. See Daniel Cosío Villegas et al., *Historia Moderna de Mexico. El Porfirato. Vol. 8, parte 2: La Vida Económica* (México, DF: Hermes, 1965), 1091, 1103, 1132–33.

42. Charles Bunker Dahlgren, *Historic Mines of Mexico: A Review of the Mines of That Republic for the Past Three Centuries* (New York: privately printed, 1883), 199.

43. R. F. Manahan, *Mining Operations of American Smelting and Refining Company in Mexico* (typeset report, 1948), pt. 2 (copy at University of Texas at El Paso library); Thomas O'Brien, "Copper Kings of the Americas: The Guggenheim Brothers," in *Mining Tycoons in the Age of Empire, 1870–1945: Entrepreneurship, High Finance, Politics and Territorial Expansion,* ed. Raymond E. Dumett (Farnham, UK: Ashgate, 2009), 195–98, 201–4.

44. Coal began to replace charcoal at Guanajuato in the 1880s. Margaret Rankine, "The Mexican Mining Industry in the Nineteenth Century with Special Reference to Guanajuato," *Bulletin of Latin American Research* 11 (January 1992): 43.

45. Marvin D. Bernstein, *The Mexican Mining Industry, 1850–1950: A Study of the Interaction of Politics, Economics, and Technology* (Albany: State University of New York, 1964), 42, 43, 159.

46. Christopher Schmitz, "The Rise of Big Business in the World Copper Industry, 1870–1930," *The Economic History Review* 39 (August 1986): 402.

47. Bernstein, *Mexican Mining Industry,* 41.

48. Helmut Waszkis, *Mining in the Americas: Stories and History* (Cambridge: Woodhead, 1993), 42–43.

49. This estimate is based on comparable figures from other smelters at the time, including, for example, the ASARCO smelters at El Paso (1,386 kg of lead [Pb] per day, 680 kg of zinc [Zn] per day, and 1.3 kg of arsenic [As] per day) and Tacoma, Washington (270 kg of Pb/day and 876 kg of As/day). See Senator Eliot Shapleigh, *ASARCO in El Paso: Moving to a Brighter Future—Away from a Polluted Past* (Austin: Senate of the State of Texas, 2007), 8, 10; Hilman C. Ratsch, *Heavy-Metal Accumulation in Soil and Vegetation from Smelter Emissions* (Corvallis, OR: National Ecological Research Laboratory, 1974), 1. The variations are mainly due to the chemical composition of the ores and not to any significant difference in emissions control techniques.

50. Leiticia Carrizales's team found Pb and As levels ten times higher than national norms, over 5,000 milligrams of arsenic per kilogram of soil and, for lead, over 4,000 mg/kg. See Carrizales et al., "Exposure to Arsenic and Lead of Children Living near a Copper-Smelter in San Luis Potosí, Mexico: Importance of Soil Contamination for Exposure of Children," *Environmental Research* 101 (2006): 1–10; and F. M. Romero, M. Villalobos, R. Aguirre, and M. E. Gutiérrez, "Solid-Phase Control on Lead Bioaccessibility in Smelter-Impacted Soils," *Archives of Environmental Contamination and Toxicology* 55 (2008): 567–68.

51. Antonio Aragón-Piña et al., "Influencia de emisiones industriales en el polvo atmosférico de la ciudad de San Luis Potosí, México," *Revista Internacional de Contaminación Ambiental* 22 (2006): 8, 10.

52. Fernando Díaz-Barriga, Lilia Batres, et al. "Exposición infantil al plomo en la zona vecina a una fundición de cobre," Unpublished ms. report (San Luis Potosí: Laboratorio de Toxicología Ambiental, Facultad de Medicina, Universidad Autónoma de San Luis Potosí, 1994), 8; Carrizales et al., "Exposure to Arsenic and Lead."

53. Walter Harvey Weed, "Notes on Certain Mines in the States of Chihuahua, Sinaloa, and Sonora, Mexico," *Transactions of the American Institute of Mining Engineers* 32 (1903): 434.

54. Arsenic concentrations in Zimapán's water supply exceed the World Health Organization drinking water standard by a factor of ten. M. A. Armienta, C. R. Rodriguez, L. K. Ongley, et al., "Origin and Fate of Arsenic in a Historic Mining Area of Mexico," *Trace Metals and Other Contaminants in the Environment* 9 (2007): 473–98.

55. A. Carrillo-Chavez, E. Gonzalez-Partida, O. Morton-Bermea, et.al., "Heavy Metal Distribution in Rocks, Sediments, Mine Tailings, Leaching Experiments, and Groundwater from the Mineral de Pozos Historical Mining Site, North-Central Mexico," *International Geology Review* 48:5 (May 2006): 466–78; I. Razo, L. Carrizales, et.al., "Arsenic and Heavy Metal Pollution of Soil, Water, and Sediments in a Semi-Arid Climate Mining Area in Mexico" *Water, Air, and Soil Pollution* 152 (February 2004): 129–52.

56. L. K. Ongley, L. Sherman, A. Armienta, et.al. "Arsenic in the soils of Zimapán, Mexico," *Environmental Pollution* 145 (February 2007): 795–96; Ana Judith Marmolejo-Rodríguez et al., "Migration of As, Hg, Pb, and Zn in Arroyo Sediments from a Semiarid Coastal System Influenced by the Abandoned Gold Mining District at El Triunfo, Baja California Sur, Mexico," *Journal of Environmental Monitoring* 13 (2011): 2182–89.

57. F. M. Romero, M. A. Armienta, M. E. Gutierrez, and G. Villaseñor, "Geological and Climatic Factors Determining Hazard and Environmental Impact of Mine Tailings," *Revista Internacional de Contaminacion Ambiental* 24:2 (2008): 52–53.

58. A wonderful history and ethnography of post-Revolutionary mining cooperatives in Guanajuato is developed in Emma Ferry, *Not Ours Alone: Patrimony, Value and Collectivity in Contemporary Mexico*(New York: Columbia University Press, 2005).

59. Servicio Geológico Mexicano, *Transición de la minería mexicana de oro y plata* (México, DF: SEG, 2011), 12.

60. Gavin Bridge, "Mapping the Bonanza: Geographies of Mining Investment in an Era of Neoliberal Reform," *Professional Geographer* 56 (2004): 407–8; Fernando Sánchez Albavera, Georgina Ortiz, and Nicole Moussa, *Panorama minero de América Latina a fines de los años noventa* (Santiago de Chile: CEPAL, 1999); David Szablowski, *Transnational Law and Local Struggles: Mining Communities and the World Bank* (London: Hart, 2007).

61. Adriana Estrada and Helena Hofbauer, *Impactos de la inversión minera canadiense en México: Una primera aproximación* (México, DF: FUNDAR, 2001).

62. André Lemieux, "Canada's Global Mining Presence," in *Canadian Minerals Yearbook, 1995* (Ottawa: Natural Resources of Canada, 1995).

63. Other players include American corporations (13%), British (1.8%), and Swiss (1.3%). Servicio Geológico Mexicano, *Anuario estadístico de la minería mexicana 2010*, 19.

64. Anthony Evans, *An Introduction to Economic Geology and Its Environmental Impact* (Malden, MA: Blackwell Science, 1997), 22, 24; MICLA, "Mexico Projects, 2008–2010," unpublished database, McGill-Queens University, 2010.

65. Alain Deneault and William Sacher, *Paradis sous terre: Comment le Canada est devenu la plaque tournante de l'industrie minère mondiale* (Montreal: Éditions Écosociété, 2012); Alain Deneault, Delphine Abade, and William Sacher, *Noir Canada: Pillage, corruption et criminalité en Afrique* (Montreal: Éditions Écosociété, 2008); Darryl Swearngin, Richard Tremblay, and Jack Silverson, "Home Base for Chilean Mining Ventures: Canada versus United States—Some Tax Considerations," *Canadian Tax Journal* 46 (1998); Bonnie Campbell, *Canadian Mining Interests and Human Rights in Africa in the Context of Globalization* (Montréal: Centre international des droits de la personne et du développement démocratique, 1999).

66. Compare this figure with the $1.6 billion raised from the London Stock Exchange or $41 million from NYSE for 2005. Market capitalization for the entire listing of mining companies on the TSX amounts to hundreds of billions of dollars.

67. Goldcorp Inc., *Peñasquito Project Technical Report: Concepción del Oro District, Zacatecas State, México* (December 2007–February 2008), 16.

68. S. G. Lasky, "How Tonnage and Grade Relations Helps Predict Ore Reserves," *Engineering and Mining Journal* (April 1950), 81–85.

69. Goldcorp Inc., *Peñasquito Project Technical Report.*

70. "Barrick Gold Bids to Acquire Cadence Energy for C$350M," CP Newswire, July 14, 2008, archived at http://www.resourceinvestor.com/2008/07/13/barrick-gold-bids-acquire-cadence-energy-c350m (accessed December 18, 2016).

71. Environment Canada, *National Inventory Report: Greenhouse Gas Sources and Sinks in Canada* (2008), http://www.ec.gc.ca/Publications/default.asp?lang=Enandxml=-492D914C-2EAB-47AB-A045-C62B2CDACC29 (accessed February 18, 2010).

72. William Sacher, *Cianuro. La cara tóxica del oro. Una introducción al uso del cianuro en la explotación del oro* (Quito, Ecuador: OCMAL, 2010), 22–25;

Robert Moran, "Cyanide Uncertainties" (Washington, DC: Mineral Policy Center, 1998).

73. H. Kempton and D. Atkins, "Delayed Environmental Impacts from Mining in Semi-arid Climates," *Proceedings from the Fifth International Conference on Acid Rock Drainage,* vol. 2 (May 2000): 1299–1308. See also H. Kempton, Thomas Bloomfield, Jason Hanson, and Patty Limerick "Guidance for Identifying and Effectively Managing Perpetual Environmental Impacts from New Hardrock Mines," *Environmental Science and Policy* 13 (October 2010): 558–66.

74. Karl Marx, *Capital: A Critique of Political Economy,* Vol. 1, trans. Samuel Moore and Edward Aveling (Moscow: Progress Publishers, 1971), 711.

CHAPTER 2

Reconstructing the Environmental History of Colonial Mining

The Real del Catorce Mining District, Northeastern New Spain/Mexico, Eighteenth and Nineteenth Centuries

ANTONIO AVALOS-LOZANO AND
MIGUEL AGUILAR-ROBLEDO

Often nothing is left of ancient societies but stones. Likewise, little is left of the social world of Mexican mining in the eighteenth and nineteenth centuries except the landscapes left behind. As this case study shows, the environmental effects of mining have indeed persisted long after mining ceased. The causes, though hidden in the past, can be inferred from methodically observing the clues that survive in landscape "palimpsests." Landscapes are dynamic systems characterized by their hysteresis,[1] which implies that their functioning and structure store information about the history that shaped them. A careful reading of the make-up of the flora, the structure of vegetation communities, the archaeological remains, the nature of local bio- and geochemical cycles, and the distribution of microclimates and eroded areas can provide information on the environmental history of Mexican mining.

This chapter offers a reconstruction of the environmental history of mining in the Real de Catorce Mining District—located in the southern part of the Chihuahuan Desert and overlapping the current Wirikuta protected area—in northeastern Mexico, from the 1770s through the 1820s, that is, from the final decades of colonial rule through the struggle for independence from Spain.[2] Based on a methodology that combines tools borrowed from paleoecology, ecology, economic history, environmental history, environmental chemistry, process engineering, and climatology, this chapter analyzes the relations between industrial metabolism and the structure, composition, and dynamics of the landscapes in this region.

More specifically, by studying the balance of matter-and-energy, or the "industrial metabolism" of the Real de Catorce Mining District, much can be recovered from the past. This method allowed estimating the amount of silver produced, the type of minerals processed, the quantities mined, and the technologies and processes used, as well as estimates of the energy sources and inputs for every facet of the production process.[3]

Further, based on documentary research, remote sensing, and field surveys, it is feasible to track commercial routes, the place of origin of inputs and materials, and the volume and destination of residual output. These data also permit inferences concerning the impacts that the metallurgical-mining system had on ecosystems and its influences on the composition and dynamics of landscapes in the southern Chihuahuan Desert. The peculiarities of the industrial metabolism in the Real de Catorce metallurgical-mining systems help us understand the complexity of Mexican metallurgical processes.

This chapter is divided into six parts: first, we sketch the history of metal mining in Mexico and in the Real de Catorce, from colonial times through the early nineteenth century; second, we offer an overview of the structure of the Mexican silver industry; third, we examine the mining operations and processing of silver ores; fourth, we turn to the formation of mining landscapes in the Real de Catorce Mining District; fifth, we offer some descriptive results from our study; and sixth, we offer our conclusions.

HISTORICAL OUTLINE OF SILVER EXTRACTION IN MEXICO

From 1561 to 1640, the production of Mexican silver increased constantly, rising in gross terms from 1,004 metric tons in the period 1561–1580 to 1,764 metric tons in the period 1621–1640.[4] After 1640, production declined as miners had to dig deeper to find rich ore. The factors that explain the fall in production can be summed up in one phrase concerning financing: "the mines, whose sterility is regretted for lack of loans and of capital."[5] Poorer ores limited investment, and limited investment further hamstrung silver production.

Fausto de Elhuyar, who served for thirty years as general director of mines in Mexico, wrote in 1825 that "in the eighteenth [century] the government began to open its eyes, and to recognize the wrong path it had until then followed."[6] According to one of the foremost historians of Mexican mining, David Brading, this new outlook on the part of Spain's imperial authorities produced a "revolution in government" that translated into a series of changes in administration known as the Bourbon Reforms (1763–1810). Brading considers the reforms to be one of the driving forces shaping mining and metallurgy in the late colonial period, one that revolutionized the structure of silver production in Mexico.[7] The innovations associated with the Bourbon Reforms spurred the economy, producing an unprecedented mining bonanza. The production of Mexican silver, which in the 1710s was 1,781 metric tons, rose to 5,985 metric tons by the 1790s.

However, this economic shake-up did not affect all the regions involved in the Mexican mining industry equally. Its effects were strongly felt in the fourteen mining towns (*asientos de minas*) that already contained 95 percent of Mexican silver production. This leads us to suggest that the ultimate causes of the great mining bonanza of Bourbon Mexico have been misidentified. It is conceivable that the Bourbon "revolution in government" was not the fundamental cause of the increase in silver production, but simply one of its consequences. The reforms certainly nurtured the bonanza in a virtuous circle, but were not the original source of the phenomenon. To attribute the enormous vigor of Bourbon mining primarily to the Crown's reforms amounts to putting the cart before the horse. It was the increasing power, negotiating skill, and volume of business of the mining aristocracy that forced the government to introduce reforms. Even though it is difficult to fathom the complexity of the processes involved, the Bourbon Reforms can hardly be envisaged in the absence of a group of entrepreneurs with sufficient clout and capital to steer imperial policy in directions they preferred. They harbored daring plans of investing in scientific and technological development, including the financing of new mining and smelting techniques out of their own pockets, as well as pursuing new prospecting explorations. This led to the discovery of Catorce, a profuse *mineral* (or ore-bearing district) that in 1803 produced 92 metric tons of pure silver, which amounted to 16 percent of the total production of silver that year in Mexico.

The powerful miner-merchant families sent various lobbying missions to Spain and came to an agreement with the minister of the Indies, José de Gálvez, regarding the restructuring of the industry. With the support of some members of the government they stripped the laborers of a substantial part of their ore share (*partido*) and obtained exemptions from the payment of various duties. In this period, defined by a veritable reconquest of Mexico undertaken against the Creoles (Mexican-born whites), the reform-minded Crown found a group of essential allies in the rich merchants of Peninsular origin (Spanish-born). By the end of Bourbon rule, a handful of families had obtained, as a reward for their loyalty, control of fourteen of the most important mining districts in the country.

Several families dominated Mexican mining, notably the Borda and Fagoaga families at the national level and the Obregon y Parrodi family in Catorce.[8] They were businessmen, founders of gigantic agro-industrial complexes.[9] These family businesses were the greatest companies of their time in Mexico, characterized by the concentration and integration of their operations in mining, agriculture, and textiles.[10] These companies were the true driving force of the Mexican economy and *provoked* crucial aspects of the Bourbon Reforms, overcoming the inertia into which mining had fallen.[11] Under their influence Mexico underwent an industrial revolution that rationalized the idiosyncrasies of mining and smelting and reconfigured the industry's social fabric. The Mining Court and School of Mines (Real Tribunal de Minería and Real Seminario de Minería), two institutions that were to become pivotal in the life of independent Mexico, were founded during this time, in the 1790s. The methods of extraction and processing of minerals were standardized. Several emblematic works of infrastructure were built. Old abandoned mining towns were reclaimed. The facilities built were so sophisticated that they allowed the reprocessing of tailings, grease, and slag.[12]

An important aspect of this revolution was the transformation of labor relations. The *partido* was reduced or eliminated,[13] while some arrangements that had not been used for a hundred years, such as the *repartimiento de indios* (forced assignment of Indian labor) and the *leva* (forced recruitment of idlers) were put back into practice.[14] The Crown had liberally resorted to forced labor throughout the long decline in population in Mexico, from the 1520s to about 1650. But with the population recovery well under way by the early eighteenth century, both practices had diminished to the point of insignificance. Then, late in the colonial era, in view of the new labor demand, they were reinstated by the Crown, which authorized them explicitly in the mining code of 1783.[15]

The main losers of this makeover were mine and hacienda laborers, as demonstrated by the multiple riots and the first recorded strike in

North America, staged by the laborers of the Vizcaína mine in 1766. But little of this wealth really benefited the country, and none was even used to alleviate the suffering of Mexican society under emergency conditions. In 1784 more than 8,000 laborers—22 percent of the national mining labor force—died of hunger in the mining town of Guanajuato, the richest in Mexico, and more than 300,000 people died throughout the country because of the loss of maize crops as a consequence of frost.[16] Alexander von Humboldt, who visited in 1803–4, was correct when he claimed that "Mexico is the country of inequality."[17]

The conditions for mine workers and mining business only worsened with the uprising of 1810, which led to the end of colonial rule. For more than a decade armies crisscrossed the country, and instability undercut the economy. In 1821, after declaring independence, Mexico was bled of capital withdrawn from the country by the Spaniards. Real de Catorce, the second most important mining district and the district that had maintained operations despite the War of Independence, was abandoned, dragging down agriculture and commerce.[18] The situation of the country was so pressing that in 1822 the Regency government issued a proclamation decreasing taxes on silver production from 13 to 3 percent. The monopoly of the colonial mint, the Casa de Apartado (facility where the gold was separated from the silver), was lifted, and the miners became free to separate gold from silver where and when they wished. Imports of quicksilver (mercury) were declared duty-free and miners were supplied with blasting powder at cost.[19] The last important shift came in 1823 when those articles of the mining ordinances of 1783 were abrogated that barred foreigners from investing in Mexican mining or metallurgy.[20] This decree allowed foreigners to form associations with Mexican citizens for the purpose of exploiting mines and smelting metal ores.

The British avidly took advantage of these reforms. They formally recognized Mexican independence and established diplomatic relations in 1825. They went on to invest more than 12 million pesos in only three years (1823–1826), through seven companies, notably the Anglo Mexicana Company in Catorce. This was the precursor to a farce that culminated in one of the worst financial disasters the City of London had faced in two hundred years.[21]

One company failed immediately, before even gaining a foothold in Mexico. The other six did establish themselves, though two failed by 1827. By the end of 1848, only the United Mexican Company survived, and in June 1849 it surrendered its interests to a company of Mexican investors and disappeared for good. The disaster was predictable, and its reasons obvious. When investing in Mexico, the English had not taken into account the effects of the prolonged War of Independence, the consequent dispersal of specialized workers, and the destruction of the country's infrastructure. While those problems could perhaps have been solved, what had no remedy was the English contempt for Mexican experience and technology, the fruit of three hundred years of mining on the part of Mexicans.[22]

THE STRUCTURE OF THE MEXICAN SILVER INDUSTRY

In the late nineteenth century, Santiago Ramírez, a mining engineer and official at Mexico's College of Mines, sketched the geology of Mexican silver. Roughly three thousand mines existed, working four or five thousand veins of silver or gold.[23] Almost all of the deposits were found in veins. Silver ores were divided according to their matrix into metallic and earthy categories. The former included minerals of lead (galena) and copper (fahlerz, arsenical copper), blendes (sphalerite), and pyrites (pyrite and chalcopyrite). The latter were all those with a matrix of quartz, clay, iron oxide, or limestone.[24]

Most Mexican mines' shallow veins had ores disseminated in ferruginous earth that the Europeans called the "iron cap," or *gossan*. The ores were also called *podridas* (rotten) because they were easy to break up. In Mexico these oxidized ores were called *colorados* (reds) and owed their name to the iron oxides that imparted a reddish color to them. Normally, the *colorados* contained chlorides, bromides, iodides, and sulfides. Native silver was also found. The *colorados* generally contained low grades of silver ore, but were profitable because they were easy to dig and amalgamated well in perhaps the most common reduction process used, the patio process—in which silver ore was crushed; mixed with water, mercury, and other substances; and then dried. In Catorce, where the amount of *colorados* was extraordinary but the weather was cold and inconvenient for the patio process, other reduction processes were deployed.[25]

At greater depth the majority of Mexican veins consisted of reduced minerals such as sulfides, in particular pyrite, and other sulfosalts, in a matrix or gangue that was hard and resistant. Their excavation was very expensive. These ores were called *negros* (black), and from a metallurgical point of view they were considered very "unruly." Ramírez states, as do other authors, that the grade of the silver ore increases with the depth of the vein.[26] For this reason, Mexican miners endeavored to establish mining operations at the greatest possible depth.

The amount of *colorados* in Mexico was enormous,[27] and in 1843, they contributed seven-eighths of the silver produced.[28] Real de Catorce had an important advantage over other mining districts in that most of its mines extracted oxidized ores, which needed less investment for extraction and refining. Smelting was unimportant in Catorce. On the whole, Mexican silver ores were of disappointing quality. José Garcés y Eguía, a nineteenth-century Mexican chemical and mining engineer, calculated that the amount of ore treated in Mexico was no less than ten million hundredweights per year (460,250 metric tons/year), yielding 690 metric tons of silver (the quantity that, with some variations, was minted in Mexico in the last years of the eighteenth century).[29] The average yield would thus have been 2.4 ounces per hundredweight, or 1.49 grams of silver per kilogram of ore.[30] Elhuyar, for his part, estimated that the Mexican ores yielded on average three to four ounces per hundredweight, or 1.87 to 2.49 grams of silver per kilogram of ore.[31] Other authors arrived at similar figures.[32] These figures show that even though the grade usually increased with depth, ordinary Mexican mines were poor on average. Garcés y Eguía asserted that rich minerals (appropiate for smelting) contain at least 60 grams of silver for every kilogram of ore.[33]

Although the grade of silver ore extracted in Mexico did not change substantially between 1570 and 1884, some mines in Catorce represented notable exceptions that contained high grades of ore. For example, the Mina del Padre Flores in its first year of operations produced as much as 46 metric tons (worth 1.6 million pesos) from rich minerals containing more than 100 grams of silver per kilogram of ore.[34]

MINING AND PROCESSING SILVER ORES

By the late eighteenth century, silver mining was an elaborate technical process. Mexican miners had little difficulty recognizing where best to look for silver. Once the veins of metal were discovered, exploitation began. Mining operations spanned everything from hacking the ore from underground veins to loading mules that would carry it to the warehouse. Mining silver required thirty different specialized tasks.[35] Considering that the grade usually increased with depth and that most Mexican veins were "inclined," the miners' work began with the construction (*cuele*) of a vertical shaft (*tiro*) that would converge with the vein at an approximate depth of 8 *varas* (6.68 m).[36] The shafts were usually rectangular, but octagonal or hexagonal ones were also found.[37] In 1732, José de Sardaneta introduced the use of *tacos* in Guanajuato for such construction activities.[38] This technique, used in Europe since 1613,[39] became indispensable and allowed the construction of shafts as deep as the one at La Valenciana, which in 1810 reached 531 meters below ground. Once the shaft reached deep enough, horizontal galleries (called *laboríos*) were built alongside the vein.[40] As mandated by the ordinances, pillars of solid rock were left in order to support tunnels. All of the operations aimed at providing access to the mine were referred to as works of *disfrute* (benefit).

As the depth of shafts increased, new difficulties and costs arose. The eighteenth-century Mexican lawyer and intellectual Francisco Javier de Gamboa noted: "It is the waters that are the major peril of the mines. [. . . W]hen the veins of the mines are picked at, the water springs as does the blood from the body. [. . .] Of this curious law of physics only the mine owners experience the effect, watching their galleries [. . .] filling with more water the deeper they are."[41] At the outset, the miners solved this problem using a mule-drawn capstan or hoist called a *malacate* to drain the mines.[42] Up to eight such capstans were placed in octagonal shafts in some mines where large volumes of water had to be removed. But some of the earliest colonial mining operations, where topography permitted, already preferred to use crosscuts, i.e., tunnels that would cut across the vein, connect the different shafts, and extract the water from the galleries by gravity.

John Percy, the father of English metallurgical literature, wrote in the preface to his *Metallurgy of Silver and Gold* (1880): "From all branches of metallurgy one that has silver as an object, is the most extensive and most varied and more complicated."[43] In the case of Mexico, the problems were even more complex, as the silver extracted from Mexican mines was of "various minerals."[44] In nature, silver can be found in native (metallic) form or combined with other substances. The silver can be an essential or a minor element in the different combinations of a mineral compound. In both cases, however, miners considered the compound a silver ore.[45]

It was crucial to establish as precisely as possible the nature of the bond that united silver with other elements, as this determined the choice of the method of processing the ore.[46] Besides the type of mineral compounds present, their assay value (*ley*) must also be considered, or the quantity of metal per unit weight that they contained, on which the commercial value of the silver would depend.

The refining of silver in Mexico involved more than eighteen different specialized tasks.[47] Different ores required different procedures.

Costly mistakes could be made at each step. In 1640, the distinguished chemist and mining engineer Alonso Barba explained: "To add quicksilver [mercury, or *azogue*] to the metal that requires fire is to lose it. To toss into the furnace what is not for smelting, is to hinder, damage, and do nothing; and even within the boundaries of what is for quicksilver, and what for fire, there are differences and degrees of ease in processing."[48] Writing 160 years later, the German metallurgist Federico Sonneschmid (a member of the late-eighteenth-century German mission sent to Mexico by the king of Spain) confirmed Barba's position on the indispensability of expertise: "Not all the minerals that contain silver are suitable for processing using the patio process of mercury amalgamation. The worker must [. . .] know [. . .] its qualities."[49] Sonneschmid's contemporary Garcés y Eguía agreed on this point while dissenting on some specifics.[50] For example, the former found pyrite-bearing minerals to "resist processing with quicksilver," while the latter found them to be suitable for the patio process, based on mercury amalgamation. In their time Sonneschmid and Garcés y Eguía were the most enlightened experts in the art of metallurgy. They perfected or invented various processing techniques. Both had gained their outstanding experience in the everyday activities of Mexican mines and haciendas.

The medley of techniques and procedures confused Humboldt to such a degree that he came to consider Mexican metallurgists and mining engineers truly inept. Actually, it was Humboldt, despite his familiarity with mining and chemistry, who did not grasp the subtleties of such a difficult art, in which one had to take into account variable climatic conditions, the great chemical and physical diversity of the ores to be processed, and the varying quality of the available supplies. The master smelters and *azogueros* (quicksilver specialists) were indeed driven by necessity to use different processing techniques to achieve the same goal, but this was not due to ineptitude. These workers drew on the experience they acquired during work, as well as the preparation they were given by the Real Colegio de Minería, which certified them. As Humboldt observed:

> The above separation of minerals [. . .] into those suitable and unsuitable for the patio process, seemed necessary to me in order to convey a general idea [. . .] but [. . .] at every step one shall find examples of unsuitable [. . .] minerals being profitably processed in the patio and [. . .] of other [. . .] suitable ones [. . .] from which not all silver content is extracted. These variable circumstances have different causes. A compact mineral [. . .] must react differently to the processing in a different way than one crushed [. . .] into fine particles [. . .] the very nature of the gravel exerts at times an influence, too.[51]

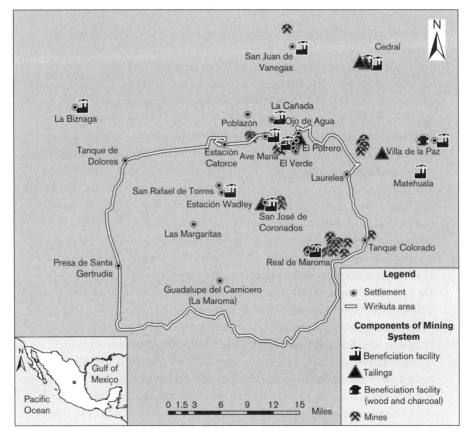

FIGURE 2.1. The Real de Catorce mining district, San Luis Potosí, New Spain/Mexico, eighteenth and nineteenth centuries. Map by José de Jesús Izaguirre Hernández.

The same types of minerals had at times such a different composition that two samples of the same ore, obtained in the same mining town, could react differently to processing.

THE FORMATION OF MINING LANDSCAPES IN THE REAL DE CATORCE DISTRICT

The mining district of Real de Catorce is located in the Altiplano region of the current state of San Luis Potosí. It includes the municipalities of Catorce, Cedral, Villa de la Paz, and portions of the municipalities of Charcas, Matehuala, and Villa de Guadalupe covered by the Natural Protected Area of Wirikuta. The study area is located in the northern part of the Altiplano, and in the southeastern quadrant of the ecological region known as the Chihuahuan Desert.[52] (See figure 2.1.)

Mining activities have been carried out in the area since the end of the eighteenth century. A large strike occurred in 1778.[53] From this moment onward, Catorce increased its production, reaching the point where it became the second most important colonial mining district, distinguished by a complex industrial metabolism that affected the composition, structure, and dynamics of the landscapes that sustained it in complicated ways.

By consulting primary sources, such as documents stored in the Mexican General Archive of the Nation and the Historical Archive of the San Luis Potosí State, we identified 124 mines that operated in the area and period under study, and the minerals obtained in each mine. The industrial facilities that operated in Wirikuta at the end of the eighteenth century and beginning of the nineteenth century were field-checked and mapped: besides the 124 mines, there were 79 processing haciendas with 644 mills (tahonas) and 248 cazos (amalgamating mills heated from below); 7 haciendas devoted to stock raising and crop cultivation; 2 estates for fattening cattle; 3 centers that supplied wood and charcoal; and 7 locales for the final disposal of residues.

In 1782, the mining town of Real de Catorce had 12 haciendas processing ore by smelting and in vats (148 tahonas, 58 vats): 4 in Ojo de Agua (70 tahonas, 25 vats); 1 in La Carbonera (24 tahonas, 13 vats); 7 haciendas processing in the patio, in vats, and by smelting in Cedral. In 1785, Matehuala had 28 haciendas processing in vats and with quicksilver, 13 by smelting, and 9 by galemes de mano (low-heat blast furnace).[54]

Sixty-six mines operated in Real de Catorce in 1794. The population of Real de Catorce rose to 15,000 inhabitants.[55] Since on-site processing was becoming difficult due to the scarcity of water at the end of the eighteenth century, the majority of processing haciendas were established nearby in the Cañada de los Catorce, Laureles, Vanegas, Potrero, Cedral, and Matehuala.[56] At that time, Real de Catorce had one smelting establishment with 9 furnaces and many small smelters (zangarros), whereas Los Catorce had 2 patio process haciendas, El Potrero had 4 and 4 zangarros, and El Cedral, the most important locality for processing, had 9 haciendas using the patio process and cazos and 11 using smelting.[57]

In 1800, the number of haciendas in Matehuala reached 14. Due to transportation costs, it was much less expensive to build haciendas near the mining town. It cost 4 pesos to transport one load of ore to

San Luis Potosí (212 kilometers), while it cost one peso to Matehuala (85 kilometers).

The twentieth-century Mexican historian Montejano calculated that mines of Real de Catorce produced 104 million Spanish ounces of silver between 1773 and 1827 (each Spanish ounce was 28.7 grams; thus 2,990 metric tons).[58] However, as in the rest of the country, the grade of the silver from Catorce mines was quite low: San Agustín produced 14 ounces per load, San Gerónimo 4, La Purísima and Santa Ana each 3.5, and the rest between 1.5 and 2. Some mines in Catorce represented a notable exception: for instance, Padre Flores produced 129 ounces per load. Considering that the average yield of Catorce ores was 7.2 ounces per load and each load was equal to 37.26 kilograms (104 million ounces divided by 7.2 ounces per load is equal to 14.44 million loads, multiplied by 37.26 kilograms), the quantity of ore extracted in Catorce between 1773 and 1827 was 538,200 metric tons.[59]

It appears that at the beginning of the nineteenth century the composition of the ores extracted in Real Catorce began to change. Between 1778 and 1803, extraction focused on the *colorados,* the "the weathered near-surface deposits, with metal in its native state (native silver and chlorargyrite)."[60] Later on, miners turned to the *negro,* or pyrite- and copper-bearing, ores. It is possible that the processing (i.e., industrial) activities—and therefore the industrial metabolism, frontier relations, and patterns of ecological disturbance—underwent modifications required by changes in the ores being processed. The shift to deeper ores, for example, likely led to an increase in charcoal demand for smelting and the formation of toxic residues such as sulfur dioxide, arsenic trioxide, and lead.

In 1803 the annual output of the mines of Real de Catorce was 92 metric tons of pure silver, which amounted to 16 percent of all Mexican silver production (575 metric tons per year). If we take into account the percentages of the silver processed in San Luis Potosí and the portion corresponding to Catorce, we arrive at the following numbers: silver obtained by amalgamation, 98.39 percent; silver obtained by smelting, 1.61 percent. Between 1773 and 1827, 529,534.98 metric tons of ore were processed by amalgamation, and 8,635.02 metric tons by smelting.

Silver mining and processing spawned a suite of associated businesses needed to provide supplies. In the mining towns of Real de Catorce and La Maroma, most of the processing was in amalgamating mills heated from below (*cazos*) and *fondón de a caballo* (a giant rotating metal plate over a fire, driven by a horse), because of the type of minerals and because the cold weather made the patio process ineffi-cient.[61] Different processes were used in Cedral, Matehuala, and San Bartolo due to the different mineral composition of the ores mined in each place. The variety of processing techniques called for various sup-plies. In addition to fuel, food, and fodder for miners and their animals, water and wood for construction were also needed. In this way, mining encouraged the development of other economic activities in Real de Catorce, such as agriculture and commerce.

Food, for example, was increasingly grown locally. The reason was simple. According to Humboldt, in 1803 the average price of a *fanega* of maize grown in Salamanca (Guanajuato state) varied as follows: where it was produced, in Salamanca it cost fourteen *reales* (one peso and six *reales*), in Querétaro (at a distance of 84 kilometers from Salamanca) two and a half pesos, and in San Luis Potosí (at a distance of 178 kilo-meters from Salamanca) four and a half pesos.[62] Humboldt did not men-tion the cost of maize in Real de Catorce, but it can be imagined, because Real del Catorce is 349 kilometers from Salamanca. Given that necessity is the mother of invention in every industry, the high price of imported food awakened the ingenuity of the inhabitants. "Men began to till the soil in the valleys and slopes of the surrounding mountains," Humboldt noted, "and everywhere where the rock was covered with mulch. Farms were established in the vicinity of mines to remedy the shortage of food and the considerable price reached by all agriculture products."[63]

Not only food became more expensive. The price of other supplies indispensable for the operation of the industrial system rose too: hides, goat tallow, *ixtle* and *pita* (fibers derived from yucca and agave), char-coal, liquor. This led to the development of various types of haciendas in the vicinity devoted to production of these and other items. They sometimes used sophisticated irrigation techniques, such as what in the Middle East are called *qanats* (known in Spanish as *galerías filtrantes*), underground channels carved through rock. These haciendas helped make the goods needed to sustain mining operations more affordable than they could have been if imported from afar. Moreover, the discov-ery in 1561 of the salt flats of Peñón Blanco, located between Zacatecas and San Luis Potosí, and their subsequent exploitation guaranteed the supply of salt at Real de Catorce at modest prices. Salt—some 78,000 metric tons between 1773 and 1827—was used in silver processing.[64]

The impact that the mining and smelting industry had on the land-scapes of the mining district of Real de Catorce in the eighteenth and nineteenth centuries has long interested observers and scholars.[65] Most

authors on the subject claim that the ecosystems suffered terrible degradation that affected the entire district and provoked the transformation of
grassland into scrubland. All of this speculation draws on a catastrophic
perspective, based on an analysis carried out in 1827 by H. G. Ward, an
agent of a British mining firm. Ward described the Sierra de Catorce in
1827 in strong terms: "There is not a single tree or a single blade of grass
in the vicinity; and yet fifty years ago the district was covered by forest.
[. . .] Whole forests were burned to clear the land and timber larger than
required for the mines is brought from a distance of twenty-two leagues
[92.18 kilometers]."[66] Ward based this assessment on reports from the
colonial commissioner Silvestre López Portillo, written in 1779 (fifty-four
years earlier) in which he described the region, in particular the land close
to the San Bartolomé River, located 30 kilometers from Real de Catorce:
"[Along the] San Bartolomé river, which springs and disappears in the
most interior and gullied parts of these mountains, [. . .] there is an
impenetrable oak- and pinewood of such imponderable corpulence [. . .]
that they may be compared with the mountains."[67]

But a careful reading of the report allows us to understand that Ward
was wrong in his interpretation of López Portillo's document. The truth
is that the commissioner described a place on the eastern slopes of the
Sierra, 30 kilometers away from Real de Catorce, which even today
supports a thriving oak and pine forest located on the eastern slope, a
wet adiabatic and more fertile zone. Ward did not even know the area
and mistook the drier northwest slope for the more humid eastern slope.

Ward was not the only one to draw hasty conclusions about vegetation history in the vicinity of Real de Catorce. More recently, scholars,
including González-Costilla, Giménez de Azcárate, García, and Aguirre
Rivera, state on the basis of secondary sources: "Thus, before the onset
of mining operations, in the last quarter of the 18th century, the Sierra
was almost uninhabited and covered by luxuriant forests, which supplied abundantly the haciendas and mines. [. . .] Fifty years on, close to
1825, there was neither tree nor scrub left."[68] However, a detailed analysis of their main source, a study by Montejano of the mines of Real de
Catorce, allows us to locate only isolated references to the subject, so
scant that they make impossible any reliable inference about the vegetation of the Sierra.[69] Montejano only cites others, who in turn repeat
Ward's statement from 1827. Those descriptions clearly impel us toward
the easy conclusion that the mining and smelting processes developed in
Catorce provoked, in only fifty-four years, a complete loss of the productive potential of the surroundings.

But the reality could have been quite different. Other sources provide more information, such as the reports of Silvestre López Portillo and Bruno de Ureña of 1779;[70] those of Ward himself on the San Cristóbal plain, written in 1826; and those of Robert Phillips from 1826.[71] From these and other sources it is possible to put together the structure and composition of the landscapes of Wirikuta at the beginning of the nineteenth century.

Four essential documentary sources, from 1779, 1779, 1822, and 1826, allow us to reconstruct local landscapes in the Real de Catorce area. Francisco Bruno de Ureña, a charted surveyor, covered the San Cristóbal plain, to the west of the Sierra de Catorce, in 1779, describing the vegetation. On the basis of the common names used in the document, it has been possible to reconstruct the vegetation communities present. From this, we infer the presence of microphyllous scrub in the San Cristóbal plain, with the presence of *Flourensia cernua* and *Larrea tridentata*.[72] In 1779, Silvestre López Portillo described the forests of the eastern slope of the Sierra up to the piedmont: "There is wood of all sizes, in some parts pines, in others oaks, as well as various other kinds, so that it is very abundant in fuelwood, and in mesquite for charcoal. Though not present in the Sierra itself, they abound on its [lower] slopes."[73]

Nearly fifty years later, Ward described the northern part of the same San Cristóbal plain, mentioning that a microphyllous scrub dominated: "[A] small shrub [. . .] which cannot be other than *la gobernadora*, and mesquites and dwarf palms (Yucca), with a fruit that is not very different from the dates."[74] Bustamante, an early-nineteenth-century Mexican scientist and influential scholar, wrote in April 1824: "An herb called *gobernadora* was found there [at Catorce]. It is abundant in the whole Sierra and is very resinous, giving a very vivid flame."[75]

These observations from men who had surveyed the region with their own eyes support the conclusion that the mining and smelting processes developed in Catorce between 1779 and 1827 did not provoke the loss of the region's productive potential. Further, the cited documentary evidence shows that the northwest slope of the Sierra de Catorce had dry-climate vegetation before the discovery of the mines, that the effects of industrial activities were superimposed on vegetation changes of a different nature provoked by a series of climatic phenomena—alternating cycles of drought and heavy rainfall, accompanied by heavy flooding. These natural changes aggravated the industrial impact. Several sources attest to some extreme climate events in the region and in Mexico as a whole toward the end of the eighteenth century.

HEALTH AND ENVIRONMENTAL PROTESTS IN THE
CONTEXT OF SILVER PROCESSING

In consulting archival documents, we have discovered social movements focused on the health threats posed by metallurgical processes. For example, in 1827, when Juan Kidell of the Catorce Company asked the mining district authorities (*diputación de minería*) to intervene because the town council of Cedral had forbidden his company to "burn" (remove sulfur from) the minerals extracted at Sereno, the council argued that smoke from his two furnaces harmed the local residents. The mining district officials appealed to the governor, accusing the council of weak-mindedness and described its position as a case of "the most vile ingratitude and shamelessness that the residents should complain about the harmful qualities of the smoke."[76] This incident ended with the shutdown of the Catorce Company.

Documents from the period reveal two opposed positions about health threats caused by metallurgic processes: an official stance held by scientists with ties to the Spanish Crown; and another, which we call "independent." One example of the official position is found in the writings of Sonneschmid, the German scientist hired by the Crown but whose research was financed by one of the great mining families of Mexico, the Fagoagas. Sonneschmid wrote: "It is noteworthy in the entire kingdom that the patio process is not a harmful operation for peons working in it, and [. . .] there would be no need for its mention, were it not for many Europeans who have let themselves be persuaded that [. . .] the refining of minerals destroys [the health of] an immense number of inhabitants."[77] However, Sonneschmid contradicted himself when he wrote about the process of distilling mercury, a necessary precursor to silver refining by amalgamation. He noted that the clay pots used to hold mercury often broke and put workers in grave danger. Referring to damage to the nervous system from mercury exposure, he wrote that he "found a number of people who [. . .] have *azogado* [shaking; original emphasis]. However from this they recovered fully, only retaining a slight tremble in their extremities."[78]

Sonneschmid, even when he admitted the existence of harmful health effects of working with mercury, sought to minimize them. Humboldt took a similar position:

> Around five to six thousand people are working in mineral amalgamation or the preceding processes. A great number of these [. . .] spend their life barefoot on mounds of ground metal [. . .] mixed with [. . .] oxidized mercury [. . .] and

it is curious to see that [. . . they] have the best health. Physicians [. . .] unanimously declare that afflictions of the nervous system are rarely seen.[79]

Other knowledgeable authors saw the risks of mercury as genuine. Gamboa wrote of "frequent diseases [. . .] poisonous foundries, and *azoguerías:* incurable, and everywhere illnesses, amidst moisture, fire and vapors."[80] José Antonio Alzate, an early-nineteenth-century Mexican scientist and influential scholar, referred to "the *azogue*'s poisonous fumes."[81] Contemporary observers remained divided about the health effects of processing silver with mercury.[82]

Notwithstanding divided opinion on mercury, authorities in the late eighteenth century understood the principle of pollution when it came to drinking water. Legislators of the time possessed enough information to point out its dangers, as proven in 1783 by the *Reales Ordenanzas* [. . .] *del Importante Cuerpo de la Minería de Nueva-España* (Royal Ordinances [. . .] of the Important Mining Corps of New Spain [Mexico]):

> Title 13. Of the supply of water and provisions for mining. Article 1. Given that drinking water is of first necessity in the mining districts [. . .] I order [. . .] that water infected with mineral particles shall not be used. Article 2. I prohibit with utmost rigor the diversion of waters from the drains [. . .] of the patios of haciendas and smelting furnaces into arroyos or aqueducts that carry it toward settlements.[83]

Humboldt's testimony implies that the law was ignored, at least in some places.[84] He relates that some of the inhabitants of Guanajuato drank the same water used for washing (in smelters) without suffering health impairment, despite the Royal Ordinances.

Although laws might be ignored, sometimes courts ruled in favor of communities claiming damage from mining operations, as demonstrated by the case recorded in Pachuca (modern-day Hidalgo State) in 1764. Pablo Aparicio was running a smelting furnace whose smoke and dust harmed his neighbors. According to the neighbors, the smoke killed their animals. On top of this, they claimed, the smelter polluted the nearby river with residues. Aparicio requested that the authorities ask the neighbors to buy back his facilities, but the final verdict obliged the accused to remove the furnaces, clean up the river, and leave the people in peace.[85]

On other occasions, however, community resistance to smelting haciendas was defeated. In a case from Chihuahua in 1732, authorities ultimately ruled in favor of the mining interests. Two mining entrepreneurs began constructing new furnaces at two sites. A group of local residents soon lodged a formal complaint before the city council (*cabildo*). The movement demanded halting the construction of the furnaces, which were located less than 170 meters from dwellings, and a commitment by the *cabildo* to prohibit the building of any furnaces in the environs of the city, because of the potential risks posed by their use. The complaint claimed: "No one should be compelled to put his life in grave danger coming [to prevent] the harm that others may suffer, and less still if it is [only] to augment someone's income and profits; for better reason still [the interests] of an individual should not be put ahead of those of the health and well-being of a community that should watch and procure its own preservation." One of the entrepreneurs appealed immediately, citing the tax revenue that his furnaces would bring to the Crown. The result, after many detours, was the approval of the haciendas and the defeat of the movement, which ended up opposing an accomplished fact, as the construction of the furnaces was furtively completed while the government consulted the opinion of physicians and scientists.[86]

THE INDUSTRIAL METABOLIC COMPLEX IN THE REAL DE CATORCE DISTRICT

A few calculations and estimates help to corroborate the impressions gleaned from the textual evidence concerning the influence of the mining and smelting industrial system in the formation of local landscapes. Let us begin with wood. Between the 1770s and 1827, a total of about 118 square kilometers (km²) of forest were cleared for the following activities: 85.39 km² were used in the making of charcoal for smelting; 10.66 km² were consumed as fuelwood in fifty-five processing haciendas; and 21.97 km² were used for domestic consumption. These calculations were carried out using the data for silver production reported by Garcés y Eguía and taking into account the yield of wood and its dry weight (830 kg/m³) according to the numbers used by Salazar.[87] The yield from a hectare of twenty-five-year old forest growing on poor soil was also factored into our calculations. This quantification does not include the wood required for the construction of ground supports, piles, and other artifacts, which from very early on was brought from distances exceeding 100 kilometers, as indicated by Ward.[88]

Ground truthing and the analysis of satellite imagery revealed that the estimated cleared area matches closely a zone that today stands out due to intensive erosion and desertification. Figure 2.1 shows the

relationships between the different vegetation communities, the identified and mapped elements of the mining and smelting system, and the
desertified areas. In 1827, Manuel Mier y Terán implied large impacts
from mining's industrial activities on the pine- and oakwood communities when he wrote that out of a total of 8,000 inhabitants of Real de
Catorce, 5,750 devoted time to hauling fuelwood and charcoal to be
sold in the city or at the silver haciendas.[89]

It is also possible to estimate the number of animals needed to produce silver in the Real de Catorce region. In an average year, some
10,000 beasts of burden, mainly mules, worked in the mines and in ore
processing—especially the patio, *fondón de a caballo,* and the rotation
of the horse capstan. These animals consumed an estimated 14,675
metric tons of maize annually. Roughly 8,000 hides of young bulls were
used annually, as well as 57.5 metric tons of tallow rendered from
100,000 goats. More than 4,000 sheep and 400 cattle were slaughtered
each year for human consumption. The herds that supplied the region,
had they roamed freely, would have required more than 5,000 square
kilometers of pasture.

Finally, in addition to the wood and animal requirements of silver
mining and processing, one can estimate the pollution legacy at Real de
Catorce.[90] As noted above, on the basis of the calculations derived from
the balances of matter and energy, we have determined that between
1773 and 1827 more than 529,534.98 metric tons of ore were processed by amalgamation and 8,635.02 metric tons by smelting.[91] Taking
as a reference the average grade of 7.2 ounces of silver per 37.26 kilogram of ore, we can estimate that a half million metric tons of tailings
containing lead, arsenic, antimony, and silver were released into local
ecosystems.[92] So were more than 125,000 metric tons of salt, between
6,000 and 37,500 metric tons of copper sulfate, and 1,250 metric tons
of mercury.[93] All of these pollutants are still lodged in different environmental matrices in the region.

CONCLUSION

From 1770 to 1827, the deforestation, erosion, and disturbance of the
Wirikuta landscape affected mostly the northern part of the Sierra de
Catorce. These disturbances were not exclusively a consequence of mining and industrial activities. Climatic anomalies (intense droughts followed by violent rainfall) occurred at the end of the eighteenth century
and magnified the industrial impact.

It is evident that the authorities and broad sectors of society were aware of the environmental and health effects of silver mining and processing. People living close to silver-refining industries complained of damage to their health, and authorities in colonial Mexico sometimes sided with them and other times with the mining interests. It is clear that many people found the evidence for ill health effects, especially of silver processing, convincing. Indeed, Mexican mining and metallurgical engineers altered technologies in ways that reduced health risks, though perhaps motivated mainly by a desire to conserve mercury. Nonetheless, power distribution inequalities allowed risky activities, ignoring demands on behalf of the possible victims.

During the late eighteenth and early nineteenth century, the emphasis of Mexican institutional policies rested mainly on selfish and short-term economic interests. Even though authorities in late colonial Mexico often observed cautionary principles when regulating the mining sector, the solution to social problems arising from the operation of potentially dangerous installations still had to answer to power relations. Support given by institutions to research caused some experts to be at the service of large capitalist interests or to be driven by cost-benefit criteria, privileging economic growth over community safety.

The incredible bonanza in Mexican silver production from the late eighteenth century to the insurrection of 1810 constituted an insurmountable barrier to the implementation of cautionary policies. These implementation efforts could not withstand the force of financial imperatives. The colonial government, and the independent one that followed after 1821, needed revenue desperately and saw the silver-mining industry as its best chance to obtain it. So both governments focused on drawing investors into mining almost at any cost. Accordingly, the health of ecosystems and human populations came in a distant second among governmental priorities. In these respects, the history of mining at Real de Catorce and in Mexico generally before 1827 resembled that elsewhere in North America in the centuries yet to come.

NOTES

The authors wish to thank the Multidisciplinary Graduate Program on Environmental Sciences, the Faculty of Social Sciences and Humanities, and the Faculty of Agronomy and Veterinary, Autonomous University of San Luis Potosí, for all their support for the research reported in this chapter; we also thank the anonymous referees for their insightful comments and suggestions; the editors, J. R. McNeill and George Vrtis, who generously helped to improve this chapter all the way from the first to its final version; Aleksander J. Borejsza, who translated the Spanish manuscript into English; and José de Jesús Izaguirre Hernández, who drafted figure 2.1. Of course, although these institutions and persons share any merit this chapter might have, the authors stand alone for its shortcomings.

1. Hysteresis is the character of a system that becomes manifest after disturbance. Even though the cause of the disturbance is no longer there, the system does not return to its original state. It stores a memory of its history, and of the path of disturbance.

2. Strictly speaking, Mexico was called the Viceroyalty of New Spain from its inception in 1521 until 1810, but for the sake of brevity we will call it Mexico throughout.

3. *Miners* refers to owners of mines; *operatives* to workers. This was the meaning of these terms in the mining industry of the study period.

4. Francisco de Elhuyar, Memoria sobre el influjo de la minería en la agricultura, industria, población y civilización de la Nueva-España en sus diferentes épocas, con varias disertaciones relativas a puntos de economía pública conexos con el propio ramo. Madrid, México: Imprenta de Amarita. Consejo de Recursos Naturales No Renovables. Ed. Facsimilar. (1964) [1825]: 47.

5. Francisco Javier de Gamboa, Comentarios a las Ordenanzas de Minas. México: Imprenta Díaz De León y White México. Miguel Angel Porrúa. Ed. Facsimilar. (1987) [1761]: 3.

6. Elhuyar, Memoria sobre el influjo de la Minería, 53.

7. D. Brading, Mineros y comerciantes en el México borbónico (1763–1810). (R. G. Ciriza, Trad.) México: Fondo de Cultura Económica. (2004): 57, 201.

8. F. Langue, Los señores de Zacatecas. Una aristocracia minera del siglo XVIII novohispano. México: Fondo de Cultura Económica. (1999): 50.

9. S. Sánchez, La minería novohispana a fines del periodo colonial. Una evaluación historiográfica. Estudios de Historia Novohispana (27), 123–64: 132.

10. F. Langue, Los señores de Zacatecas, 114.

11. Elhuyar, Memoria sobre el influjo de la Minería, 49.

12. Sánchez, La minería novohispana, 131–32.

13. "In Guanajuato, Catorce, Zacatecas and Real del Monte, once the worker completed his daily quota of mineral, called *tequio,* he would get from 50 to 30% of the additional ore extracted during the rest of the day" (Brading, Mineros y comerciantes, 202). The proportion varied in other mining towns. Debt peonage was common and prompted several laws that limited loans to four to eight months of wages (P. J. Bakewell, Minería y sociedad en el México Colonial Zacatecas, 1546–1700. México: Fondo de Cultura Económica. [1997]: 177).

14. S. Sánchez, La minería novohispana, 131–32.

15. Brading, Mineros y comerciantes, 201.

16. Alejandro de Humboldt, Ensayo Político sobre el Reino de la Nueva España (Séptima ed.). México: Editorial Porrúa. (2004): 48–49, 251.

17. Ibid., 68.

18. H. G. Ward, México en 1827. México: Fondo de Cultura Económica. (1995): 346.

19. Ibid.

20. R. Randall, Real del Monte. Una empresa minera británica en México. México: Fondo de Cultura Económica. (2006): 41.

21. Ward, México en 1827, 357–58.

22. Ibid., 357.

23. Santiago Ramírez, Noticia Histórica de la Riqueza Minera de México. México: Oficina Tipográfica de la Secretaría de Fomento. (1884): 65.

24. L. Carrión, Metalurgia por vía seca del plomo, plata, cobre, mercurio y oro. Pachuca: Tipografía del Gobierno del Estado. (1900): 35.

25. Ramírez, Noticia Histórica de la Riqueza Minera de México, 79.

26. Ibid., 80.

27. Humboldt, Ensayo Político sobre el Reino de la Nueva España, 341.

28. S. Duport, Métaux Précieux au Mexique Considérée Dans ses Rapports Avec La Géologie. La Métallurgie et L'Économie Politique. Paris: Chez Firmin Didot Frères, Libraires. (1843): 29.

29. Joseph Garcés y Eguía, Nueva Teórica y Práctica del Beneficio de los Metales de Oro y Plata. México: Imprenta de Díaz de León y White. (1873): 105–8.

30. Humboldt, Ensayo Político sobre el Reino de la Nueva España, 341.

31. Ramírez, Noticia Histórica de la Riqueza Minera de México, 36.

32. Duport, Métaux Précieux au Mexique, 143–44.

33. Garcés y Eguía, Nueva Teórica y Práctica del Beneficio de los Metales de Oro y Plata, 78.

34. Humboldt, Ensayo Político sobre el Reino de la Nueva España, 359.

35. Doris Ladd, The Making of a Strike. Mexican Silver Workers' Struggles in Real del Monte, 1766–1775. Lincoln: University of Nebraska Press. (1988): 7.

36. One *vara* equals 0.836 meters (Humboldt, Ensayo Político sobre el Reino de la Nueva España, CXLIV).

37. Francisco de Sarría, Ensayo de Metalurgia o descripción por mayor de las catorce materias metálicas, del modo de ensayarlas, del laborío de las minas, y del beneficio de los frutos minerales de la plata. México: Impreso por D. Felipe de Zúñiga y Ontiveros. (1784): 86.

38. *Tacos* in this context were paper cartridges filled with black gunpowder and tied together with *ixtle* fiber and sealed with bentonite. Ladd, The Making of a Strike, 10.

39. Daubuisson, cited in Humboldt, Ensayo Político sobre el Reino de la Nueva España, 366.

40. Sarría, Ensayo de Metalurgia, 86.

41. Gamboa, Comentarios a las Ordenanzas de Minas, 353.

42. Ramírez, Noticia Histórica de la Riqueza Minera de México, 622.

43. Percy quoted in H. Collins, Metallurgy of Lead & Silver, Part I. Lead. Edited by W. C. Roberts-Austen, K. C. B., D. C. L., F. R. S. London: Charles Griffin & Company, Limited. (1900): v.

44. Humboldt, Ensayo Político sobre el Reino de la Nueva España, 337.

45. Collins, Metallurgy of Lead & Silver, Part I, 19.

46. Ibid.

47. Ladd, The Making of a Strike, 7.

48. Barba, Arte de los Metales en que se Enseña el Verdadero Beneficio de los de Oro y Plata por Azogue. El modo de Fundirlos Todos, y como se han de

refinar, y apartar unos de otros. Madrid: En la oficina de la Viuda de Manuel Fernández, Casa C. Bermejo. Ed. Facsimilar. (1932) [1770]: 73.

49. F. Sonneschmid, Tratado de la amalgamación de Nueva España. México: Sociedad de Exalumnos de la Facultad de Ingeniería, UNAM. Ed. Facsimilar. (1983) [1825]: 54–56.

50. Garcés y Eguía, Nueva Teórica y Práctica del Beneficio de los Metales, 77–78.

51. Sonneschmid, Tratado de la amalgamación de Nueva España, 54–57.

52. Onésimo González-Costilla, Relación entre Bioclima y Vegetación en la Sierra de Catorce y Territorios Adyacentes (Altiplano Norte del Estado de San Luis Potosí, México), Tesis doctoral. Universidad Computlense de Madrid, Facultad de Farmacia, Departamento de Biología Vegetal II. (2005): 1.

53. Humboldt, Ensayo Político sobre el Reino de la Nueva España, 359.

54. G. Palmer, Real de Catorce: Articulación Regional, 1770–1810. San Luis Potosí, SLP: El Colegio de San Luis. (2002): 84.

55. Rafael Montejano y Aguiñaga, El Real de Minas de la Purísima Concepción de los Catorce, SLP. (Tercera ed.). San Luis Potosí, SLP: Academia de Historia Potosina, AC. (1974): 43.

56. Ibid., 173.

57. Ibid., 174.

58. Ibid., 172.

59. Humboldt, Ensayo Político sobre el Reino de la Nueva España, 341.

60. Modesto Bargalló, La química inorgánica y el beneficio de los metales en el México prehispánico y colonial. México, DF: Facultad de Química de la Universidad Nacional Autónoma de México. (1966).

61. A variation on the method of hot (vat) amalgamation developed in Catorce in the last decade of the eighteenth century by Miguel de Aguirre. It employed an enormous copper disk that rotated on a set of ovens, propelled by a horse.

62. Humboldt, Ensayo Político sobre el Reino de la Nueva España, 252.

63. Ibid., 238.

64. Montejano y Aguiñaga, El Real de Minas de la Purísima Concepción de los Catorce, 4.

65. Ibid.; Onésimo González-Costilla, Joaquín Giménez de Azcárate, José García Pérez, and Rogelio Aguirre Rivera, Flórula Vascular de la Sierra de Catorce y Territorios Adyacentes, San Luis Potosí, México. Acta Botanica Mexicana (78): 1–38. (2007); Palmer, Real de Catorce: Articulación Regional, 1770–1810. (2002).

66. Ward, México en 1827, 587.

67. Silvestre López Portillo, Documento de peritaje en el Real de Nuestra Señora de la Concepción de Guadalupe de Alamos, en catorze de Agto de mis setecientos setenta y nueve años, quoted in Primo Feliciano Velázquez, Colección de documentos para la historia de San Luis Potosí. (Prima Feliciano Velázquez, Comp.) (1987) [1779]: tomo 3, p. 480.

68. González-Costilla, Giménez de Azcárate, García, and Aguirre Rivera, Flórula Vascular de la Sierra de Catorce (2007): 3.

69. Montejano y Aguiñaga, El Real de Minas de la Purísima Concepción de los Catorce.

70. Velázquez, Colección de documentos para la historia de San Luis Potosí, 395, 481, 489.

71. Robert Phillips, Detalles de un viaje desde Altamira a Catorce (Serie Cuadernos 28 ed.). San Luis Potosí, SLP: Biblioteca de Historia Potosina. (1973).

72. Velázquez, Colección de documentos para la historia de San Luis Potosí, 395.

73. López Portillo, Documento de peritaje en el Real de Nuestra Señora de la Concepción de Guadalupe de Alamos, tomo 3, p. 481.

74. Ward, México en 1827, 607.

75. Carlos M. Bustamante, Carlos María, Diario Histórico de México, 1822–1848. México: Centro de Investigaciones y Estudios Superiores en Antropología Social, Colegio de México, 2003.

76. Archivo Histórico de San Luis Potosí, Secretaría General de Gobierno, 1826, February 14, 1827.

77. Sonneschmid, Tratado de la amalgamación de Nueva España, 94.

78. Ibid., 51.

79. Humboldt, Ensayo Político sobre el Reino de la Nueva España, 49.

80. Gamboa, Comentarios a las Ordenanzas de Minas, 463.

81. Alzate, Elogio Histórico del Sr. D. Francisco Javier de Gamboa Regente que fue de esta Real Audiencia de México. Gaceta de Literatura de México. 373–84. (1831): 380.

82. Luis Chávez, La situación del minero asalariado en la Nueva España a fines del siglo XVIII. México: UCPEET/STPS. (1987): 48–49.

83. Reales Ordenanzas para la Dirección, Regimen y Gobierno, del Importante Cuerpo de la Minería de Nueva España y de su Real Tribunal General. De orden de su Mejestad. México: Sociedad de Exalumnos de la Facultad de Ingeniería. Ed. Facsimilar. (1979) [1783]: 134–35.

84. Humboldt, Ensayo Político sobre el Reino de la Nueva España, 49.

85. Ladd, The Making of a Strike.

86. B. Hausberger, Una iniciativa ecológica contra la industria minera en Chihuahua. Separata de: Estudios de Historia Novohispana. (1993) [1732]: vol. XIII, pp. 116–34: 4–5.

87. Garcés y Eguía, Nueva Teórica y Práctica del Beneficio de los Metales; Guadalupe Salazar, Las haciendas en el siglo XVII en la Región Minera de San Luis Potosí. San Luis Potosí, México: Universidad Autónoma de San Luis Potosí. (2000): 38.

88. Ward, México en 1827, 587.

89. Manuel Mier y Terán, 1827, as quoted in Montejano y Aguiñaga, El Real de Minas de la Purísima Concepción de los Catorce.

90. Ward, México en 1827, 338, 375, 601.

91. Elhuyar, Memoria sobre el influjo de la Minería, 90.

92. Humboldt, Ensayo Político sobre el Reino de la Nueva España, 341, 372.

93. Garcés y Eguía, Nueva Teórica y Práctica del Beneficio de los Metales, 108.

PART TWO

Industrial Catalysts

CHAPTER 3

A World of Mines and Mills

Precious-Metals Mining, Industrialization,
and the Nature of the Colorado Front Range

GEORGE VRTIS

One of the more striking and difficult things about seeing a mine—or studying the environmental history of mining—is trying to grasp the sheer environmental tumult involved. In the largest precious-metals mines operating today, vast open pits can reach more than a mile in width and a half mile in depth. Along the edge of the open pits are waste-rock piles that can tower thousands of feet above the surrounding landscape and can grow by hundreds of thousands of tons a day. These are the newest mountain chains to be found anywhere on earth. Depending on the mineral reduction and concentration processes used, settling ponds (or tailing dams) are also often found nearby. These water bodies can take on the dimensions of small lakes, but beyond that superficial likeness, they bear little resemblance to their naturally occurring cousins. They are routinely filled with the toxic by-products generated during mineral processing, making their water so polluted that they have been known to poison the unlucky waterfowl that land on them. Less obvious but perhaps even more complicated to discern is the wind- and water-driven dispersal of pollutants, which can carry the effects of mining far beyond any mining district. Whether we turn to images, figures, or analytical assessments, the environmental repercussions involved in contemporary precious-metals mining are not easily comprehended.[1]

Nor were they easily grasped when industrial mining first began sinking its roots into the mountainsides of the American West in the middle of the nineteenth century. Though historical western mining districts were different in many ways from contemporary mines, numerous visi-tors from the 1840s onward struggled to come to terms with what they saw. One of those visitors was James Meline. In 1866, the former New York City journalist and Union Army colonel wandered his way west-ward from Fort Leavenworth to Colorado and then on to New Mexico, writing thirty-six letters along the way. Six of those letters focused on Colorado, and nothing captured Meline's imagination more than the high-country mining districts lying just west of Denver in the Front Range of the Rocky Mountains. There Meline gazed upon the ecologi-cal remnants of placer mining, noting the seemingly endless number of prospector's holes, trenches, and mounds of debris that marked every stream. "[N]ot one stone left upon another," Meline wrote, "not one where Nature put it." Scanning up the surrounding mountainsides where lode mining was now the center of attention, Meline described the mountains as being "in a shockingly bad state of affairs." "Trees and vegetation," Meline observed, "have long since disappeared. Holes, shafts, and excavations almost obliterate the original surface." A few days later, Meline seemed to try to sum up his views on what he had seen all across the Front Range: "Mining, from prospecting to smelting, is here, directly or indirectly, the 'all in all' of everyone's existence."[2]

Like Meline, environmental historians have been struggling to under-stand the world that mining created in Colorado and many other cor-ners of North America and the world. As the essays in this volume and earlier work in the emerging field of mining environmental history reveal, scholars have begun unearthing the complicated stories that weave together human societies, our growing commitment to mineral-dependent cultural formats, and the way these developments interface with the natural world. From these works, we have a growing sense of the interaction among some of the key scientific and technological developments, political and economic forces, and the cultural and eco-logical dynamics that have been shaping our extraordinary relationship with minerals and mines since the advent of the industrial age.[3]

In the essay that follows, I draw on and extend aspects of the enviro-technical perspective that some scholars have recently used to analyze one of the largest and most valuable precious-metals mining regions in nineteenth-century North America, the Colorado Front Range (see figure 3.1).[4] Between the mid-1860s and the end of the century, the Front Range was transformed from a peripheral frontier outpost into a modern industrial mining region in ways that dramatically recast the natural world. The often degrading environmental results, I argue, were

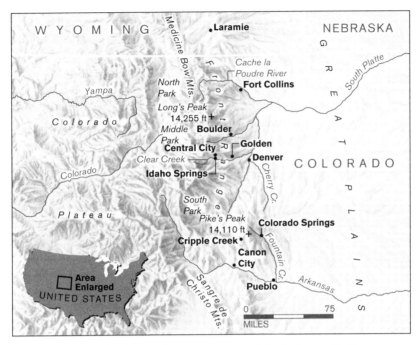

FIGURE 3.1. The Colorado Front Range, 1860–1900. Map by Jerome Cookson.

manifestations of interlocking scientific, technological, economic, and cultural processes that were just then coming together in the form of industrialization and reshaping mining landscapes all across North America. But to see only environmental change and ruin is to miss a critical feature of this story. These same dynamics, I further suggest, began to seep into social and political consciousness in Colorado and at the national level, giving rise to an emerging eco-cultural pensiveness, to legislative and cultural developments, and to new concerns about the superabundance of the earth and scarcity that can further deepen our understanding of the American conservation movement.[5]

To pursue this argument, I begin with a brief assessment of Front Range precious-metals mining as it grew increasingly industrialized following the gold rush years from 1858 to 1864. Then, I turn to a detailed assessment of the enormous environmental transformations these developments set loose, using both traditional historical and scientific sources. In the chapter's final section, I examine the way these environmental changes stirred sociopolitical concerns and reactions during the American conservation movement.

THE SHAPE OF INDUSTRIAL MINING

Beginning in 1868, mining entered a new, more complex phase of development across the Front Range as the region was pulled into the orbit of the industrializing world. Among the clearest signs of the new era were smelters, improved stamp mills, and railroads. They stood as both the symbols and the agents of the industrial world, and together signaled an important and complex break with the past. During the three decades following their nearly simultaneous introduction into the Front Range in the late 1860s and early 1870s, the spread of these and other scientific and technological developments—and the ways they interlocked with market dynamics, cultural changes, and the shift toward a new coal-powered energy regime—helped miners intensify production and environmental change on a scale never before seen in the region nor in much of North America at the time. Although the conventional view of America's nineteenth-century industrialization is often linked to eastern factory life, workplace conflict, new technologies, and the well-worn path from New England's textile mills to Andrew Carnegie's sprawling Pittsburgh steelworks, industrial activity also exerted a powerful influence on the natural world, especially in the resource-rich American West.[6] As smelters, stamp mills, and railroads led the Front Range down the industrial path, they helped miners fashion a radically different environment for all who followed.

The first decade of Front Range mining activity followed a pattern typical of frontier mining regions in North America. During the early boom years, from 1858 to about 1864, the easily worked surface deposits were exhausted, and the region gradually slipped into a depression as the world of the small-scale, largely independent gold seekers gave way to the much larger, more heavily capitalized corporate interests that began to control the industry after the mid-1860s. Although the passage between these two eras was marked by a massive influx of eastern and British capital, a consolidation of properties, new labor relations, the discovery of silver lodes, and a multitude of economic and social problems unleashed by both the Civil War and Great Plains Indian wars, its most important and difficult hurdle lay hidden deep within the ground.[7] Often less than ninety feet below the surface of the ground, miners began encountering the Front Range's so-called "rebellious ores." As existing stamp-milling and other reduction processes failed to recover an economically viable percentage of the gold and silver contained in these ores, production plummeted, mines and mills sat idle, and miners began a desperate search for new reduction methods that eventually led to one of the great watersheds in the region's nineteenth-century mining and environmental history.

In his 1870 report to Congress on the condition of the western mining industry, U.S. commissioner of mining statistics Rossiter Raymond characterized the years from 1864 to 1867 as dominated by a sort of "process-mania." "Upon the first failure of the stamp-mills," Raymond wrote, "people came to the conclusion that the ores must be roasted before the gold could be amalgamated. One invention for this purpose followed another; *desulphurization* became the Abracadabra of the new alchemists; and millions of dollars were wasted in speculations, based on the sweeping claims of perfect success put forward by deluded or deluding proprietors of patents."[8] Raymond then assessed twelve of the more familiar processes that had been or were currently being used in the Front Range. Most of them, he thought, were essentially useless and one—the Bartola process—was deemed so impracticable that he found it "difficult to reconcile the history of this invention with honesty on the part of the inventor." The last two on the list, however, were Nathaniel Hill's Boston and Colorado smelter and a more advanced version of the common stamp mill, both of which, Raymond observed, had already achieved some success and were busy revitalizing the industry.[9]

A former professor of chemistry at Brown University, Hill established the Front Range's first successful smelter at Black Hawk—the Boston and Colorado Smelting Company—based on the best scientific knowledge available in Europe. After examining the ores around Central City and making two trips to consult with metallurgists and mining engineers at the world-renowned metallurgy centers at Swansea, Wales, and Freiburg, Germany, in the mid-1860s, Hill returned to Colorado in 1867 and began constructing a small smelting operation modeled on the famous Swansea process he observed in Wales. At the center of the process were a series of reverberatory furnaces where Hill melted the ore in order to separate the precious metals from the worthless country rock that encased it. Although the process was too expensive for treating anything but the mines' first-class ore (the richest part of any vein), Hill's operations were extremely effective. Estimates suggest that from the time he began operations in 1868, Hill was able to save at least 95 percent of the assay value of the ore, an extraordinary accomplishment for the day since yields prior to Hill's smelter had dropped to as little as 10 percent of the ore's gold content.[10] As the editor of Denver's *Rocky Mountain News* reflected just four years after Hill had begun operations, the

Boston and Colorado Smelting Company had "done more, probably, than any one thing to re-establish confidence and build up our yield of the precious metals."[11]

At the same time that Hill was organizing his smelting operations, changes in stamp-milling technologies provided a more effective and economical method for treating the great bulk of the gold ore mined: the second-class ore. After several years of experimenting with California and Nevada prototypes—each "hailed," as Raymond put it, "as another Moses to lead us out of the wilderness"[12]—Front Range mill men turned sharply away from the advice of miners in other western regions and developed their own unique milling processes. In addition to lining the amalgamation tables surrounding the stamp batteries with quicksilver-coated copper plates, the real innovations of Front Range mill men focused on the production of a much more finely crushed ore. To this effect, they increased the depth of mortars, decreased the size of discharge screens, and slowed down the entire milling process. For example, compared to California stamp mills, which ran at about 90 to 105 drops per minute and crushed between 2.5 and 3 tons of ore a day, Front Range mills averaged just 30 drops per minute and crushed only about 1 ton of ore per day. The result of all these changes was promising. As the more finely crushed ore passed more slowly over the amalgamated copper plates, the loss of gold dropped substantially. From the estimated 75 to 90 percent thought lost by stamp mills in 1863, Raymond determined, the loss had been reduced to somewhere between 30 and 70 percent by 1869, while another government mining authority narrowed his estimate of the loss to between 40 and 50 percent.[13]

Though crucial developments, it is important to keep in mind that Hill's Boston and Colorado Smelting Company and the improved stamp mill were only the tip of a huge industrial mining technology iceberg that took shape after 1867. Rival smelters were soon erected in nearly every mining district, and each employed increasingly well trained European and American metallurgists and mining engineers who calibrated their processes to the region's various geological complexities. For instance, while Hill's Swansea process worked well on the copper and iron pyrites found across much of Gilpin and lower Clear Creek Counties, it was inadequate for reducing the rich silver-bearing lead deposits found near Georgetown and Empire in upper Clear Creek County in the mid-1860s. To process these ores, miners turned to blast furnaces and still other technologies. And after 1874, miners and mill men also began relying on the scientific and technological advances developed at the Colorado

School of Mines in Golden. Innovation was almost continuous. Pneumatic drills, electricity, more powerful steam engines, a better understanding of ores—over and over again, as a number of mining historians and historians of science have emphasized, advances in mining technology, metallurgy, and engineering allowed miners to overcome some difficult barrier and push the industry forward.[14]

Stitching together these new reduction processes and underpinning their long-term success was the nearly concurrent development of railroads. Between 1867 and 1874, six important railways began to serve the greater Front Range region. The first three created links to the East, effectively shrinking the 690 miles separating the Front Range from the nearest Missouri River valley settlements, thereby making an arduous monthlong journey into a relatively comfortable two-day train ride. In 1867, the Union Pacific reached Cheyenne, Wyoming, 106 miles north of Denver. That advent was soon followed by the completion of two other lines in 1870: the Denver Pacific, which connected Denver to the Union Pacific at Cheyenne; and the Kansas Pacific, which negotiated the Smoky Hill River valley from Kansas City to Denver. Taken together, these three lines linked the Front Range directly to the East and to national markets by rail for the first time.[15]

The other three railways were all principally local, narrow-gauge lines that originated in Denver and penetrated the mountains at various points. Constructed during the early 1870s, the Colorado Central, the Denver and Rio Grande, and the Denver, South Park and Pacific were engineering marvels that knitted the region together like modern highways. In addition to making transportation more feasible and reliable year-round, all of these railways combined to slash the cost of all commodities (from labor to machinery, food, and fuel) and reduced the price of moving ore around, thereby recasting the economics of mining and stimulating a reopening of shuttered mines.[16]

ENVIRONMENTAL UPHEAVAL:
MOUNTAINS INTO MINES

As smelters, more advanced stamp mills, and railroads combined to lift the region out of the depression that had gripped it since 1864, they mediated an even more complex relationship between miners and Front Range ecosystems than the one that had emerged during the gold rush era.[17] To the earlier worlds of rather simplistic placer and lode mining were added these more powerful engines of change, which accelerated the scale and rate of some environmental changes while initiating and modifying the course of others. As in the gold rush era, miners, mill men, and their supporting casts refashioned vast stretches of the mountains in their search for precious metals. Although placer mining would never again achieve a fraction of the importance it had between 1858 and 1863, small-scale operations continued to emerge from time to time and replumb waterways and destroy riparian habitat.[18] Far more significant, however, was lode mining. From 1868 to the end of the century, it contributed more than 99 percent of the total value of precious-metals production, and numerous contemporary observers wrote lucid descriptions detailing its far-reaching impact on the mountains, including this succinct assessment by a visitor to the Central City region in 1873: "Everything betokens the industry of the district. The mountains are spotted with dump piles and prospect holes [. . .] The people are a mining people, earning their worldly wealth, for the most part, by delving in the bowels of the earth."[19]

Much the same could have been written about any of the Front Range's other major mining districts in the later nineteenth century. As the newspaperman and traveler Robert Strahorn observed in the mid-1870s, more than 60,000 mines had already been opened in the region and "the din of the quartz mill, the drill and the blast echo night and day from a thousand mountainsides and mountain depths."[20] One way of gaining a general, though clearly imperfect, sense of the mining-related environmental changes that Strahorn and others observed during the post–gold rush era is by examining the production of precious metals. Although few nineteenth-century miners would have conceived of production statements in these terms, they can be used as a basic index of environmental change. With each bar of gold or silver added to a mining company's financial statements came a rearranging of local ecosystems, the effects radiating outward from the centers of production. In general, these changes became most pronounced around the oldest and most productive mining districts located in Gilpin, Clear Creek, and Boulder Counties; and after 1891, when a rich gold deposit was discovered in the throat of an extinct volcano in Teller County's Cripple Creek District, that region too was transformed (see table 3.1).

Next to the countless mine shafts, prospect holes, and mill buildings that littered every mining district, perhaps the most readily apparent environmental change that took shape during the industrial mining era was the proliferation of increasingly vast tailing piles and debris dumps (see figure 3.2). As new mines were opened and others were pushed to

TABLE 3.1   PRODUCTION OF PRECIOUS METALS (GOLD AND SILVER), COLORADO
FRONT RANGE MINERAL-PRODUCING COUNTIES, BY DECADE, 1858–1900

	1858–1869	1870s	1880s	1890–1900	Total
Arapahoe	—	—	$741	$6,453	$7,194
Boulder	$349,700	$5,080,690	4,820,420	6,944,976	17,195,786
Clear Creek	2,882,321	15,700,057	18,902,899	20,251,156	57,736,433
Douglas	—	—	1,495	779	2,274
Gilpin	14,318,130	21,246,416	18,602,357	18,672,447	72,839,350
Grand	—	—	—	6,913	6,913
Jefferson	—	—	4,495	39,837	44,332
Larimer	—	—	—	18,426	18,426
Park	2,580,000	3,794,706	3,591,570	2,065,778	12,032,054
Pueblo	—	—	—	652	652
Teller	—	—	—	76,909,834	76,909,834
Total	$20,130,151	$45,821,869	$45,923,977	$124,917,251	$236,793,248

NOTE: The data in this table is drawn from Charles W. Henderson, *Mining in Colorado: A History of Discovery, Development, and Production*, U.S. Geological Survey Professional Paper 138 (Washington, DC: U.S. Government Printing Office, 1926), 88–96, and compiled in George Vrtis, "The Front Range of the Rocky Mountains: An Environmental History, 1700–1900" (Ph.D. dissertation, Georgetown University, 2006), table A.1. All of the figures are unadjusted for inflation.

greater and greater depths, the amount of material removed from sub-surface shafts and tunnels and piled onto slopes and valleys increased enormously. Although the scarcity of finely detailed mining records makes it impossible to determine with precision the total amount of rock excavated, we can at least work backward and make a rough estimate. In 1875, for instance, as many of the Front Range's most productive mines were pushing beyond 500 feet in depth,[21] the region's lode mines produced a total of $1,641,402 (about 83,109 ounces) in gold, $2,506,841 (2,021,646 ounces) in silver, and $63,745 (280,815 pounds) in copper.[22] Since one-half to two-thirds of the total amount of rock excavated was gangue (the waste rock surrounding enriched veins and fissures) and even the richest first-class ores seldom assayed over 40 ounces of gold and silver and 60 pounds of copper per ton, that meant somewhere on the order of 1,968 pounds of waste rock (or 98 percent of every ton) were added to tailing piles for every ton excavated and subsequently treated by smelters, stamp mills, or other reduction processes.[23] In terms of the total production figures for 1875, then, the amount of waste rock amounted to at least some 104,000 tons for just one average year in the 1870s and 1880s.[24] From the 1860s onward, the shear mass of all of this material being redistributed around mountainsides and valleys destabilized numerous slopes, increasing the

FIGURE 3.2. The Silver Plume mining region, Clear Creek County, Colorado, 1870s. The discolored areas above the town are tailing piles and waste-rock dumps. Courtesy of William Henry Jackson Collection (Scan 10025699), History Colorado.

number of mass movements around mining districts and adding yet more sediment and debris to stream channels and aquatic habitats that had already been choked with waste rock during the gold rush era.[25]

The tons of ore that were piled onto slopes and precipitated landslides had their counterpart in the warren of shafts, drifts, and tunnels that took shape beneath the mountains' surface. In this underground world, miners altered the region's geology and hydrology in a number of significant ways. Digging subterranean caverns removed support from the overlying strata, sometimes causing them to sag or, in extreme cases, collapse into the underlying void. Though much more common in coal-mining operations, due both to the extraction methods used and the sedimentary geology involved, ground subsidence also occurred in hard-rock mining areas.[26] Mining also rearranged groundwater flows and even shifted flows between watersheds. The impressive four-mile-long Argo Tunnel (completed between 1893 and 1904) dug between Idaho Springs and the more elevated Central City, for instance, reconnected hydrological zones that had been separated for millions of years. And much as with surface water diversions, the Argo Tunnel reconfigured downstream ecological communities and habitats dependent on earlier flow regimes. Scores of other tunnels presumably produced comparable, if less well documented, hydrological and geographical changes.[27]

As miners constructed new geographies above and below ground, they also extended the process of deforestation begun during the gold rush era. Deeper mines meant more and heavier mine props, strengthened headframes and shafts, improved mill buildings, and an even larger consumption of wood for fuel. Below about 200 feet in depth, miners turned to wood-powered steam engines to power hoisting works, ventilation systems, and the huge pumps necessary to keep deep mines from filling up with groundwater. As the part-time historian and editor of the *Central City Herald,* Frank Fossett, observed, "No heavy mining work can be carried on without steam-power."[28] And, under the egalitarian prescriptions of local mining district codes and the 1872 federal mining law, each of these timber demands was multiplied thousands of times over.[29] Although some miners managed to consolidate their claims into larger companies and gain economies of scale by sinking a single shaft, most sunk and timbered their own shafts and outfitted their own mines, further accelerating local deforestation. In 1872, for instance, six companies owned and operated claims on the original 800-foot-long Bobtail lode in Gilpin County, the smallest being just 33 feet, 4 inches long. The well-known Burroughs lode was similarly crowded. Its 2,347 feet were divided into seventeen distinct claims, each unleashing a seemingly insatiable demand for timber.[30]

Smelting operations also propelled regional deforestation. For instance, an 1876 study of the Boston and Colorado Smelting Company shows that it consumed an estimated 37.8 cords of wood in its furnaces and other reduction processes every working day. Allowing a month of downtime for repairs or weather-related problems, that amounts to an annual consumption rate of about 12,660 cords, which can be visualized as a stack of wood four feet high and four feet wide, extending the nineteen miles from Black Hawk all the way down twisting Clear Creek Canyon to the piedmont city of Golden. Obtaining that much cordwood meant clear-cutting some 160 to 210 acres of forested mountainside.[31] When the timber consumption rates of all the other Front Range smelters, blast furnaces, and steam engines are considered, it is likely that miners consumed more wood for fuel during the last three decades of the nineteenth century than for any other reason.[32]

To these demands on the forest, however, must still be added those required by the construction of railroads. Beginning in the late 1860s, the construction of the region's railroads required massive amounts of timber for trestles, station buildings, and most of all, for the crossties that supported iron rails. On average, every mile of standard-gauge track required around 2,400 crossties, each measuring eight feet long, seven inches on the face, and at least seven inches in depth. The narrow-gauge railways that ran into the mountains used slightly smaller crossties, but more of them, sometimes laying as many as 3,000 per mile. Constructing the 106-mile-long Denver Pacific Railway between Denver and Cheyenne, for example, required 254,000 to 260,000 crossties. On a more comprehensive scale, one study of the crosstie industry estimates that Front Range forests supplied approximately 7 million ties in its first ten years alone. Considering that hundreds of miles of track were laid in the greater Front Range region between 1868 and 1900, that ties had to be replaced every four to seven years, and that the region's forests were also tapped to construct and maintain distant lines, it is clear that tie-cutting operations required vast numbers of mature trees.[33]

Not surprisingly, such enormous demands for timber leveled vast tracts of forested mountainside, creating islands of deforestation around mining districts, along waterways, and in other areas where small armies of lumbermen turned trees into crossties. In the early 1870s, for instance, a correspondent from the *New York Tribune* described the widespread deforestation from atop a summit overlooking Central City: "All the slopes and mountain tops were once covered with a heavy growth of pine timber, but for several miles around they have been cut away for the smelting of ores in Black Hawk and Central."[34] On his way north toward Boulder, he added a rare description of the relentless march of the lumbermen: "After five miles of travel we passed thousands of cords of wood by the roadside, which had been hauled from the mountain slopes on either hand, and we could see vast blocks taken out of the solid forest. Not many years can pass before all the timber within ten or fifteen miles of Central will be gone, and then ores must be taken to the coal at the foot of the mountains."[35] The correspondent's prophecy was soon proven true but not without some important variations. Although the forests surrounding Central City and other major mining districts were rapidly cleared during the mid-nineteenth century, more distant forests continued to thrive, and some even managed to survive the miners' onslaught altogether.

In the southern and central portions of the range, deforestation reached its greatest extent in the mixed ponderosa pine–Douglas fir forests that dominated the montane forest region (6,000–9,300 feet), where most mines, smelters, lumber mills, and crosstie-cutting camps were located. Since hauling wood up and down steep mountain slopes was an expensive and difficult proposition, lumbermen naturally focused their efforts on either the nearest or most easily transported sources. For the crosstie cutters, this meant harvesting stands as close as possible to the region's major waterways. The eight-foot-long crossties weighed about 100 pounds each, making river transport the most economical method for moving them downstream to the piedmont, where oxen teams hauled them to rail lines. Until railroads reached well into the mountains in the 1870s, most crossties were moved by water and oxen teams. In some areas, such as along Clear Creek or practically anywhere below about 8,000 feet, the combined effects of the lumbermen and crosstie cutters cleared nearly all of the timber. At higher elevations, however, the dense stands of Engelmann spruce and subalpine fir that stretched across the subalpine forest region (9,300–11,400 feet) attracted far less attention. Except for those near major mining districts, these vast forests largely escaped widespread logging and today contain stands characterized by three-hundred- and four-hundred-year-old trees.[36]

Further north, beyond Boulder County, the pattern changed significantly. Since the northern portion of the Front Range lies outside the Colorado mineral belt and far from major mining districts, most of the logging in this region was associated with the production of crossties and thus concentrated along waterways. The heaviest logging occurred along St. Vrain Creek and the Big Thompson and Cache la Poudre Rivers, waterways that had high enough spring flows to float crossties downstream with the annual snowmelt. The crosstie drives were often immense operations that choked waterways with tens of thousands of ties. During the winter of 1868–69, for instance, more than 200,000 crossties were cut and floated down the Cache la Poudre alone. The drives continued in the Cache la Poudre and other northern watersheds until the mid-1880s, when most of the desirable trees had been cut and the industry migrated across the Continental Divide to plunder the vast forests that clothed Colorado's western slope.[37]

As local deforestation ensued and railroads began to crisscross the mountains in the 1870s, Hill and other smelter men began to look more and more to relocating their operations down along the eastern piedmont where they could take advantage of railroad connections and distant ore markets, expand operations, slash labor and transportation costs, and make the long-desired switch in fuel from wood to coal. In Hill's case, the decision to relocate finally came to a head in 1877 when Secretary of the Interior Carl Schurz persuaded the U.S. Justice Department to file suit against the Boston and Colorado Smelting Company for allegedly possessing 50,000 cords of timber illegally harvested from the public domain.

Supported by a jury that upheld the public's right to the productive resources of the public domain, Hill prevailed in the trial. Still, he must have recognized that the lawsuit was a harbinger of changing times, and it likely helped him decide to move his operations downslope to the piedmont. In 1878, after a decade of plundering the forests around Black Hawk, Hill established the sprawling new Argo plant alongside the Colorado Central's tracks just north of Denver and turned to coal.[38]

Located along much of the Front Range's eastern edge and western parks were vast coal deposits that smelter men had eyed since they first established operations in the high country. Since the cost of oxen- or horse-drawn wagon freight was generally too high to haul large amounts of coal to mountain smelters in the late 1860s, the earliest coal mines were generally small, seasonal operations that supplied local household markets in Denver, Golden, and Boulder. Beginning in the 1870s, the development of the region's railroads and the relocation of mountain smelters changed all this. Every Colorado railway ran a line to a Front Range coalfield, and energy-intensive industries, such as smelting and iron production, increasingly began to cluster along the sides of their tracks. By the mid-1880s, the symbiotic relationship between railroads, coal, mining, and other industrial developments had led to the opening of major coal mines near Cañon City, Colorado Springs, and Boulder, as well as the establishment of some of the nineteenth-century West's largest smelting operations in Denver and Pueblo, including the Boston and Colorado Smelting Company's Argo plant, the Holden Smelting Company, and the Pueblo Smelting and Refining Company.[39]

The combined effects of the development of railways, the shift from a predominantly wood-powered to a coal-powered energy regime, and the various ways these intermingled with economic activity were part of a larger pattern then engulfing the industrializing world, and they contributed to environmental change in ways that few fully appreciated. Before the advent of the railroad and the use of fossil fuels, Front Range miners were beholden to the small network of heavily traveled dirt and plank roads and renewable energy sources (essentially muscle, biomass, and flowing water) that, like all transportation systems and energy regimes, set boundaries on resource strategies and economic development. This meant, for instance, locating stamp mills and smelters as close to mines and fuel sources as possible, since moving ore and fuel around was difficult and expensive. The railroad and coal broke these constraints, much as the wheel, horse, and sail had done for earlier human societies. Although the first locomotives burned wood, railways had begun the gradual shift toward coal in the 1830s as high-pressure, coal-burning boilers became available. By harnessing the stored photo-synthesis of millions of years ago and feeding it to locomotives and smelters, Front Range miners were able to free themselves from the confines of earlier transportation and energy regimes, create ever more effective and elaborate economic linkages with the wider world, and send mining—as well as the whole process of industrialization—hurtling forward at an even more furious pace.[40]

The environmental ramifications of all of these developments were profound. At the most basic level, the shift from wood to coal merely transferred some of the local fuel demand from the mountain forests to the piedmont coal deposits. This had the twin effects of reducing the pressure on Front Range forests and narrowing the immediate ecological degradation caused by fuel extraction, since coal mines were much more concentrated operations than logging concerns.[41] At a more complex level, the network of rails and the commitment to coal can also be seen as crucial factors in transforming the Front Range from a peripheral frontier mining society into a modern industrial one, with all the attendant and far-flung ecological consequences. In the increasingly integrated and diversified economy that emerged after the 1870s, mining entrepreneurs were able to expand their tributary spheres and tap distant ore markets from Montana to northern Mexico, while blast furnace operators working with silver-bearing lead ores could rely on high-grade coke produced in southern and western Colorado, southern Wyoming, and northern Mexico. Similarly, Front Range ores were shipped to smelters in Omaha, Chicago, and Pittsburgh, and its coal was used to power mines, smelters, and other industrial activities in Kansas City, Butte, and Salt Lake City.[42] As the century wore on, the effects of mining became ever more diffused across far-distant landscapes.

While the coal mines and passing railcars signaled extensive environmental consequences, other less clearly visible ecological changes were also taking place. Throughout the nineteenth century, precious-metals mining relied on one of the most toxic heavy metals, mercury (quicksilver), to help amalgamate gold in virtually all placer- and early lode-mining devices. It was added to the riffle bars of rockers and sluices, as well as to the stamp mortars and amalgamation tables of stamp mills. Though present throughout the biosphere and in all living organisms, mercury is normally found in tiny amounts counted in parts per million (PPM) or parts per billion (PPB). For example, soils usually contain between 0.01 and 0.5 PPB; food crops may hold as much as 1.0 PPM;

and human body tissue normally has between 0.2 and 0.7 PPM. In terms of aggregate amounts, the total amount of mercury commonly found in an average adult human body is just six milligrams, and the amount absorbed—through the digestive tract, skin and lungs—on a daily basis is about three micrograms.[43]

While the toxicity of mercury varies across species and environmental conditions, the liquid form used by nineteenth-century miners was particularly lethal since it could be volatilized and absorbed by the lungs. To cite just one well-documented example that highlights its poisonous effects, a single flask containing some 76 pounds of mercury broke open aboard the British sloop *Triumph* in 1810. The accident affected the entire crew, killing three, as well as all the cattle and birds aboard.[44] Exactly how high mercury concentrations rose in Front Range mining districts in the nineteenth century is unknown, but they likely exceeded modern safety standards many thousandfold. According to an 1870 engineering analysis of thirty-five Gilpin County stamp mills, between 423 and 438 pounds of mercury were lost during normal operations *every month*. One mill alone, the Black Hawk, was reportedly losing 60 pounds a month.[45] The losses, as Rossiter Raymond explained, occurred during nearly every step of the reduction process: "Every piece of wood that has come in contact with quicksilver, the canvas straining-sacks, the worn-out pan-shoes and dies, even after careful washing and breaking, the thoroughly washed and shaken quicksilver-flasks, the used up kettles and dippers, the floors, & c., all have quicksilver sticking to them; the men carry quicksilver on their boots and clothes, and it is found scattered in very small quantities outside of the mill. It goes everywhere."[46] And, Raymond might have continued, it often stayed there too. Like the sites of countless other nineteenth-century gold mines in the American West, mercury concentrations still remain highly elevated around many former Front Range mines.[47]

In addition to mercury, many other toxic heavy metals and hazardous substances contaminated the environment around mining districts. Stamp mill operators customarily disposed of the various concoctions of cyanide of potassium, ammonia, lime, lye, and nitric acid that they used to clean their amalgamated copper plates by dumping the leftover solution into the nearest gulch or creek.[48] Smelter men pumped countless tons of sulfur dioxide, lead, arsenic, and other volatile elements into the atmosphere, coating the surrounding mountainsides with hazardous substances and, likely, sulfuric acid rain.[49] In places such as Black Hawk, where Hill's Boston and Colorado Smelting Company was located until 1877, the billowing fumes became hemmed in by Clear

FIGURE 3.3. The Boston and Colorado Smelting Company,
Gilpin County, Colorado, 1870s. Courtesy Subject File
Collection, Hill Smelter (Scan 10052994), History Colorado.

Creek Canyon's high walls and gave the town a well-deserved reputation, as one visitor put it in 1873, for its "sulphurous vapors."[50] Another
visitor, a correspondent for the *Engineering and Mining Journal*,
described the smoke emanating from Hill's smelters as filling "the
atmosphere with coal dust and darkness [. . .] volumes of blackness
from *seventeen* huge smoke stacks."[51] Although similar problems
accompanied smelting operations wherever they located, they were particularly severe in places like Black Hawk that were located on the floor
of a narrow canyon (see figure 3.3).

Similarly pernicious and lingering environmental problems stemmed from chemical reactions taking place within the massive tailing piles and numerous shafts and tunnels that marked every mining district. Since most Front Range gold and silver deposits were locked within complex sulfides, the ongoing oxidation and weathering of these ores released varying concentrations of acidic water and toxic trace elements into surrounding watersheds. For example, as mining exposed the abundant pyrite and country rock found near Central City to water and dissolved oxygen, chemical reactions took place that left behind elevated concentrations of many hazardous metals, including arsenic, copper, cadmium, iron, lead, manganese, and zinc. Even today, nearly a century after precious-metal mining collapsed along the Front Range, the soils, groundwater, and surface water around many mining districts remain heavily polluted. Among the most seriously contaminated areas is the Clear Creek basin. The concentration of heavy metals at a number of old mining sites in the basin remains so high that they have been designated federal Superfund sites, and the cleanup continues.[52]

Polluted streams, soils and air, the widespread decline in forestland, and the ongoing transformation of valleys and mountainsides into mining districts further reduced wildlife populations dependent on those habitats. Already pushed to the edges of the region's expanding mining districts and towns during the gold rush era, deer, elk, buffalo, antelope, bighorn sheep, bear, and wild turkey all but disappeared from the vast cutover regions that had been rapidly expanding since the gold rush years. Below the streams that flowed through mining districts, bottom-dwelling invertebrate communities, mollusks, crustaceans, and fish populations were also decimated and had begun to shift toward more metal- and pollution-tolerant species. By 1872, the decline in wildlife had become so great that the Colorado Territorial Assembly passed its first comprehensive wildlife protection act to establish hunting seasons and levy fines to protect many species of birds and animals. Fifteen years later, the numbers of bighorn sheep and buffalo, in particular, had plunged so sharply that the state closed the season on bighorn for eight years and on buffalo for an entire decade. For the once-numerous buffalo, the act was more of an early epitaph than anything else. The last wild buffalo in the Front Range was believed shot in South Park in 1897, leaving only place-names, such as Buffalo Peak and Buffalo Pass in North Park, to mark their former gathering places.[53]

Just as mining was tough on land and wild creatures, it also took a brutal toll on miners and all who lived and labored in its shadow.

Mining was a dangerous occupation, and miners faced a plethora of perils in their poorly lit underground chambers. Falling rocks, collapsing shafts, earth-rattling explosions, noninsulated electrical wires, whirling machinery, pockets of groundwater, unseen holes, diseases— each of these and others disabled, debilitated, or killed countless miners.[54] "The mining section," as one Colorado resident recalled after touring the region, "is full of men with but one or no eye, and with fingers missing, while hundreds are cut down in their prime, by twos and threes, every decade."[55] If miners somehow managed to avoid all these hazards, they still had to contend with the most insidious danger of them all. The millions of microscopic silica particles produced by blasting and drilling through quartz lodes gradually accumulated in the lungs, creating tiny incisions that formed scar tissue and slowly impaired their ability to function properly. Known as silicosis or "miner's consumption," it frequently suffocated its victims or led to other occasionally fatal pulmonary illnesses, such as pneumonia or tuberculosis. In the early twentieth century, silicosis was eventually recognized as the leading cause of death among nineteenth-century hard-rock miners.[56]

Once miners left work and returned home, they, like all those who resided near mining districts, still had to contend with the smelter smoke, layers of soot, poisoned water, landslides, avalanches, floods, and other dangers peculiar to mining environments. Even a simple walk down a street, as one woman found out in Central City, could turn hazardous. According to the *Daily Central City Register,* a woman was seriously injured after falling ten to fifteen feet into an abandoned shaft that had been covered over and used as a cross street.[57] Others were less fortunate. Patrick Ryan broke his neck and died after falling into a shaft near Central City, and a small child died after tumbling into another shaft filled with water.[58] The direct effects of mining created a seemingly endless number of hazards for all who worked in or lived near the mines. As the influential editor of the *Springfield (Mass.) Republican,* Samuel Bowles, explained after touring the Central City mining district in 1868, the gulch is "torn with floods, and dirty with the debris of mills and mines that spread themselves over everything."[59]

Writing early on in the Front Range's industrial development, Bowles was more perceptive than he probably realized. Hill's Boston and Colorado Smelting Company had just begun operations; new and retrofitted stamp mills were only then starting to demonstrate their worth; and the railroad era still lay two years in the future. Each of these, and the various ways they combined with other scientific, technological, and cultural developments to stimulate precious-metals production and integration into the emerging national and international marketplace, further deepened the environmental transformations that Bowles seemed to lament. In removing millions of dollars' worth of gold and silver, miners created new landscapes and altered primary ecological processes, leaving behind a new world of gutted mountainsides, poisoned watersheds, denuded forests, and decimated wildlife populations that affected all life in the region. These two accounts are, in reality, opposite sides of the same coin and together reveal the essence of precious-metals mining in the early industrial age. Scientific, technological, capital-intensive, economically integrated, environmentally rapacious—industrial precious-metals mining was a wonder of productivity and environmental change. As Front Range mines passed from the periphery toward the core of the industrial age, the region and its creatures were profoundly transformed.

CONSERVATION-MINDED STIRRINGS

Those environmental changes were never idle developments. Over time, as Bowles and others noted, the environmental effects of precious-metals mining gave rise to anxieties and misgivings over the course of the industry's relationship with the natural world. Many of these concerns are well known, and the powerful responses they generated in the Front Range and elsewhere in the United States have effectively made them pillars of the American conservation movement that emerged during the Progressive era. But precious-metals mining also raised important questions that do not fit neatly into the existing contours of conservationist thought in the nineteenth century and that have received too little attention from environmental historians. Minerals are fundamentally different from trees and other living things, and when miners dug into the earth in search of them, they also uncovered pathways toward a deeper and perhaps more worrisome engagement with the idea of conservation and Americans' relationship with the natural world than has been traditionally recognized.

The historiography on the American conservation movement has some very well established grooves. Although scholars have recently pushed the chronology back into the eighteenth century and expanded the scope beyond its traditional figures and concerns, the movement's major benchmarks continue to revolve around a series of environmental threats that gathered national attention between about 1890 and 1920: dwindling forests and wildlife, polluted cities and unhealthy workplaces, wasteful extraction methods and production practices, and the decline of wilderness and the general cornering-up of nature. In response to these developments, early conservationists—such as the diplomat and philologist George Perkins Marsh; the first chief of the U.S. Forest Service, Gifford Pinchot; President Theodore Roosevelt; and the preservationist and founder of the Sierra Club, John Muir—pushed for the creation of the nation's first national forests, first wildlife refuges, and first national parks, as well as new game protection laws, new sanitation regulations, and new and expansive public health initiatives. As scholars have shown, many of these foundational features of the American conservation movement were influenced by western precious-metals mining and the social and political responses that followed.[60]

In the Front Range, for instance, the decline in forests led concerned members of Colorado's state constitutional convention in 1876 to push successfully for Article 18, which pledged the General Assembly to "enact laws in order to prevent the destruction of, and to keep in good preservation, the forests upon the lands of the state." This was followed in the mid-1880s by the establishment of the Colorado State Forestry Association (1884) and the Office of the State Forest Commissioner (1885), which sought to prevent forest fires, preserve forests, and encourage silviculture. These initiatives eventually mingled with the concerns of others across the country and helped propel passage of the 1891 Forest Reserve Act at the national level, creating the National Forest Reserve system. Within two years of its passage, President Benjamin Harrison had established fifteen forest reserves, including three in the Front Range that amounted to just over 1 million acres.[61]

Similarly, the billowing fumes, airborne silica particles, diseases, and other hazardous environmental effects of precious-metals mining in the Front Range led to the formation of local benevolent associations and fraternal lodges to care for miners and their families who had been injured, fallen sick, or been killed. Miners also formed labor unions to carry forward the work of the benevolent associations and fraternal lodges, and to press mine owners and the Colorado legislature for mine safety and mine inspection laws, the eight-hour workday, a workmen's compensation system, and other initiatives. The first hard-rock mining union formed anywhere in the American West was organized in the Front Range at Central City in 1863. From these beginnings, miners' unions became a regular feature of Colorado's mining industry for the rest of the century. By 1900, at least half of Colorado's miners are thought to have joined the Western Federation of Miners, arguably the most powerful miners' union in the United States at the time.[62]

The environmental transformation of the Front Range also played a role in the establishment of the region's first national park, Rocky Mountain National Park. In 1915, Congress set aside 231,000 acres of land along the western edge of the Front Range, extending from Estes Park west to the Continental Divide. As with other national parks, the campaign for Rocky Mountain reflected concerns about the beauty and wonder of nature that have long been associated nationally with John Muir, and in Colorado with one of Muir's chief disciples, the naturalist Enos Mills. In 1909, when Mills was just beginning the campaign that would lead to the establishment of Rocky Mountain, he worried aloud about what was being lost: "Extensive areas of primeval forest have been misused and ruined; the once numerous big game has been hunted almost out of existence, the birds are falling, the wild flowers vanishing, and the picturesque beaver, except where protected, are almost gone." In 1915, those imperiled landscapes were set aside by congress "for recreation purposes by the public and for the preservation of the natural conditions and scenic beauty thereof."[63]

Each of these responses to the Front Range's changing environment—the creation of forest reserves, the establishment of benevolent associations and unions, the push for mine safety laws, and the establishment of Rocky Mountain National Park—ties neatly into the historiography of the American conservation movement. But there is also evidence to suggest that some public officials carried their concerns about mining and the environmental changes it precipitated beyond the conservation movement's well-known focal points. In 1868, for instance, the nation's first U.S. commissioner of mines and mining statistics, J. Ross Browne, included the following passage in his report to Congress on the condition of the western mining industry:

> No country in the world can show such wasteful systems of mining as prevail in ours. At a moderate calculation, there has been an unnecessary loss of precious metals since the discovery of our mines of more than $300,000,000, scarcely a fraction of which can ever be recovered. This is a serious consideration. The question arises whether it is not the duty of government to prevent, as far as may be consistent with individual rights, this waste of a common heritage, in which not only ourselves but our posterity are

interested. The miner has a right to the product of his labor, but has he a right to deprive others of the benefits to be derived from the treasures of the earth, placed there for the common good?[64]

Although the thrust of Browne's comments was clearly directed at increasing mining and metallurgical efficiencies in order to reduce mineral losses, his views are unmistakably tinged with notions of restraint, equity, or what we might today call sustainability. While Browne did not try to answer the far-reaching questions he posed, they stand like signposts to the age, pointing the way toward a deeper engagement with the heart of a conservation ideology that was just beginning to be pieced together in various corners of the country.

Like Browne, his successor as U.S. commissioner of mines and mining statistics and one of the nation's leading authorities on mining, Rossiter Raymond, also grappled with weighty questions not customarily seen as part of the conservation movement. In 1885, for instance, Raymond published an ambitious article on the history of mining law. Reaching from the Phoenicians, Greeks, and Romans to modern European states, Mexico, Canada, and the United States, Raymond discussed mining rights, patents, and royalty systems. But the concern that seems to animate Raymond's inquiry is what he called "the unique character of the industry itself." As he explained, "The deterioration of the soils through ignorant or reckless agriculture may be cured in time by wiser methods; forests wantonly destroyed may be replanted; fishing grounds left to themselves or restocked artificially may recover their prolific abundance of supply, but coal, iron, copper, lead, petroleum, gold, and silver will not come again within the history of the human race into the places from which they have been extracted. A waste of them is a waste forever." Within the same paragraph, Raymond underscored these ideas, characterizing the mining of minerals "as constantly tending towards a permanent exhaustion of the natural resources of the land."[65]

Raymond's and Browne's concerns about mineral exhaustion and a sense of environmental limits were unusual for the age. Throughout much of the nineteenth century, American attitudes toward mineral stocks were far more ebullient, far closer to the one expressed in 1869 by the geologist Ferdinand Hayden, whose famous surveys of the American West included the Front Range. Hayden described the mines around South Clear Creek as "very rich and practically inexhaustible."[66] His sense of the region's seemingly endless mineral abundance drew on a particular view of mineral veins that was ubiquitous across many levels of society in the 1860s and well into the later nineteenth century. It was evident in the views of geologists and miners, as well as in those of boosters and speculators. As a professor of geology at Brown University, George Chase, put it:

> The universal experience in Colorado is that the lodes increase in richness as they are followed downwards. In this respect they are remarkably distinguished from those of Australia and most other gold districts, where the richest ore is found in the first hundred feet from the surface. The deeper workings are less productive; and at length become unprofitable, and are abandoned. The apparent exception to the general law governing gold-bearing lodes—an exception in favor of those in Colorado—adds greatly to their desirableness and value.[67]

Others echoed Chase's position. Writing for an unspecified mining committee report in 1864, the mining attorney William Rockwell described Colorado's "true fissure veins [as] having no termination downwards attainable by human ingenuity." In fact, Rockwell summed up, "no well-developed and defined vein has ever been found entirely terminating in depth."[68] Another mining attorney, J. Weatherbee Jr., characterized Colorado's mineral veins similarly, noting in an 1863 brief on the Colorado Territory that "if the vein pays at the blossom, it will pay better as you descend." And, as if to ward off unease, Weatherbee went on to state that even if a vein reached a cap, "it is sure to open again, and the richer for it, if your courage holds."[69]

And yet, even as such hopeful views of Colorado's veins were carrying forward an older cornucopian vision of nature, experiences in the Front Range and elsewhere in the American West were beginning to tilt opinion in Raymond's and Browne's direction. As vein after vein pinched out into cap, and as mine after mine exhausted its paying ore and shuttered its tunnels, anxieties like Raymond's and Browne's were being levered into new conceptual, social, and political consciousness.[70] We see this most clearly in 1908, when President Theodore Roosevelt invited all of the state governors to the White House to discuss the idea of conservation. The proceedings of that conference and the *Report of the National Conservation Commission* that followed are filled with warnings about limited and declining stocks of minerals, and about how wasteful mining and production practices threaten America's industrial prosperity.[71] Although Congress failed to maintain the National Conservation Commission and its work on coordinating and assessing resource exploitation for the country, the commission—and particularly its focus on minerals—raised striking new concerns about the earth's superabundance and the depths that mining pushed those concerns during the Progressive era.[72]

CONCLUSION

When James Meline wandered into the Front Range in 1866, he found himself in the midst of multiple revolutions. The industrialization of mining, the dramatic evolution of science and technology, the rise of new energy sources, the expansion and integration of economic markets—all of these and other developments were just then becoming entangled in the Front Range and beginning to recast the world around him in fundamental ways. As these processes unfolded, they also gave rise to an emerging eco-cultural pensiveness and to new understandings about mineral deposits, which expressed themselves in the form of environmental and cultural misgivings, anxieties, and eventually, legislative and cultural measures that looked to counter some of the most destructive trends. From this viewpoint, the history of the American conservation movement takes on new meaning. Not only did western precious-metals mining contribute to well-known features of the conservation movement such as deforestation and the creation of national parks, but it also seeded larger questions—stubborn, difficult questions that merit further research—about mineral exhaustion, scarcity, progress, and the superabundance of the earth in ways that forests and other renewable resources never did, or perhaps could. As we think our way through this past, widening our understanding of the scope of the conservation movement and blurring the sharp lines that once separated earlier ideas about conservation from our own today, we might just find some fresh new insights and a historical forum for thinking about our current environmental circumstances and the challenges they pose.

NOTES

1. For two recent articles on how environmental concerns tie up with contemporary gold mining, see Edwin Dobb, "Alaska's Choice: Salmon or Gold," *National Geographic* 218:6 (December 2010): 100–125; and Brook Larmer, "The Real Price of Gold," *National Geographic* 215:1 (January 2009): 34–61.

2. James F. Meline, *Two Thousand Miles on Horseback, Santa Fe and Back: A Summer Tour through Kansas, Nebraska, Colorado, and New Mexico in the Year 1866* (New York: Hurd and Houghton, 1867), 63–65.

3. Foundational works in this area include Thomas G. Andrews, *Killing for Coal: America's Deadliest Labor War* (Cambridge, MA: Harvard University

Press, 2008); Andrew C. Isenberg, *Mining California: An Ecological History* (New York: Hill and Wang, 2005); Kathryn Morse, *The Nature of Gold: An Environmental History of the Klondike Gold Rush* (Seattle: University of Washington Press, 2003); and Duane Smith, *Mining America: The Industry and the Environment, 1800–1980* (Niwot: University Press of Colorado, 1993).

4. On the use of enviro-technical perspectives in environmental history, see the pathbreaking work done by Timothy J. LeCain, in *Mass Destruction: The Men and Giant Mines That Wired America and Scarred the Planet* (New Brunswick: Rutgers University Press, 2009), and the collected essays in Susan R. Schrepfer and Philip Scranton, eds., *Industrializing Organisms: Introducing Evolutionary History* (New York: Routledge, 2004). Though not consciously framed as enviro-technical, also very useful is Paul R. Josephson, *Industrialized Nature: Brute Force Technology and the Transformation of the Natural World* (Washington, DC: Island Press/Shearwater Books, 2002). In addition, for deeper theoretical considerations, see Rosalind Williams, *Notes on the Underground: An Essay on Technology, Society, and the Imagination,* new ed. (Cambridge, MA: MIT Press, 2008); Bruno Latour, *Science in Action: How to Follow Scientists and Engineers through Society* (New Haven: Yale University Press, 1988); and the now-classic work of Lewis Mumford, *Technics and Civilization* (1934; reprint, New York: Harcourt Brace, 1963).

5. For recent scholarship that focuses on the early history of the American conservation movement, see Richard W. Judd, *The Untilled Garden: Natural History and the Spirit of Conservation, 1740–1840* (New York: Cambridge University Press, 2009); Aaron Sachs, *The Humboldt Current: Nineteenth-Century Exploration and the Roots of American Environmentalism* (New York: Penguin Books, 2006); and Richard W. Judd, *Common Lands, Common People: The Origins of Conservation in Northern New England* (Cambridge, MA: Harvard University Press, 1997). On the relationship between developments in the nineteenth-century American West and the conservation movement, see, for instance, Stephen Fox, *The American Conservation Movement: John Muir and His Legacy* (Madison: University of Wisconsin Press, 1981); Samuel P. Hays, *Conservation and the Gospel of Efficiency: The Progressive Conservation Movement, 1890–1920* (1959; reprint, Pittsburgh: University of Pittsburgh Press, 1999); G. Michael McCarthy, *Hour of Trial: The Conservation Conflict in Colorado and the West, 1891–1907* (Norman: University of Oklahoma Press, 1977); and Ronald C. Brown, *Hard-Rock Miners: The Intermountain West, 1860–1920* (College Station: Texas A&M University Press, 1979).

6. For recent works that look at industrialization and environmental change within the confines of the nineteenth-century American West, see David Igler, "Engineering the Elephant: Industrialism and the Environment in the Greater West," in *A Companion to the American West*, ed. William Deverell (Malden, MA: Wiley-Blackwell, 2004), 93–111; David Igler, *Industrial Cowboys: Miller & Lux and the Transformation of the Far West, 1850–1920* (Berkeley: University of California Press, 2001); David Igler, "The Industrial Far West: Region and Nation in the Late Nineteenth Century," *Pacific Historical Review* 69 (May 2000): 159–92; and Steven Stoll, *The Fruits of Natural Advantage: Making the Industrial Countryside in California* (Berkeley: University of California Press, 1998).

7. For a comparative examination of western mining regions, see Rodman Wilson Paul and Elliott West, *Mining Frontiers of the Far West, 1848–1880*, revised and expanded ed. (Albuquerque: University of New Mexico Press, 2001). For the relationship among eastern and British financiers, economic conditions, and Front Range mining developments in the nineteenth century, see Joseph E. King, *A Mine to Make a Mine: Financing the Colorado Mining Industry, 1859–1902* (College Station: Texas A&M University Press, 1977), and Clark C. Spence, *British Investments and the American Mining Frontier, 1860–1901* (Ithaca, NY: Cornell University Press, 1958).

8. Rossiter W. Raymond, *Statistics of Mines and Mining in the States and Territories West of the Rocky Mountains*, U.S. Department of the Treasury (Washington, DC: U.S. Government Printing Office, 1870), 347–48. Though imprecisely labeled, these annual reports to Congress also cover developments in the Rockies. Raymond issued eight annual reports in this series (1869–1877); all but the first share the same title and are hereafter cited by annual report year.

9. Raymond, *Statistics of Mines and Mining in the States and Territories West of the Rocky Mountains* (1870), 356–65. See also James W. Taylor, *Report of James W. Taylor on the Mineral Resources of the United States East of the Rocky Mountains*, U.S. Department of the Treasury (Washington, DC: U.S. Government Printing Office, 1868), 21.

10. James D. Hague and Clarence King, *Mining Industry, by James D. Hague, with Geological Contributions by Clarence King; Submitted to the Chief of Engineers and Published by Order of the Secretary of War Under Authority of Congress*, in *Report of the Geological Exploration of the Fortieth Parallel* (Washington, DC: U.S. Government Printing Office, 1870), 3:577–88; Ovando J. Hollister, *The Mines of Colorado* (Springfield, MA: Samuel Bowles and Co., 1867), 134; Frank Fossett, *Colorado: Its Gold and Silver Mines, Farms and Stock Ranges, and Health and Pleasure Resorts* (1879; reprint, New York: Arno Press, 1973), 143, 226; Henry Dudley Teetor, "Refining and Smelting in Colorado: Prof. Richard Pearce, F.G.S., Manager of the Boston and Colorado Smelting Works," *Magazine of Western History* 11:3 (January 1890): 278–82. For secondary works that treat the significance of Nathaniel Hill and the Boston and Colorado Smelting Company, see James E. Fell, *Ores to Metals: The Rocky Mountain Smelting Industry* (Lincoln: University of Nebraska Press, 1979), 11–38; Jesse D. Hale, "The First Successful Smelter in Colorado," *Colorado Magazine* 13:5 (September 1936): 161–67; C.H. Hanington, "Smelting in Colorado," *Colorado Magazine* 23:2 (March 1946): 80–84; and Charles W. Henderson, *Mining in Colorado: A History of Discovery, Development, and Production*, U.S. Geological Survey Professional Paper 138 (Washington, DC: U.S. Government Printing Office, 1926), 9, 27–32, 36–40.

11. *Rocky Mountain News*, 27 November 1872.

12. Raymond, *Statistics of Mines and Mining in the States and Territories West of the Rocky Mountains* (1870), 365.

13. Ibid., 364–65; Raymond, *Statistics of Mines and Mining in the States and Territories West of the Rocky Mountains* (1872), 339–73; Hague and King, *Mining Industry*, 3:548–73; A.N. Rogers, "The Mines and Mills of Gilpin County," *Transactions of the American Institute of Mining Engineers* 11

(1883): 29–55; T. A. Rickard, "Limitations of the Gold Stamp Mill," *Transactions of the American Institute of Mining Engineers* 23 (1893): 137–47; Edson S. Bastin and James M. Hill, *Economic Geology of Gilpin County and Adjacent Parts of Clear Creek and Boulder Counties, Colorado,* U.S. Geological Survey Professional Paper 94 (Washington, DC: U.S. Government Printing Office, 1917), 154–56; Samuel Cushman and J. P. Waterman, *The Gold Mines of Gilpin County, Colorado: Historical, Descriptive and Statistical* (Central City, CO: Register Steam Printing House, 1876), 99–102.

14. The evolution and spread of the science and technology of mining and mineral reduction processes are nicely covered in Paul and West, *Mining Frontiers of the Far West,* 123–34, 270–77; Fell, *Ores to Metals*; Crane, *Gold and Silver,* 496–552; Thomas Tonge, "Smelting Gold and Silver Ores in Colorado: The Development of the Industry and Its Effects on the Mining Business," *Mines and Minerals* 19:3 (October 1898): 97–100; and Clark C. Spence, *Mining Engineers and the American West: The Lace-Boot Brigade, 1849–1933* (New Haven: Yale University Press, 1970). On developments at the Colorado School of Mines, see Kathleen H. Ochs, "The Rise of American Mining Engineers: A Case Study of the Colorado School of Mines," *Technology and Culture* 33:2 (April 1992): 278–301. Firsthand observations and illustrations of many of these processes (including Brückner's revolving cylinders and the Washoe pan process), as well as the intellectual advances that propelled them, can be found in Raymond, *Statistics of Mines and Mining in the States and Territories West of the Rocky Mountains* (1870), parts 4–5; Raymond, *Statistics of Mines and Mining in the States and Territories West of the Rocky Mountains* (1874), 407–98; Hague and King, *Mining Industry,* 3:547–88, 606–16; and Richard Pearce, "Progress of Metallurgical Science in the West," *Transactions of the American Institute of Mining Engineers* 18 (1889): 55–72. On Front Range ore types, see Bastin and Hill, *Economic Geology of Gilpin County and Adjacent Parts of Clear Creek and Boulder Counties*; T. S. Lovering and E. N. Goddard, *Geology and Ore Deposits of the Front Range, Colorado,* U.S. Geological Survey Professional Paper 223 (Washington, DC: U.S. Government Printing Office, 1950); and Waldemar Lindgren and Frederick Leslie Ransome, *Geology and Gold Deposits of the Cripple Creek District, Colorado,* U.S. Geological Survey Professional Paper 54 (Washington, DC: U.S. Government Printing Office, 1906).

15. On the importance of railroads in shaping western development, see Richard White, *Railroaded: The Transcontinentals and the Making of Modern America* (New York: W. W. Norton, 2011); Carlos A. Schwantes and James P. Ronda, *The West the Railroads Made* (Seattle: University of Washington Press, 2008); William Cronon, *Nature's Metropolis: Chicago and the Great West* (New York: W. W. Norton, 1991), 63–81; and Duane A. Smith, *Rocky Mountain West: Colorado, Wyoming, and Montana, 1859–1915* (Albuquerque: University of New Mexico Press, 1992), 54, 98–120.

16. The development of Colorado's early railway networks is covered in E. O. Davis, *The First Five Years of the Railroad Era in Colorado* (Golden, CO: Sage Books, 1948); Tivis E. Wilkens, *Colorado Railroads: Chronological Development* (Boulder, CO: Pruett, 1974); R. A. LeMassena, *Colorado's Mountain Railroads* (Golden, CO: Smoking Stack Press, 1968); Kenneth Jessen, *Railroads*

of Northern Colorado (Boulder, CO: Pruett, 1982); and Robert G. Athearn, *Rebel of the Rockies: A History of the Denver and Rio Grande Western Railroad* (New Haven: Yale University Press, 1962). For the precise routes traveled by these lines in Colorado, see Kenneth A. Erickson and Albert W. Smith, *Atlas of Colorado* (Boulder: Colorado Associated University Press, 1985), 28–29.

17. On the environmental history of Colorado during the gold rush years, see George Vrtis, "Gold Rush Ecology: The Colorado Experience," *Journal of the West* 49:2 (Spring 2010): 23–31.

18. For instance, see F. V. Hayden, *Third Annual Report of the United States Geological Survey of the Territories, Embracing Colorado and New Mexico* (Washington, DC: U.S. Government Printing Office, 1869), 225; F. V. Hayden, *Ninth Annual Report of the United States Geological and Geographical Survey of the Territories, Embracing Colorado and Parts of Adjacent Territories; Being a Report of Progress of the Exploration for the Year 1875* (Washington, DC: U.S. Government Printing Office, 1877), 425; Raymond, *Statistics of Mines and Mining in the States and Territories West of the Rocky Mountains* (1873), 296; and Raymond, *Statistics of Mines and Mining in the States and Territories West of the Rocky Mountains* (1874), 298–99.

19. On the relative contributions of placer and lode mining, see the detailed calculations in George Vrtis, "The Front Range of the Rocky Mountains: An Environmental History, 1700–1900" (Ph.D. dissertation, Georgetown University, 2006), table A.1. For the visitor quotation, see Charles Harrington, *Summering in Colorado* (Denver: Richards and Co., 1874), 31–32.

20. Robert E. Strahorn, *To the Rockies and Beyond; or, A Summer on the Union Pacific Railroad and Branches: Saunterings in the Popular Health, Pleasure, and Hunting Resorts of Nebraska, Dakota, Wyoming, Colorado, Utah, Idaho, Oregon, Washington and Montana,* 2nd ed. (Omaha: New West Publishing Co., 1879), 34. See also Fossett, *Colorado,* 58, 62, 386.

21. Cushman and Waterman, *The Gold Mines of Gilpin County, Colorado,* 43–98; Raymond, *Statistics of Mines and Mining in the States and Territories West of the Rocky Mountains* (1877), 290–321, esp. 291.

22. For 1875 production figures, see Vrtis, "The Front Range of the Rocky Mountains," table A.1. For the conversion of the total value of gold into ounces, I have used $19.75 per ounce (the official price of gold paid by the U.S. Mint throughout the later nineteenth century).

23. Raymond, *Statistics of Mines and Mining in the States and Territories West of the Rocky Mountains* (1872), 347; and Bastin and Hill, *Economic Geology of Gilpin County,* 109–20.

24. This is a conservative estimate of the amount of waste rock produced by milling processes since (1) it uses the minimum value of one-half for the amount of waste rock produced from initial sorting operations, and (2) it assumes that all of the precious metals were recovered from a single type of ore, which occurred, but was certainly not the norm for the region.

25. Ellen E. Wohl, *Virtual Rivers: Lessons from the Mountain Rivers of the Colorado Front Range* (New Haven: Yale University Press, 2001), 78–81; Raymond, *Statistics of Mines and Mining in the States and Territories West of the Rocky Mountains* (1877), 291–92.

26. Vicki Cowart, "When the Ground Lets You Down: Ground Subsidence and Settlement Hazards in Colorado," *Rock Talk* (Colorado Geological Survey) 4:4 (October 2001): 1–12; Jeffrey L. Hynes, ed., *Proceedings of the 1985 Conference on Coal Mine Subsidence in the Rocky Mountain Region,* Colorado Geological Society Special Publication 31 (Denver: Colorado Geological Society, 1986).

27. The Argo and other tunnels are discussed in Bastin and Hill, *Economic Geology of Gilpin County,* 303–6 (Argo), 177–367 (other tunnels). On the downstream effects of water diversion projects, see Wohl, *Virtual Rivers,* 72, 117–25.

28. Fossett, *Colorado,* 205, quote on 290.

29. On the Mining Law of 1872, see Gordon Morris Bakken, *The Mining Law of 1872: Past, Present, and Prospects* (Albuquerque: University of New Mexico Press, 2008).

30. Raymond, *Statistics of Mines and Mining in the States and Territories West of the Rocky Mountains* (1873), 288.

31. The details used in determining these timber consumption rates are drawn from Thomas Egleston, "Boston and Colorado Smelting Works," *Transactions of the American Institute of Mining Engineers* 4 (1876): 276–98. The cords per acre of forested mountainside conversion factor of 60 to 80 used here is based on Raymond, *Statistics of Mines and Mining in the States and Territories West of the Rocky Mountains* (1877), 330.

32. For instance, according to Raymond's 1874 annual report, the number of smelters and blast furnaces operating in the Front Range amounted to fifteen. He does not provide a listing for stamp mills, but unlike smelting enterprises, which tended to be organized as independent companies, most of the hundreds of mines also had their own stamp-milling operations. See Raymond, *Statistics of Mines and Mining in the States and Territories West of the Rocky Mountains* (1875), 388. For an analysis of the timber consumption rates of steam engines operating in Gilpin County in 1871, see "Report on the Committee on Statistics," *Daily Central City Register,* 7 May 1871. Based on this report, the total number of steam engines then in use required 170 cords of wood per day to operate, an amount four times greater than the daily amount used by the Boston and Colorado Smelter.

33. William H. Wroten Jr., "The Railroad Tie Industry in the Central Rocky Mountain Region, 1867–1900" (Ph.D. dissertation, University of Colorado at Boulder, 1956), 4, 65–156, esp. 124.

34. *New York Tribune* correspondent quoted in Raymond, *Statistics of Mines and Mining in the States and Territories West of the Rocky Mountains* (1872), 325. For similar observations, see F. V. Hayden, *Annual Report of the United States Geological and Geographical Survey of the Territories, Embracing Colorado; Being a Report of Progress of the Exploration for the Year 1873* (Washington, DC: U.S. Government Printing Office, 1874), 280; *Daily Central City Register,* 10 August 1874; and George A. Crofutt, *Crofutt's Grip-Sack Guide of Colorado: A Complete Encyclopedia of the State* (Omaha: Overland Publishing Co., 1881), 42.

35. *New York Tribune* correspondent quoted in Raymond, *Statistics of Mines and Mining in the States and Territories West of the Rocky Mountains* (1872), 325.

36. Some of the best comprehensive data on the extent of nineteenth-century deforestation comes from modern scientific studies. The following studies of Front Range forest dynamics include quantitative and repeat photographic data on stand characteristics, which can, by working backward, provide an indication of the composition of earlier time periods: John W. Marr, *Ecosystems of the East Slope of the Front Range in Colorado,* University of Colorado Studies Series in Biology No. 8 (Boulder: University of Colorado Press, 1967), esp. 25–75; Robert K. Peet, "Forest Vegetation of the Colorado Front Range: Composition and Dynamics," *Vegetation* 45 (1981): 8–9, 37–75; Thomas T. Veblen and Diane C. Lorenz, *The Colorado Front Range: A Century of Ecological Change* (Salt Lake City: University of Utah Press, 1991), esp. 18–9, 173–76; and Thomas T. Veblen and Diane C. Lorenz, "Anthropogenic Disturbance and Recovery Patterns in Montane Forests, Colorado Front Range," *Physical Geography* 7:1 (1986): 1–22. On tie-cutting operations, see Wroten, "The Railroad Tie Industry in the Central Rocky Mountain Region," 65–156, 264–87. For firsthand observations on the extent of the regional deforestation, see Fossett, *Colorado,* 240, 284; Strahorn, *To the Rockies and Beyond,* 58; John G. Jack, *Pikes Peak, Plum Creek and South Platte Forest Reserves, Showing Density of Forests,* U.S. Geological Survey Twentieth Annual Report, part 5 (Washington, DC: U.S. Geological Survey, 1898), plate 8; and, more generally, Enos A. Mills and W. G. M. Stone, *The Forests and Exotic Trees of Colorado* (Denver: Colorado State Forestry Association, 1905), 5–17.

37. Marr, *Ecosystems of the East Slope of the Front Range in Colorado,* esp. 25–75; Peet, "Forest Vegetation of the Colorado Front Range," 8–9, 37–75; Veblen and Lorenz, *The Colorado Front Range,* 18–9, 173–76; Wroten, "The Railroad Tie Industry in the Central Rocky Mountain Region," 65–156, 264–87; Howard Ensign Evans and Mary Alice Evans, *Cache la Poudre: The Natural History of a Rocky Mountain River* (Niwot: University Press of Colorado, 1991), 201.

38. Fossett, *Colorado,* 240; Fell, *Ores to Metals,* 52–54, 133–38; Teetor, "Refining and Smelting in Colorado," 280; Hale, "The First Successful Smelter in Colorado," 163–64; Case Declaration and Plea Documents, *United States v. Nathaniel P. Hill and the Boston and Colorado Smelting Company,* File Folder 121, Box 12, Civil Case Files for 1878, Records of the United States District Courts–Colorado, Record Group 21, National Archives and Records Administration, Rocky Mountain Region, Denver, Colorado.

39. On the development of the Colorado coal industry and its relationship to railroads and mining, see Lee Scamehorn, *High Altitude Energy: A History of Fossil Fuels in Colorado* (Boulder: University Press of Colorado, 2002), 1–12; Charles W. Henderson, "Mining in Colorado," in *History of Colorado,* ed. James H. Baker and LeRoy R. Hafen (Denver: Linderman Co., 1927), 1:547–51; Andrews, *Killing for Coal,* chs. 1–3; and Fell, *Ores to Metals,* 141–200. For firsthand observations on early coal-mining operations, see the following government reports and Fossett's carefully detailed tourist guide: Hague and King, *Mining Industry,* 3:475–84; Raymond, *Statistics of Mines and Mining in the States and Territories West of the Rocky Mountains* (1870), 378–79; Raymond,

Statistics of Mines and Mining in the States and Territories West of the Rocky Mountains (1872), 293, 379; F.V. Hayden, *Annual Report of the United States Geological and Geographical Survey of the Territories, Embracing Colorado and Parts of Adjacent Territories; Being a Report of Progress of the Exploration for the Year 1874* (Washington, DC: U.S. Government Printing Office, 1876), 33–34; and Fossett, *Colorado*, 539–40.

40. On the complex relationships among technology, energy regimes, economic activity, and environmental change, see Martin V. Melosi, *Coping with Abundance: Energy and Environment in Industrial America* (New York: Alfred A. Knopf, 1985), 17–34; J.R. McNeill, *Something New Under the Sun: An Environmental History of the Twentieth-Century World* (New York: W.W. Norton, 2000), 296–324; Clive Ponting, *A Green History of the World: The Environment and the Collapse of Great Civilizations* (New York: Penguin Books, 1991), 267–94; and Vaclav Smil, *Energies: An Illustrated Guide to the Biosphere and Civilization* (Cambridge, MA: MIT Press, 1999), esp. 105–73.

41. On coal-mining methods and their effects, see Scamehorn, *High Altitude Energy*, 9–11.

42. Raymond, *Statistics of Mines and Mining in the States and Territories West of the Rocky Mountains* (1875), 359; Fell, *Ores to Metals*, 71–75, 139–41, 155–57, 163–64; Scamehorn, *High Altitude Energy*, 6–42.

43. John Emsley, *Nature's Building Blocks: An A–Z Guide to the Elements* (New York: Oxford University Press, 2001), 254–61.

44. Ibid., 256.

45. Hague and King, *Mining Industry*, 3:574–75.

46. Raymond, *Statistics of Mines and Mining in the States and Territories West of the Rocky Mountains* (1874), 415. See also Hague and King, *Mining Industry*, 3:557; Raymond, *Statistics of Mines and Mining in the States and Territories West of the Rocky Mountains* (1872), 345.

47. Jerome O. Nriagu and Henry K.T. Wong, "Gold Rushes and Mercury Pollution," in *Metal Ions in Biological Systems*, vol. 34, *Mercury and Its Effects on Environment and Biology*, ed. Astrid Sigel and Helmut Sigel (New York: Marcel Dekker, 1997), 147; Charles R. Goldman and Darell G. Slotton, "Mercury Contamination in California: A Mining Legacy," in *Proceedings of the 20th Biennial Ground Water Conference*, Water Resources Center Report No. 88, ed. J.J. DeVries and J. Woled (Davis: University of California Center for Water and Wildland Resources, 1996), 145–54; Wohl, *Virtual Rivers*, 76.

48. Raymond, *Statistics of Mines and Mining in the States and Territories West of the Rocky Mountains* (1872), 339, 367.

49. Fell, *Ores to Metals*, 49–51.

50. Harrington, *Summering in Colorado*, 24. See also Smith, *Mining America*, 11–12; and Fell, *Ores to Metals*, 49–51.

51. *Engineering and Mining Journal*, 23 September 1877, quoted in Smith, *Mining America*, 12 (original emphasis).

52. Kathleen S. Smith, "Generation and Interaction of Mine Drainage at the Argo Tunnel," in *Guidebook on the Geology, History, and Surface-Water Contamination and Remediation in the Area from Denver to Idaho Springs,*

Colorado, ed. K. C. Stewart and R. C. Severson, U.S. Geological Survey Circular 1097 (Washington, DC: U.S. Government Printing Office, 1994), 39–40; Walter H. Ficklin and Kathleen S. Smith, "Influence of Mine Drainage on Clear Creek, Colorado," in *Guidebook on the Geology, History, and Surface-Water Contamination and Remediation,* 43–48; Wohl, *Virtual Rivers,* 72–77; and U.S. Environmental Protection Agency, *EPA Superfund Program: Central City, Clear Creek, Idaho Springs, CO,* cumulis.epa.gov/supercpad/cursites/csitinfo .cfm?id=0800257 (accessed July 2016).

53. Peter Hansson, Charles Hjelte, and George Feltner, eds., *A Look Back: A 65-Year History of the Colorado Game and Fish Department,* 1961 Annual Report of the Colorado Game and Fish Department (Denver: Colorado Game and Fish Department, 1961), 6–10. On invertebrates, mollusks, crustaceans, and fish, see Wohl, *Virtual Rivers,* 74–75. On the last stand of the Front Range buffalo, see F. M. Fryxell, "The Former Range of the Bison in the Rocky Mountains," *Journal of Mammalogy* 9:2 (1928): 129–35.

54. For accounts of common mine accidents, see *(Denver) Rocky Mountain News,* 8 January 1869; *Rocky Mountain News,* 6 December 1869; Brown, *Hard-Rock Miners,* 75–98; and Duane A. Smith and Ronald C. Brown, *No One Ailing except a Physician: Medicine in the Mining West, 1848–1919* (Boulder: University Press of Colorado, 2001), 67–76.

55. John A. Hitching Diary, 60, Hitching Collection, Western History Collections, University of Colorado Libraries, Boulder, quoted in Smith and Brown, *No One Ailing except a Physician,* 70.

56. Although no accurate statistics on silicosis exist before the twentieth century, the following two studies rely on adjusted sex ratio analysis and an early-twentieth-century U.S. Bureau of Mines investigation, respectively, to make their points: Leanne Louise Sander, "'The Men All Died of Miners' Disease': Women and Families in the Industrial Mining Environment of Upper Clear Creek, Colorado, 1870–1900" (Ph.D. dissertation, University of Colorado at Boulder, 1990), 180–88; Brown, *Hard-Rock Miners,* 80–81, 93–94. See also Smith and Brown, *No One Ailing except a Physician,* 75–76, 116–19.

57. *Daily Central City Register,* 10 August 1874.

58. *Rocky Mountain News,* 20 July 1868, 18 March 1870.

59. Samuel Bowles, *The Switzerland of America: A Summer Vacation in the Parks and Mountains of Colorado* (Springfield, MA: Samuel Bowles and Co., 1869), 151.

60. See note 5. See also Smith, *Mining America;* Kathleen A. Brosnan, *Uniting Mountain and Plain: Cities, Law, and Environmental Change along the Front Range* (Albuquerque: University of New Mexico Press, 2002); and the following works on urban America in the Progressive era: Martin V. Melosi, *Garbage in the Cities: Refuse, Reform, and the Environment, 1880–1980,* rev. ed. (College Station: Texas A&M University Press, 2005); Martin V. Melosi, *The Sanitary City: Urban Infrastructure in America from Colonial Times to the Present* (Baltimore: Johns Hopkins University Press, 2000); and Joel A. Tarr, *The Search for the Ultimate Sink: Urban Pollution in Historical Perspective* (Akron, OH: University of Akron Press, 1996).

61. McCarthy, *Hour of Trial*, 37–41; Michael Williams, *Americans and Their Forests: A Historical Geography* (New York: Cambridge University Press, 1989), 409–11; William Wyckoff, *Creating Colorado: The Making of a Western Landscape, 1860–1940* (New Haven: Yale University Press, 1999), 92. On the Colorado State Forestry Association and the Office of the State Forest Commissioner, see Colorado State Forestry Association, *Colorado State Forestry Association: Organized November 19th, 1884* (Denver: Republican Publishing Co., Printers, 1884), 2; and Edgar Ensign, *Forestry in Colorado: First Annual Report of the State Forest Commissioner* (Denver: Collier and Cleaveland, 1885), 1–15.

62. Mark Wyman, *Hard Rock Epic: Western Miners and the Industrial Revolution, 1860–1910* (Berkeley: University of California Press, 1979), ch. 6, esp. 151–52, 172–79; Richard E. Lingenfelter, *The Hardrock Miners: A History of the Mining Labor Movement in the American West, 1863–1893* (Berkeley: University of California Press, 1974), 103–5, 219–28.

63. Thomas Andrews, *Coyote Valley: Deep History in the High Rockies* (Cambridge, MA: Harvard University Press, 2015), ch. 6, quotes on 146, 157.

64. J. Ross Browne, *Report of J. Ross Browne, on the Mineral Resources of the States and Territories West of the Rocky Mountains* (Washington, DC: U.S. Government Printing Office, 1868), 9.

65. Rossiter W. Raymond, "Historical Sketch of Mining Law," in *Mineral Resources of the United States, Calendar Years 1883 and 1884* (Washington, DC: U.S. Government Printing Office, 1885), 2:988.

66. Hayden, *Third Annual Report of the United States Geological Survey of the Territories, Embracing Colorado and New Mexico*, 188.

67. George I. Chase, "Report of George I. Chase, Professor of Geology at Brown University, Providence, R.I.," in S. F. Tappan, *Colorado Territory* (Washington, DC: n.p., 1866), 11.

68. William Selden Rockwell, *Colorado: Its Mineral and Agricultural Resources* (New York: n.p., 1864), 13.

69. J. Weatherbee Jr., *A Brief Sketch of Colorado Territory and the Gold Mines of That Region* (Boston: Wright and Potter, Printers, 1863), 15–16. Accounts similar to those of Hayden, Chase, Rockwell, and Weatherbee are numerous: see, for example, Henry van Schaick, H. Fales, and Willis Gregory, *Report to the Board of Trustees of the American Flag Gold Mining Co.* (New York: n.p., 1864).

70. On changing geological understandings of mineral deposits in the mid-nineteenth century, see Charles Lyell, *Principles of Geology* (1830–33; reprint, Chicago: University of Chicago Press, 1990); Louis Agassiz, *Geological Sketches* (Boston: Ticknor and Fields, 1866); R. M. Macbrair, *Geology and Geologists; or, Visions of Philosophers in the Nineteenth Century* (London: Simpkin, Marshall, and Co., 1843); and Mott T. Greene, *Geology in the Nineteenth Century: Changing Views of a Changing World* (Ithaca, NY: Cornell University Press, 1982).

71. On the White House conference of governors and the report of the National Conservation Commission, see W. J. McGee, ed., *Proceedings of a Conference of Governors in the White House* (Washington, DC: U.S. Govern-

ment Printing Office, 1909); and G. R. Green, ed., *Report of the National Conservation Commission* (Washington, DC: U.S. Government Printing Office, 1909).

72. On the history of Americans' relationship with abundance and scarcity, see and ponder Donald Worster's provocative new book, *Shrinking the Earth: The Rise and Decline of American Abundance* (New York: Oxford University Press, 2016).

CHAPTER 4

Consequences of the Comstock

The Remaking of Working Environments on America's Largest Silver Strike, 1859–1880

ROBERT N. CHESTER III

On the evening of Wednesday, June 28, 1870, a tunnel collapsed at the Yellow Jacket mine in Gold Hill, Nevada, killing Manuel Alameda, Patrick Doherty, John Kennedy, and Ralph Hanson. The four miners had been at work more than 800 feet underground when the roof caved and tons of debris buried them. According to the *Alpine Chronicle*, "All four were new hands in the mine and are believed to have been single men." The next morning, reporter Alfred Doten ran down to the scene of the accident. He recorded the grisly details of the recovery effort in his journal: "They got out one body about 1 oclock this morning and after working all day got out another about 5PM—I saw both—The first was not much distorted, but the last was—His bowels were crushed out of the lower part of his belly—& he was black & blue & stunk very badly—2 more left in the mine." The next morning they pulled the third body out of the mine, and workers recovered the fourth corpse later that same day. As this incident suggests, a small cave-in could have a large impact in the death toll it delivered.[1] Hundreds of other men mining the Comstock also suffered death or sustained catastrophic injuries that left them permanently disabled.

As social historians Roy Rosenzweig and David Thelen have remarked, "The meaning of the large lay in *the stories of the small*."[2] So what do these deaths reveal about the miners' working environments on the Comstock Lode? How did environmental forces in northwestern Nevada shape demands for labor and the occupational hazards encountered by workers underground? How in turn did subterranean environments contribute to the rapid expansion of technological systems and the rise of the corporation? As the largest silver strike in the history of the United States and the nation's most dangerous mining district from 1860 to 1880, the Comstock Lode proved hard-rock mining's earliest industrial crucible in the nineteenth-century American West.

The vein's rich concentrations of ore occurred within a matrix composed of both silver and gold. The abundance of both metals made this mining district extraordinarily productive and valuable. Within the first two years of production, eighty-six corporations formed to establish mines on the Comstock. The combined capitalization of these companies exceeded $61 million. By 1880 more than one hundred mines still operated on the Comstock. However, very few Comstock corporations ever returned profits to their investors. In part, these failures reflected both the unpredictable and finite character of mineral veins and the manipulation and mismanagement of these companies by unscrupulous financiers. Equally important, though, were the immense costs of operating these mines. The many hundreds of millions of dollars that mining companies spent on labor and highly elaborate technological systems laid the foundation for the rapid expansion and enormous scale of hard-rock mining for the remainder of the nineteenth century. Ultimately, between 1859 and 1880 the mines of Gold Hill and Virginia City produced silver and gold worth more than $300 million.[3] Opportunities for profit appeared real enough, but the risks for workers often exceeded the rewards.[4]

Between 1863 and 1880 approximately nine hundred men died or suffered severe injuries while working on the Comstock. The gruesome fashion in which most of these men died reveals the dynamic and dangerous character of the working environments they helped to build but could never fully control. A diverse array of unpredictable and overlapping forces made the Comstock Lode's subterranean environments expensive and dangerous to mine: flooding, extreme heat, cave-ins, mechanized drills, explosions, the depth of shafts, the lowering and hoisting of cages, steam engines, pumping machinery, and elaborate ventilation systems. Corporations, engineers, and hundreds of laborers worked together to construct new and elaborate technological systems to stabilize loose earth, remove immense quantities of water, and mitigate the intense heat encountered during extraction. Industrial managers repeatedly expressed great frustration with these environmental conditions.[5]

Death records, newspaper articles, and the reports of superintendents provide historians rich insights into the work routines, occupational hazards, and costs of doing business on the Comstock Lode.
Although many scholars have examined the ways that capital flows,
technology, community formation, and unionization both shaped and
reflected conditions and opportunities on the Comstock, the central role
played by environmental forces in the industrialization of this region
remains underdeveloped. Large-scale, industrialized precious-metals
mining on the Comstock unfolded according to a series of dialectical
relationships among the imperatives of corporate capitalism, enormous
demands for labor, and the unpredictable forces encountered in subterranean working environments. The technological systems that mining
companies introduced on the Comstock Lode made mining more feasible in the face of treacherous natural forces, but some new technologies
also introduced hazards to the miners' working environment.

The expansion of industrial technologies and the adoption of the corporate form help explain why the Comstock Lode dwarfed all other mining operations elsewhere in the 1860s and most of the 1870s.[6] Thanks to
the capital generated by the California Gold Rush and the subsequent
emergence of hydraulic mining, investors increasingly exploited a favorable legal environment that allowed them to create new mining corporations. In response to a boom in mining stocks, enterprising investors
established the San Francisco Stock and Exchange Board of Trade in
1862. The board's centralization and standardization of transactions dramatically increased the volume of commerce and facilitated the flow of
capital across the Sierra Nevada into the mines of Nevada's Comstock.[7]

Some close observers believed the costs of labor and technological
systems on the Comstock were unsustainable for corporations trying to
make profits.[8] By the 1870s, the Comstock increasingly witnessed more
prominent patterns of corporate consolidation, as fewer syndicates controlled an ever-larger share of both mines and mills. In addition to the
particular financiers and investors who stood to profit from such trends,
Rossiter Raymond, the U.S. commissioner of mining statistics, also
favored these developments. He believed that competition between
companies impeded coordination and standardization in the construction of large, expensive systems for drainage, ventilation, and hoisting.
The diffusion of capital created massive administrative inefficiencies in
the Comstock's mining economy. Writing in 1869, he insisted that
"foremost among the necessary reforms is the consolidation of companies, and the consequent reduction in the cost of administration."[9]

Although corporate consolidation did occur, it never approached the level of uniform coordination and control Raymond thought necessary for the maximization of profit. Environmental obstacles underground and corporate competition increased the costs of expansion, maintenance, and machinery. In Raymond's view, without more systematic and rational improvements offered by consolidation, the methods employed by many competing companies were "fast approaching their economical limit of application to these mines."[10] In the abstract, Raymond's conclusions logically followed from the evidence he presented concerning potential profits and projections of the costs of operations at ever-lower depths. Despite the inability of most of these corporations to deliver consistent dividends to their shareholders, mining continued as it had, and companies survived these inefficiencies far longer than Raymond thought possible. Obstacles to extracting ore and making money on the Comstock remained numerous and severe over the course of its development.[11]

OBSTACLES BELOW

Placer miners first discovered the surface deposits of the Comstock Lode in June 1859 while digging a reservoir. The massive deposits they soon encountered overwhelmed the earliest prospectors, who lacked adequate capital and the technical expertise to extract and process these ores. The earliest and most dangerous obstacle encountered by miners on the Comstock Lode was cave-ins. Because of the large size of the ore bodies extracted and the geological peculiarities of the lode, mining companies faced entirely new challenges concerning the timbering of mines. Wet, porous, and crushed quartz led ceilings to sag and collapse. In December 1860, less than 200 feet from the surface of George Hearst's Ophir mine, workers uncovered an ore body more than 40 feet wide. More than a decade later, reporter William Wright, writing under the penname Dan De Quille, described the new challenges in his hyperbolic style: "No such great width of ore had been seen before and the miners were at their wits' end to know how to keep up the superincumbent ground." Workers refused to keep excavating until they believed the mine was again safe.[12] The shortage of large timbers in the immediate vicinity of the mines prompted these companies to use composite support beams created by splicing separate lengths of wood together using iron fasteners. These longer, thinner supports failed to withstand the weight and pressure exerted by the relatively loose rock that constituted much of the mountains.

To prevent catastrophic cave-ins and loss of life and capital, Hearst and his associates hired an imaginative engineer from California to solve the problem. A Hessian Jew who had left Europe for California at nineteen, Philip Deidesheimer arrived in the goldfields of the Sierra Nevada foothills in 1851. A decade later, he had accumulated a great deal of knowledge and experience in California's early hard-rock mining operations. Applying lessons he had learned in the young quartz-mining operations in El Dorado County, California, and adjusting his plans to the particular demands made by the Comstock's geology, he devised a system of timbering that became known as square-sets. These hollow rectangular frames were typically 4 feet by 5 to 6 feet and held together by mortise-and-tenon joints at the corners of both top and bottom. The tenon—a notch protruding from either end of the beams—fit into the mortise joint chiseled into the inside ends of beams, thus creating a system of interlocking frames.[13]

This design used the extremely snug joints at the corners of the square-sets as a way to help redistribute weight more evenly over space. The new invention stabilized the spatial voids produced by ore removal. Workers filled these empty cubes with waste rock to provide more solid foundations for the mines as they continued to expand. Deidesheimer's innovation allowed the Ophir to expand operations and made Hearst's first fortune possible. It also made mining the rest of the lode viable.[14] As a result, mining companies had a much easier time convincing investors to purchase their stock, and workers to go back underground. Companies began to expand their operations at rates unprecedented in the history of hard-rock mining. Square-sets became a common feature of the underground landscape in working environments throughout the West for the rest of the century.

Although this innovative response to nature's subterranean obstacles facilitated the success of mining and made working conditions safer, Deidesheimer's ingenuity and artifice could only delay gravity's will. Charles Bonner, superintendent of the Savage Mining Company, recognized these persistent obstacles, as he wrote about the difficult task of preventing cave-ins. In early April 1868, he complained that "the ground is very soft & so full of water that it is very [difficult] to keep from caving." Thus, Deidesheimer's square-sets may have been the best answer available to the uncertainties encountered underground, and they undoubtedly saved many lives, but cave-ins remained a common occurrence and would persist as long as people and companies chose to chase the treasures inside Nevada's Virginia Mountains.[15]

Long after the development of the square-sets, dozens of miners continued to die each year due to cave-ins and falling rock and timber. Nearly half of all those miners killed underground between 1869 and 1870 died from causes directly attributed to collapsing tunnels or falling matter.[16] In October 1877 a cave-in occurred on the 1,700-foot level of the Savage mine. The accident killed Richard Kitz—a twenty-nine-year-old miner from Kentucky.[17]

On the morning of September 11, 1874, foreman Jerry Cross knelt down on the 1,100-foot level of the Yellow Jacket mine. As the *Territorial Enterprise* recounted, "He was in the process of leveling a sill, which was being put down as the foundation for a tank. While he was engaged [in this task], two or three carloads of rock suddenly fell from the top of the chamber [. . .] crushing him to the ground and covering him up." Although his co-workers immediately pulled him from the rubble, "it was ascertained that he was much bruised in the region of the abdomen, that he had received a frightful gash across the temple, another on the top of his head, and that one of his ankles was badly lacerated if not fractured." The reporter concluded, "His recovery is considered doubtful."[18]

Men would often hear the groaning strain of crumbling earth and cracking timbers prior to the collapse of sinking floors and falling roofs, but such audible omens did not necessarily allow workers to escape. Sometimes these warnings saved lives, and other times they simply foretold a fatal end. But more than simply cave-ins threatened their lives and health. Rock could also fall through the square-sets without crushing the timbers or the tunnel, and sometimes fell upon workers. Jerry Cross had been an experienced foreman working in a "place [that . . .] appeared perfectly safe, but, as was afterwards seen, there was a clay seam above the flake of rock which fell, therefore there was nothing to hold it in place."[19]

Randomly distributed seams of pliable clay further contributed to the unpredictable movements of the Comstock's subterranean geology. As William Wright observed, "The whole body of the clay appears to be creeping. It has the almost imperceptible motion of the glacier. [. . .] Its action is so mysterious that some of the miners [. . .] explain it by saying that the clay comes out and fills up the drifts because 'nature abhors a vacuum.'" Though hyperbolic, stories recounting the destruction wrought by geological blobs reveal the exceedingly unpredictable and volatile nature of the miner's working environment. A number of unseen variables complicated attempts to mine in a safe and predictable fashion.[20]

In his misanthropic style, reporter J. Ross Browne anthropomor-phized the battle waged between miners on the Comstock and the mountains they attempted to exploit:

> Perhaps there is not another spot upon the face of the globe that presents a scene so weird and desolate in its natural aspect, yet so replete with busy life, so animate with human interest. It is as if a wondrous battle raged, in which the combatants were man and earth. Myriads of swarthy, bearded, dust-covered men are piercing into the grim old mountains, ripping them open, thrusting murderous holes through their naked bodies; piling up engines to cut out their vital arteries; stamping and crushing up with infernal machines their disemboweled fragments [. . .] while the mighty earth, blasted, barren, and scarred by the tempests of ages, fiercely affronts the foe—smiting him with disease and death.[21]

Although Browne's rhetoric may have rung hollow for most people, the miners themselves could appreciate such metaphors. Nature in the mines was to be respected. This respect came from fear and the wisdom of lived experience that instilled poignant lessons about the earth's power to crush even the most cautious miners.

In addition to the settling and shifting of crushed quartz and creeping seams of clay, the Comstock's peculiar geology had also created throughout the lode large, hidden deposits of water. Pumps more than once failed to keep pace with the flooding caused by the puncturing and release of underground aquifers. The challenges confronted in the Ophir mine provide an instructive example of the disasters wrought by flood-ing in the mines and demonstrated how imperfect pumping remained. Eliot Lord from the U.S. Geological Survey recounted how workers cut a hole in the rock face in the fall of 1869 only to find to their horror "a flood [that] poured so violently that it rose irresistibly in the shaft, though the pumps were worked at their full capacity and discharged 20,000 gallons hourly. [. . . T]he water was 270 feet deep in the shaft." The flooding overtaxed pumping machinery on the lower levels, causing the equipment there to seize up and stop working. It was not until April 1870 that the company finally succeeded in draining the mine to the point where the flood had started.[22]

The more flooding that occurred and the longer the water filled tun-nels, the faster timbers would decay and collapse, which in turn endan-gered more men and cost companies more capital. Moreover, because the water increasingly strained the machinery as the pumps lifted it ever higher, mining companies typically set up a series of pumping stations every couple of hundred feet to work pumps more efficiently and to keep them in good repair, which increased operating costs even more. Nature on the Comstock would not yield silver and gold easily or cheaply; rather, it imposed huge costs on corporations that built and maintained these mines, and these companies repeatedly turned to their shareholders to bear these expenses. Investors thus assumed added risks in pursuit of potential profits. The worker who labored underground, however, confronted both economic and physical risks in the form of injury and death.[23]

THE DEPTHS OF HELL

The Comstock Lode's enormous veins formed in response to rupturing mountains pulled apart by tectonic extension. Plates in the Great Basin have continued to diverge from one another over the course of millions of years. In the process, this movement has slowly wrenched the earth's crust into pieces, leading to the deformation of mountain ranges generated by earlier geological epochs.[24] This process eventually allowed the mantle to rise closer to the earth's surface and created enormous fissures surrounded by extraordinarily hot rock. Over time, rainwater accumulated in these fissures. Minerals embedded in hot rocks dissolved into the water. The water then continued to lift and deposit much of this matter in the fissures, which ultimately created the veins of the lode. The same sort of processes that created gold and silver deposits also led to both unstable matrices of decomposing rock and the formation of thermal hot springs throughout Nevada and eastern California. Where these types of ore deposits existed in the Great Basin, mineral springs commonly occurred. Some of these springs remained trapped in the subterranean depths of the mountains until a miner's drill or a blast of black powder released their contents.[25]

In addition to their volume and the flooding they caused, many of these waters reached scalding-hot temperatures. In early February 1878 Ophir superintendent James Fair reported "a strong flow of hot water from the top of the drift which is very hot rendering it almost [im]possible to work in the drift [even] with the aid of two strong blowers." This problem continued to trouble the Ophir for the following month.[26] Mining engineer and professor John A. Church examined this phenomenon as part of his report on the occupational hazards caused by the Comstock's extreme depths and peculiar geological features: "[W]e have some of the most singular occurrences known in mining. The injuries by scalding were occasioned entirely by falling into hot mine waters. Their temperature varies by locality, but the maximum which I have observed 156°F. [. . .] The water is hot and gaseous, and the unfortunate man who falls in it sinks deeply and probably finds it difficult to regain the surface."[27] As the Comstock mines pushed ever deeper into Mount Davidson, the temperature of water, air, and rock continued to rise. "One of the most striking phenomena [. . . in] the mines on the Comstock lode," Church observed, "is the extreme heat encountered in the lower levels. The heat [. . .] proceeds from the rock" and its "temperature [stays] very much higher than the average of the atmosphere in Nevada."[28]

Church argued that the intense heat of the local rock and abundant water distinguished the Comstock from other mining districts not only in the United States but throughout the world as well. He believed that "the heat of the mines is a matter of more than usual interest, for they are the only hot ones now worked in the United States, and both in the present temperature encountered and in the increase which is to be expected as greater depths are reached, they appear to surpass any foreign mines of which we have a record."[29] The heat exuded from the rocks in the lower levels of the mines raised the temperature of the air and contributed to a stifling working environment. Even with the aid of blowers that fed cooler air through thousands of feet of ventilation pipes and tons of ice stacked against the walls, the temperature in the mines often remained well above 110°F. Many workers encountered great difficulty laboring in such conditions. James Galloway reported in 1875 that the heat got to him while working in the Consolidated Virginia mine. He wrote in his diary that the heat inside the mines at the 1,400-foot level made him sick.[30]

Fierce heat, shifting rocks, and immense depths increasingly drove the rapid innovation of new technologies and made the Comstock Lode a laboratory for modern industrialism. By transforming inhospitable, alien spaces into habitable work zones, mining companies encouraged the implementation of more capital-intensive technological systems to surmount ever-more extreme environmental conditions. For example, many companies began using larger ventilation pipes to distribute more compressed air to lower levels.[31]

The very task of extending ventilation pipes, however, was itself fraught with peril, especially for newer employees unaccustomed to working in such extreme heat. In the first week of March 1877, Thomas Wilson, a thirty-eight-year-old miner from Edinburgh, Scotland, was working deep in the Imperial mine. Workers were extending an air pipe and had only been at work for about fifteen minutes when Wilson began to suffer from heat exhaustion and slipped while trying to climb up the incline to the "cooling off room." He then tumbled backward, falling down a series of steps. Co-workers carried Wilson back to the cooling station, but he died before they got him near the surface. Nearly four years later, Phillip Harrington, a twenty-four-year-old Irishmen, also died from the effects of extreme heat in the mines.[32] Sometimes the very technological systems designed to make the working environment cooler and safer could fail and cause deadly accidents. In August 1879, the *Territorial Enterprise* reported the death of P. E. Woodward and the injury of John Allen, noting that a flywheel on an air blower nearly sliced Woodward in half and that shrapnel gashed Allen's hip.[33]

Ever-deeper mines demanded ever-larger ventilation pipes. The longer and wider iron pipes allowed engineers to conduct more air from compressors with greater force to the work sites below. On September 28, 1872, Superintendent Smith reported progress in the Belcher mine. Workers put air boxes into tunnels more than 1,000 feet below to improve ventilation to a level at which miners could resume their work. By December the company had begun to see results in its efforts to create a safer and cooler environment.[34] To the modern reader, large pipes may not seem an advanced technology, but placed in the context of environmental engineering in the 1870s, these systems were revolutionary and can be compared to the elaborate systems of hydraulic engineering and energy-intensive air conditioning in the rapidly growing cities of the American Southwest after World War II. The scale of environmental hazards and the expenses incurred in implementing these technological remedies only further amplified the unparalleled power of corporations to overcome physical constraints more rapidly than other forms of economic organization could.[35]

Even with the vast corporate expenditures on machinery for ventilation, cooling, and drainage, the heat, moisture, and work on these lower levels continued to tax the health and endurance of miners. The wet, poorly ventilated, and exceedingly close quarters in which miners worked combined with the intense heat to create an optimal breeding ground for the spread of respiratory and infectious diseases such as pneumonia, tuberculosis, and typhoid fever. And the transition from this oppressively hot, moisture-laden environment to the cold mountain air on the surface throttled the breathing capacity of even the most acclimated of miners. This problem especially assaulted miners during the early 1870s when mining companies first reached the hotter levels.[36]

Another lethal hazard was gravity. The deeper mine shafts went, the likelier it became that falls were fatal. In late August 1873, the *Virginia*

Evening Chronicle reported one such horrible accident: "James Stile [. . .] working in the Savage [mine . . .] falling from an ascending cage. [. . .] about 600 feet [. . .] was instantly killed." The poor lighting and fierce noise of explosions and drills often distorted worker perceptions. Occasionally men would walk into the shaft without first making certain that the cage had not left his level.[37]

Such accidents happened throughout the Mining West, but because the Comstock expanded so rapidly and became the first truly massive hard-rock mining district in the American West, contemporaries alleged that the Comstock suffered more such casualties than did other mining districts. John Church also repeated the claim: "The most appalling accident which can occur in mining work, the falling of men down a deep shaft, repeats itself in the Comstock mines with a frequency which I believe is unknown elsewhere."[38] Since the Comstock dwarfed all other hard-rock mining operations in the American West in the late 1870s, this was certainly true for the preceding two decades, but there is no reason to believe that other districts that achieved equal or greater depths in later years did not suffer just as many or more such casualties.

Deep shafts exacerbated the risk of fatal falls. The movement of shifting rocks also warped timbers and bent tunnels into less symmetrical configurations. Offering a complex analysis of how a diverse array of environmental factors may combine to cause more falls on the Comstock than elsewhere, Church again looked to heat as a probable co-conspirator, but he pinpointed another variable to account for the number of men who fell down shafts and winzes: "[T]he Comstock rocks are forever moving, swelling, and forcing the shafts out of line. [. . .] The work of repair is necessarily dangerous and more hazardous in a hot steaming upcast than in a cool one, and it is to the frequency with which the miner is called upon to perform this work that the number of falls in the shaft is attributed." Church also emphasized "the unfavorable local conditions [that] may contribute essentially to the result. But of the two causes I consider that frequent opportunity is greater." Church failed, however, to connect all the dots, as the "opportunity" to repair the shafts remained inseparable from the very local conditions he cited.[39]

One could also fall down a shaft while still on a cage. A falling cage could crush workers at the shaft's lowest levels, especially if a cable broke or an inattentive engineer failed to stop winding the hoisting mechanism. In late August 1870, Alfred Doten reported a horrible accident in the Hale & Norcross mine in which "bolt of brake strap broke and let cage fall 1150 feet to bottom—killed two men working there & hurt another so bad it is thought he will also die." Four years later, Joseph Farnsworth was the lone victim in a less deadly cage accident. The *Territorial Enterprise* reported: "[H]e was severely injured while at work in the bottom of the main shaft of the Sierra Nevada Company by the cage being lowered upon him. [. . . T]he weight of the cage [. . .] crushed him down to the ground. [. . . H]e was insensible when taken out of the shaft." As a result of the accident, Farnsworth sustained severe injuries to the head and spine.[40]

In addition to workers and cages plunging down these shafts, the act of raising men on cages also led to a series of horrific accidents. Eliot Lord recalled one of the worst hoisting accidents ever on the Comstock. In December of 1879 the engineer of the Union mine began to lift a cage carrying seventeen men from the bottom of the shaft. Lord recounted how the engineer mistakenly sped up the ascending cage when he had actually intended to slow it down. Instead of slowing and stopping, "the cage shot upward with a sudden bound under the increased pressure. [. . .] in a moment the cage was torn out of the shaft as if shot from a catapult [. . .] and the men on the cage [. . .] were thrown sprawling over the floor of the shaft-house." Two miners died and six more were permanently disabled as a result of the accident. Ultimately, authorities deemed the engineer at the Union mine sober and attentive. Instead, they attributed his mistake to a fit of absentmindedness.[41]

Some hazards reflected the peculiarities of the Comstock's subterranean environment. Others derived from the dangers of new technologies and human error. Of course, even in environments as dangerous as those on the Comstock, on most days, most engineers and other employees regularly performed their duties without incident, but danger and death lurked in every mine.

THE RHYTHMS OF LABOR

The scale of hard-rock mining on the Comstock called upon each company to employ several score to hundreds of miners. With fewer available workers, a cost of living that was inflated due to the geographic isolation, and competition between employers, labor proved very costly at more than $3 per day per miner. Occupational hazards eventually helped unify workers in their efforts to organize. With unionization in the mid-1860s, the daily wage for all men who toiled underground eventually climbed to $4 per day. By comparison, most common laborers in industrial manufacturing firms during the 1870s averaged a daily wage just over $1 per day. In the West the wage scale was typically higher, at approximately $2 per day for a common laborer.[42] Because of the number of miners needed, labor quickly became the most expensive cost of doing business on the Comstock. In 1860 alone, nearly 2,000 miners already worked on the lode. That number rose slowly to approximately 2,700 by 1870. However, after the discovery and development of the famed Big Bonanza ores starting in 1873, the number of miners working on the Comstock shot up and peaked in 1875 at more than 4,900. With major declines in production in the second half of the 1870s, the workforce shrank rapidly thereafter, but as late as 1880, it still stood at more than 2,400.[43]

By the 1870s, miners typically worked one of three eight-hour shifts: morning, afternoon, or night. Most mining companies rotated these shifts among their workers, so that each man would work each shift for a series of weeks. Miners showed up at 7 A.M., 3 P.M., or 11 P.M. every day of the week. However, such structured schedules were often the ideal, not the reality. Employment was frequently erratic. James Hezlep Galloway worked for at least three mines—the Consolidated Virginia, Ophir, and Consolidated Union—between 1875 and 1877. He recorded in April 1876, "Did not work machinery broke at Ophir."[44] Although wages were high at $4 per day, frequent disruptions of the work routine could undermine financial security. The very technological systems that required so much labor and capital to build could and did fail. Technological interventions intended to improve natural environments could also result in unexpected outcomes. The costs sunk into infrastructure and equipment could not be recouped when these systems stopped working and halted ore production. Both workers and corporations suffered adverse effects. For some, these delays would prove costly but temporary setbacks; for others, depending on the length of work stoppages, such obstacles might be devastating.

The management of the labor force reflected the increasingly corporatized structure of work. The superintendent represented the highest link on the chain of authority in the shaft house. Beneath him a shift boss monitored and directed work and attempted to solve any problems that might occur on his watch. He also recorded a summary of his crew's progress. A clerk or timekeeper kept track of each and every miner who entered and left the mine. Men starting or ending their shift checked in at a window. Mechanics, engineers, blacksmiths, and carpenters also worked in the shaft house. Mechanics installed and repaired the equipment, and engineers operated the hoisting works that lowered and lifted the cages. Blacksmiths kept tools sharp, and carpenters performed a variety of tasks in the shaft house and down in the mines.[45]

The hierarchy and delegation of specialized duties, along with the indispensability of constantly evolving technologies, demonstrate both the corporate and industrial character of hard-rock mining on the Comstock Lode. Upon entering a shaft house, reporter William Wright inventoried the equipment and noted both the sophistication and scale of the machines: "In the mass of buildings before us we see nothing to cause us to think of a mine. What we have before us more nearly resembles a large iron-foundry or big manufactory of some kind." The new, massive industrial machinery led people to start referring to these large operations as "extractories." This term revealed the hybrid quality of mining economies in the mid-nineteenth century. Though an extractive enterprise, the techniques employed to remove, transport, and reduce ore involved a series of highly industrial processes. Moreover, the most modern and sophisticated of new technologies—such as elevators (cages and hoisting works), ventilation systems, and mechanized drills—often emerged as a result of the immediate needs of the mining industry.[46]

Upon descending a shaft, one found a working environment that appeared far less industrial than the world above. The orderly appearance of each station, with its stocks of supplies, hid the tedious backbreaking labor that occurred further down the drift. In every level of the mine, a foreman supervised a crew of miners, muckers, and timbermen. Miners stoped—excavated steps or layers of rock in—the breast or face of the drifts and crosscuts and drilled holes into which they inserted charges of blasting powder with fuses they typically ignited from relatively safe distances. Muckers filled carts with waste rock and ore, being sure to keep them separate when possible. It fell to these men to clean up the rubble and knock down any loose debris that hung from the blasted areas. One mucker might serve as many as a dozen miners and assumed a position akin to an apprentice. Timbermen floated from place to place in order to insert new square-sets, replace rotting timbers, and inspect the general health and stability of the mine's supports.[47]

Each miner received three candles per shift. Employers expected that he would make them last throughout his eight-hour shift. Due to foul air quality, these candles frequently flickered out. Some miners used a steel holder with a sharp end to stick these lights into the walls of the mine, while others mounted a candle atop their hats. Both strategies allowed the miners to keep their hands free and illuminate their workspace. The dim light made the miner's work all the more difficult when drilling—swinging a hammer against a hand drill. Sometimes a man did this by himself, and in other instances two men worked in tandem, as one held the drill and the other swung the hammer.[48]

INDUSTRIAL INSTRUMENTS

If mechanization made the work of the miner more productive, it did not necessarily follow that the new technologies were safer. Ventilation and pumping equipment typically improved working conditions and made them safer. But other technologies created new hazards, such as the corrosive effects of increased particulate matter and violent explosions from drilling and blasting rock. Machine drills and dynamite increased levels of productivity, and workers on the Comstock, unlike miners in other mining districts, did not actively oppose these changes. Instead, miners quickly and efficiently integrated these new tools into their age-old rhythms.

Mining companies began employing machine drills on the Comstock during the early 1870s. Typically, two men operated these new drills, which sat atop tripods. The same compressed air blown into the mines for purposes of ventilation also served to power these heavy machines. Compared to the rate of drilling by hand, the machine drills accelerated the rate at which miners could move through rock by three to four times. But these advances came with a price, one borne most heavily by the miners' lungs. As the drills burrowed into walls, they ground the rock and released immense amounts of rock dust into the air. After working for a sustained period of years in such conditions, many miners developed a series of respiratory ailments, the most famous and controversial of which was silicosis, popularly known as "miner's consumption." The constant blasting of charges released additional matter that also strained and damaged lungs, but the drill men disproportionately contracted respiratory diseases and suffered their debilitating symptoms.[49]

A few of the miners on the Comstock Lode may have suffered from silicosis, but the local rock contained much lower amounts of silica than mines in the Rockies. And the wet conditions of the Comstock mines reduced the friction between drill and rock. Nor did the humidity allow the dust particles to move as liberally through the air. However, miners frequently contracted a variety of respiratory illnesses other than silicosis. Pneumonia and tuberculosis, in particular, plagued miners' lungs and exhibited symptoms similar to those manifested in silicosis. Even when very little silica existed in the rock, mechanical drills still produced corrosive dust that could cut lung tissue and make workers more susceptible to other respiratory ailments. Although precise figures remain elusive, silicosis was eventually thought to be the leading cause of death among hard-rock miners in the nineteenth-century American West. And the limited data on diseases on the Comstock suggest that death rates from respiratory illness remained alarmingly high throughout the second half of the nineteenth century.[50] One study of coroner and burial records for Gold Hill and Virginia City between 1863 and 1904 attributes more than 40 percent of deaths to infectious and respiratory diseases. Tuberculosis was the cause of death in no less than a third of cases linked to "infectious and parasitic reasons." Among respiratory illnesses pneumonia killed more people than any other disease. Such high incidence of pneumonia and tuberculosis also makes sense given the thousands of coughing miners who repeatedly exchanged microbes in the cramped, damp, and hot spaces of the mines.[51]

The diverse environmental conditions on the Comstock and period death records suggest that prevalent causes of mortality varied from location to location. Where the mines were extraordinarily hot and wet, both pneumonia and tuberculosis prevailed, and the dust that machine drills and explosions produced abraded lung tissue and made miners more susceptible to a variety of respiratory ailments. In drier, cooler portions of the district, where heavier concentrations of silica might have existed, silicosis may have worked in tandem with tuberculosis to claim the health and lives of miners. Historian Christopher Sellers points out that it remains difficult to assess "the biological impact of the workplace." Silicosis in particular generated a great deal of confusion and controversy. The disease shared symptoms with many other illnesses common outside industrial working environments, making it difficult for many miners to prove their cases in a court of law.[52]

Black powder and dynamite posed yet other health risks and occupational hazards. Some of the risks in deploying such tools were more insidious than first appearances would indicate. After drilling holes in the rock face, miners would insert a series of charges. They then attached these volatile capsules to fuses that could be ignited from a distance. After igniting the fuses, men scrambled away and awaited the blast(s) that followed. Initially, mines employed black powder to blast the rock, and some companies briefly experimented with nitroglycerine. Although black powder was, due to its superior stability, less powerful than nitroglycerine, it remained the explosive of choice until the arrival of dynamite. Nitroglycerine's volatility had proved unpredictable and incredibly dangerous. Though more powerful than black powder and less volatile than nitroglycerine, dynamite could explode prematurely or much later than expected. Dynamite also could fail to ignite on schedule and remain inert until someone ignited the hidden charge by accident hours, days, or months later.[53]

A lack of communication between co-workers could exacerbate the risk of mining accidents. Miners who broke for lunch or ended their shift occasionally left undetonated cartridges in the rock face. Sometimes men on the following shift came in and started drilling only to ignite the explosives. The concussion caused by triggering these cartridges with a machine drill was especially fierce, as the compressed air powering the tool could combine with the hidden explosives to create terribly violent accidents. Charges could also be ignited by the movement of rock caused by work on the levels below and above. However, sometimes simple recklessness on a miner's part could be his own undoing. In 1880, Charles Richards was working in the Forman shaft and used the handle on the end of his shovel to tamp down a fuse onto a charge of "giant powder." The cartridge exploded, killing both Richards and a co-worker.[54]

Charles Richards's story, though tragic, also offers evidence of improved vigilance on the part of *most* miners and their employers. Those who testified at the coroner's inquest confirmed that three of twenty-five charges had failed to explode during the previous shift. These details were passed on from one shift boss to the next in order to prevent any accidents. In response, shift bosses halted all drilling to avoid unintentional explosions. Miner George Greening provided the most pointed eyewitness testimony when he described watching Richards from the other end of the shaft. He recounted that Richards used his shovel handle to force the fuse into the hole containing the charge of powder. A perplexed Greening explained that most veteran miners would never even consider such a dangerous technique. "In putting a cartridge into a hole, if we find it tight, we don't use force in pushing it down. I would press the powder down with a stick. I think the tamping caused the explosion. The caps are exploded by fire and not by concussion." The jury decided that no one but Richards could be held responsible for the accident.[55]

Explosions and corrosive dust were part of the Faustian bargain generated by the increased productivity of labor-saving devices. On the Comstock at least, the injuries miners sustained and diseases they contracted as a result of the mechanization of their profession were not misfortunes that better medical knowledge or militant protests were likely to prevent. Mines on the Comstock adopted these new technologies rather late in the history of this mining district but very early in the history of these technologies. When combined with the handsome wages they earned and the host of other geological and hydrothermal hazards they confronted underground, it is possible that the dangers posed by mechanized drills still remained poorly understood in the early 1870s and that such threats would be considered lesser evils on the Comstock.[56]

CONCLUSION: CONSEQUENCES OF THE COMSTOCK

Economic and environmental forces converged on the Comstock Lode to propel hard-rock mining onto a trajectory dominated by large corporations and highly industrial technological systems that allowed miners to work at ever-deeper levels, in ever–more inhospitable places. Not until 1877–78, with the development of large silver deposits in Leadville, Colorado, and gold deposits in the Black Hills of South Dakota, did any industrial precious-metals mining operations in the United States begin to rival the scale of the Comstock Lode's workforce or elaborate technological systems. By then, the Comstock had begun its steady decline.[57]

Miners, managers, and investors carried extremely valuable knowledge, technologies, and capital with them from the industrialized hard-rock operations of the Comstock Lode to other mining regions across the North American West. These men and the resources and experience they took from the Comstock contributed to the development of many of the largest mining operations in the world for the remainder of the nineteenth century. From the Hearst syndicate's Homestake Mining Company in South Dakota to Marcus Daly's Anaconda Copper Mining Company in Montana, men who made millions from investments in the Comstock and those who accumulated valuable experience working in its subterranean environments would continue to shape much of the industry for the rest of the nineteenth century. In the case of Daly and copper, trends of capital-intensive production and rapid corporate expansion demonstrated that many of the patterns that emerged on the Comstock Lode would also play a large role in shaping the growth patterns in the industrial mining of base metals.

Similar to the ways that men who cut their teeth in the California Gold Rush brought capital and technical knowledge to northwestern Nevada in the early 1860s, many rank-and-file Comstock veterans brought their experiences to bear in scores of mining districts across the North American West in the 1870s and 1880s. The Comstock's influence radiated across the industry and shaped large-scale flows of both capital and labor.

Mining on the Comstock Lode witnessed the first successful attempts to organize western miners into large industrial unions of "nonskilled" laborers. Successful unionization relied on a solidarity forged by the shared risks of death and injury, which seemed to lurk everywhere underground. Both geological and hydrological features contributed to the instability and unpredictability of the subterranean environment. In response to these challenges, corporate capital enabled the construction of massive technological systems to mitigate environmental hazards and facilitate expansion, extraction, and maintenance. These systems played a large role in making mining on the Comstock more productive and in some ways, perhaps, safer. However, cages, steam engines, mechanized drills, blasting powder, and compressed air and ventilation systems introduced new variables into an extremely volatile work environment.

Finally, what went on in the heart of the Comstock's working environments also shaped the capital-intensive technologies and high labor costs needed to expand operations underground. In an effort to uncover and extract subterranean deposits, corporations and workers constructed new underground environments that strained the reach of technical innovation, human endurance, and financial solvency. Through the sale of stock, corporations accumulated sufficient capital to flex the financial muscle necessary for the operation of mines on the Comstock, but profitable and unprofitable companies alike still relied overwhelmingly on the human muscle power of scores to hundreds of workers to perform the strenuous and debilitating labor required underground. The same volatile environmental forces of heat, water, explosions, and shifting and falling rock that made mining so expensive for corporations made working conditions extremely dangerous for miners. Hundreds lost their lives, and hundreds more suffered horrific accidents that left them permanently disabled.

Environmental forces underground posed grave threats to both physical safety and large capital investments. These obstacles demanded technical innovation, thousands of laborers, and huge sums of capital to defray the costs of doing business on the Comstock Lode. These relationships ultimately made corporations the most expedient business organizations for meeting persistent economic, environmental, and technical challenges. Although scores of corporations failed to achieve the efficiencies found in consolidation, and although few actually reaped the profits hoped for by Rossiter Raymond, they nevertheless contributed to the expansion of an industrialized business paradigm and the creation of working environments that would soon dominate mining.

Such forces further facilitated the rapid spread of corporations across the American industrial landscape. The same environmental forces and technological systems that contributed to the growth of corporate power also structured the daily lives, labor, and safety of the thousands of men who worked underground. Their individual and collective stories, both in life and in death, provide us insights into how working-class Americans experienced the transformative impacts of industrialization and the rise of corporate capitalism in the United States in the second half of the nineteenth century.

NOTES

1. *(Markleeville, CA) Alpine Chronicle,* July 2, 1870. Alfred Doten, *The Journals of Alfred Doten* (Reno: University of Nevada Press, 1973), 1097.

2. Roy Rosenzweig and David Thelen, *The Presence of the Past: Popular Uses of History in American Life* (New York: Columbia University Press, 1998), 93. For an example of the ways that death records can help historians illuminate and partially reconstruct the lives, routines, and material worlds of working people, see Barbara Hanawalt, *The Ties That Bound: Peasant Families in Medieval England* (Oxford: Oxford University Press, 1989).

3. Maureen Jung, "The Comstocks and the California Mining Economy, 1848–1900: The Stock Market and the Modern Corporation" (Ph.D. dissertation, Department of Sociology, University of California, Santa Barbara, 1988), 69.

4. Production figures for the Comstock remain contested and somewhat unclear. What is clear, however, is that the district's mines produced bullion that was worth well over $300 million and, according to some estimates, may have come close to nearly $400 million. Of course, the Comstock was the largest silver strike in the history of the United States, but the gold extracted from these mines generated far more profits in relative terms because of the higher price fetched by this precious metal when compared to silver. For estimates of production totals, see Eliot Lord, *Comstock Mining and Miners* (Washington, DC: U.S. Government Printing Office, 1883), 407–35; Grant H. Smith, *The History of the Comstock Lode, 1850–1997,* with new material Joseph V. Tingley (Reno: Nevada Bureau of Mines and Geology; University of Nevada Press, 1998), 310–14; and Rodman Wilson Paul, *Mining Frontiers of the Far West, 1848–1880,* revised and expanded by Elliot West (Albuquerque: University of New Mexico Press, 2001), 56–86.

5. Death Records Entered by Storey County Recorder, April 30, 1879, to December 10, 1880, Storey County Recorder's Office, Virginia City, NV; Richard E. Lingenfelter, *The Hardrock Miners: A History of the Mining Labor Movement in the American West, 1863–1893* (Berkeley: University of California Press, 1974), 23–30; Gunther Peck, "Manly Gambles: The Politics of Risk on the Comstock, 1860–1880," in *Across the Great Divide: Cultures of Manhood in the American West,* ed. Matthew Basso, Laura McCall, and Dee Garceau (New York: Routledge, 2001), 73–96.

6. Paul, *Mining Frontiers*, 57–58, 128–29, 192–96.

7. Joseph L. King, *History of the San Francisco Stock and Exchange Board, by the Chairman, Jos. L. King* (San Francisco: J.L. King, 1910); Michael J. Makley, *The Infamous King of the Comstock: William Sharon and the Gilded Age in the West* (Reno: University of Nevada Press, 2006), 14, 214.

8. Many historians have argued for hard-rock mining's inherently industrial character. Regrettably, the industry still seems relegated to a separate, regional category that obscures the far more compelling connections between eastern and western patterns of industrialization in the United States in the second half of the nineteenth century. For the best treatment of mining as an industry, see Mark T. Wyman, *Hard Rock Epic: Western Miners and the Industrial Revolution, 1860–1910* (Berkeley: University of California Press, 1979).

9. Rossiter W. Raymond, *Statistics of Mines and Mining in the States and Territories West of the Rocky Mountains* (Washington, DC: U.S. Government Printing Office, 1870), 98–99.

10. Ibid.

11. For more on debates over the advantages and inefficiencies of corporations in a relatively new and rapidly expanding economy, see Richard White, *Railroaded: The Transcontinentals and the Making of Modern America* (New York: W.W. Norton, 2011); Charles Perrow, *Organizing America: Wealth, Power, and the Origins of Corporate Capitalism* (Princeton: Princeton University Press, 2002); William G. Roy, *Socializing Capital: The Rise of the Large Industrial Corporation in America* (Princeton: Princeton University Press, 1997); Gerald Berk, *Alternative Tracks: The Constitution of the American Industrial Order, 1865–1917* (Baltimore: Johns Hopkins University Press, 1994); Olivier Zunz, *Making America Corporate, 1870–1920* (Chicago: University of Chicago Press, 1990); Naomi R. Lamoreaux, *The Great Merger Movement in American Business, 1895–1904* (New York: Cambridge University Press, 1985); Alfred D. Chandler, Jr., *The Visible Hand: The Managerial Revolution in American Business* (Cambridge, MA: Belknap Press, 1977).

12. Dan De Quille, *The Big Bonanza: An Authentic Account of the Discovery, History, and Working of the World-Renowned Comstock Lode of Nevada* (Hartford, CT: American Publishing, 1876), 90. For a discussion of the climatic, hydrological, and geological obstacles that threatened the lives and fortunes of miners in the Yukon Gold Rush, see Kathryn Morse, *The Nature of Gold: An Environmental History of the Klondike Gold Rush* (Seattle: University of Washington Press, 2003), 89–114. Also see Thomas Andrews, *Killing for Coal: America's Deadliest Labor War* (Cambridge, MA: Harvard University Press, 2008).

13. Dan De Quille, "Notes on System of Timbering Mines in Square-Sets," Box 3, Dan De Quille Papers, Bancroft Library, University of California, Berkeley.

14. Ibid.; Paul, *Mining Frontiers*, 64.

15. Charles Bonner to unknown recipient, April 4, 1868, Box 11, Folder 202, Savage Mining Company and Associated Records, Beinecke Library, Yale University, New Haven, CT.

16. 1870 U.S. Census, Storey County, NV (Mortality Schedule). Following are the names and ages of all the miners who died, by date. August 1869: Henry Melchan (47), E. Caine (52) and J. L Roach (50). October 1869: Pat Dalley

(38). December 1869: Daniel White (36) and C.E. Brightmore (38). February 1870: J. Monagere (45). April 1870: John (33) and Walter (35) Tregallis. May 1870: Andrew Johnson (33) and Samuel Roberts (29). Others also may have been killed by caving accidents or fallen timbers or rock during these months, but many causes of death as listed in the census data are attributed to unspecified categories such as "Accident-Mining" or "Killed in mine."

17. Testimony of B.G. Williams, Foreman of Savage Mine, Coroner's Inquest, attached to Certificate of Death for Richard S. Kitz, written by J.D.F. Hodges, Public Administrator and Ex-officio Coroner, Storey County, Virginia City, Nevada, October 5, 1877, Storey County Clerk's Office, Virginia City, Nevada.

18. *(Virginia City, NV) Daily Territorial Enterprise* September 12, 1874.

19. Ibid.

20. De Quille, *The Big Bonanza,* 391.

21. J. Ross Browne, *Crusoe's Island, California and Washoe* (New York: Harper and Brothers, 1864, reprinted in *A Peep at Washoe and Washoe Revisited* (Balboa Island, CA: Paisano Press, 1959, 179.

22. Lord, *Comstock Mining and Miners,* 296–97.

23. Adolph Sutro, *The Advantages and Necessity of a Deep Drain Tunnel, for the Great Comstock Ledge* (San Francisco: n.p., February 1865), 16.

24. John McPhee, *The Annals of the Former World* (New York: Farrar, Strauss and Giroux, 1998), 46–50, 101–5.

25. Ibid.

26. James Fair to Weller, February 2, 9, 16, and 23, and March 2, 1878, Box 10, Records of Ophir Silver Mining Company, Special Collections, Getchell Library, University of Nevada, Reno.

27. John A. Church, *Accidents in the Comstock Mines and their Relation to Deep Mining,* paper presented at the Pittsburgh Meeting of the American Institute of Mining Engineers, May 1879 (printed for the author), Beinecke Library, Yale University, New Haven, CT.

28. John A. Church, *The Heat of the Comstock Mines* (Washington, DC: U.S. Government Printing Office, 1878.)

29. Ibid.; D.B. Barton, *A History of Tin Mining and Smelting in Cornwall* (Truro, Cornwall, UK: privately printed, 1967), 217.

30. James Hezlep Galloway, "Diary," entry for March 6, 1875, Bancroft Library, University of California, Berkeley.

31. H.H. Smith to J.D. Fry, "Weekly Report," August 31, 1872, Belcher Silver Mining Company, Box 1, Special Collections, Getchell Library, University of Nevada, Reno.

32. Testimony of William Burns and Neil McFarland, Coroner's Inquest, Certificate of Death for Thomas Wilson, written by J.D.F. Hodges, Public Administrator and Ex-officio Coroner, Storey County, Virginia City, NV, March 3, 1877, Storey County Clerk's Office, Virginia City, NV; Death Record, Phillip Harrington, August 13, 1880, Storey County Recorder, Virginia City, NV.

33. *Daily Territorial Enterprise,* August 1, 1879.

34. H.H. Smith to J.D. Fry, "Weekly Report," September 28, October 5, and December 7, 1872, Belcher Silver Mining Company, Box 1, Special Collections, Getchell Library, University of Nevada, Reno.

35. H.H. Smith to J.D. Fry, "Weekly Report," October 5, 1872, Belcher Silver Mining Company, Box 1, Special Collections, Getchell Library, University of Nevada, Reno; Records of Ophir Silver Mining Company, Box 10, Special Collections, Getchell Library, University of Nevada, Reno.

36. Elizabeth Baines, "Mortality and Migration: A Study of the Comstock Lode" (Master's thesis, University of Nevada, Reno, 2000), 50–51; Lord, *Comstock Mining and Miners*, 374; De Quille, *The Big Bonanza*, 150.

37. *Virginia Evening Chronicle*, August 29, 1873.

38. Church, *Accidents*, 7.

39. Ibid.

40. Doten, *Journals*, 1103; *Daily Territorial Enterprise*, September 12, 1874.

41. Lord, *Comstock Mining and Miners*, 402–3.

42. U.S. Census Bureau, *Wages of Mechanics Employed in the Manufacturing and Mining Industries*, United States Census, 1880, Volume 20 (Washington, DC: U.S. Government Printing Office, 1885), 544–63.

43. For the size of the workforce and the best social history and general overview of the Comstock Lode, see Ronald M. James, *The Roar and the Silence: A History of Virginia City and the Comstock Lode* (Reno: University of Nevada Press, 1998), 26, 92, 139; Baines, "Migration and Mortality."

44. Galloway, "Diary," entries for March 1875 through 1877.

45. De Quille, *The Big Bonanza*, 244.

46. Ibid., 221.

47. Ibid.

48. Lingenfelter, *Hard Rock Miners*, 17–18.

49. Wyman, *Hard Rock Epic*, 84–93; Baines, "Mortality and Migration," 15, 50–65. Historian Ronald James argues that the Comstock miners as a rule did not suffer from silicosis, because the local rock lacked silica. While there is no consensus, contemporary Dan De Quille noted the frequency of the disease. Some historians of hard-rock mining have emphasized the ubiquity of silicosis, while others have pointed to the uneven distribution of high and low rates across mining districts in the West. Whatever the reality, one estimate claims that between 1890 and 1930 more than half of the hard-rock miners working in the West died from the disease. James C. Foster, "The Western Dilemma: Miners, Silicosis, and Compensation," *Labor History* 26:2 (Spring 1985), 268–87; James, *The Roar and the Silence*, 121.

50. Ronald C. Brown, *Hard-Rock Miners: The Intermountain West, 1860–1920* (College Station: Texas A&M University Press, 1979); Duane A. Smith and Ronald C. Brown, *No One Ailing except a Physician: Medicine in the Mining West, 1848–1919* (Boulder: University Press of Colorado, 2001).

51. Wyman, *Hard Rock Epic*, 84–93; Baines, "Mortality and Migration," 15, 50–65.

52. Baines, "Mortality and Migration," 15, 50–65; Foster, "The Western Dilemma," 268–87; James, *The Roar and the Silence*, 121; Christopher C. Sellers, *Hazards of the Job: From Industrial Disease to Environmental Health Science* (Chapel Hill: University of North Carolina Press, 1997), 21. For more on the history of occupational illness, see Paul Weindling, ed., *The Social History of Occupational Health* (London: Croom Helm, 1985).

53. Popularly known as giant powder, dynamite was invented by Alfred Nobel in 1866. See Mark Wyman, "Industrial Revolution in the West: Hard-Rock Miners and the New Technology," *Western Historical Quarterly* 5:1 (January 1974): 41–42.

54. *Daily Territorial Enterprise,* December 11, 1880.

55. Ibid.

56. De Quille, *The Big Bonanza,* 145–48; Wyman, *Hard Rock Epic,* 104–6.

57. Paul, *Mining Frontiers,* 57–58, 128–29, 192–96.

CHAPTER 5

Dust to Dust

*The Colorado Coal Mine Explosion
Crisis of 1910*

THOMAS G. ANDREWS

"MORE MEN AND MACHINERY FOR C.F. & I. MINES": The *Trinidad Chronicle-News* headline seemed to herald welcome news to readers in Colorado's Las Animas County, heart of the most productive coal-mining region west of the Mississippi River. Surging demand for coal, the *Chronicle-News* explained in late January 1910, "made it necessary" to increase production. Colorado Fuel & Iron (CF&I), the Rockefeller-controlled behemoth that dominated the western fuel trade, was consequently scrambling to hire workers "from every employment agency in the state." CF&I managers had especially high ambitions for the company's Primero mines, which plunged beneath the Rocky Mountain foothills a dozen miles north of the New Mexico border. The coal giant announced that Primero, which already boasted "the largest producing capacity of any [coal mines] in Colorado," would be expected to produce "all that it can give up."[1]

Just a week later, another *Chronicle-News* headline carried altogether darker tidings: Primero had exploded for the second time in three years, "claim[ing] an awful toll of death." Seventy-seven men and adolescent boys were toiling underground on January 31 when a "ball of flame" ripped through the mine around 4:30 P.M. Newspaper reporters invoked natural phenomena to describe the blast. "The explosion," one wrote, "came ruthlessly and like a thunderbolt from a clear sky." "Like a volcano," another claimed, "fire and debris and smoke and dead men's bodies spouted out the entrance to the workings." As the explosion snuffed out dozens of lives, it propelled materials generally safely confined beneath the earth—"debris" and parts of "dead men's bodies"—out of Primero's depths and onto the surface above. In the process, this eruption of profound and unsettling potency upset the ostensibly clear and stable boundaries that ordinarily segregated the terrestrial from the subterranean, the natural from the artificial, and the Promethean from the infernal.[2] The Primero disaster thus offered a revealing case study in the complicated and sometimes violent interconnections between environmental, social, and political dynamics that had long prevailed in and around the world's coal mine workscapes.[3]

Although the explosion killed 76 men, it turned turn out to be but the first of three blasts to strike Las Animas County miners that year, claiming a grand total of more than 200 lives. The Starkville mine exploded in October, killing 56, followed a month later by the deaths of 79 more at Delagua.[4] Never before in the annals of world mining history—and never since—would three mines lying in such close proximity unleash so much death and destruction in such short succession.[5]

A highly contingent conjuncture of interactions between people and nonhuman nature caused all three blasts. When sparks or flames loosed the vast potential energy latent in extraordinarily volatile mine atmospheres, they catalyzed two additional material processes: first, the toxification of mine atmospheres as the sudden ignition of coal dust, methane, and other substances saturated mine workscapes with a deadly stew of carbon monoxide and other combustion by-products known to Colorado miners as afterdamp; and second, the rapid decomposition of the men and mules killed, whether by the initial blast, the inhalation of afterdamp, or other causes. Together, combustion, toxification, and decomposition not only drove the course of rescue and recovery efforts but also mobilized popular discontent, legislative reform, and labor militancy.

Environmental histories of mining have rightly emphasized the intensive, far-reaching, and often long-lasting ecological consequences of human efforts to unearth, transport, process, and consume mineral resources. Most retrospective assessments of mining's environmental consequences, however, adopt a bird's-eye view of terrestrial landscapes while ignoring subterranean mine workscapes. This essay, by contrast, maintains a tight spatial and temporal frame on the tunnels and chambers of the Primero, Starkville, and Delagua mines, showing how disasters animated by sudden and extraordinarily volatile reactions between labor and nature reenergized long-festering conflicts among mine workers, mining companies, and the broader public. The resulting struggles over who should bear the mounting costs that coal-powered industrialism inflicted upon workers' bodies and Colorado's democratic institutions laid the groundwork, in turn, for the great coalfield war of 1913–14, the deadliest strike in U.S. history.[6]

Wherever coal was mined and consumed, vast and capricious energies threatened to erupt. Between the Civil War and the early twentieth century, coal-powered industrialization transformed the United States into what legal historian John Fabian Witt rightly calls "the accidental republic." As railroad derailments, streetcar collisions, boiler explosions, smelter accidents, and other coal-fired calamities filled newspaper columns and cemetery rows, Americans suffered "an accident crisis like none the world had ever seen."[7]

The burdens of this crisis, as Colorado's coal miners knew only too well, fell disproportionately on working-class Americans. In the five years preceding the Colorado mine tragedies of 1910, American coal miners perished at three times the rate of their U.K. counterparts, leading one British expert to scoff that the United States "enjoy[ed] the unenviable reputation of being the most backward of the civilized nations" in terms of coal-mine safety.[8] Colorado's coalfields figured among the most "backward" and dangerous in the nation, with a fatality rate roughly double the national average for the period from 1884 to 1912.[9] Between 1884 and 1907, a total of ten large blasts killed 239 Colorado colliers.[10] Contemporaries often likened the appalling toll of underground labor to industrial tribute, with the bodies of dead miners serving as sacrifices to King Coal. The Western Federation of Miners (WFM) expressed a similar idea even more sharply. "Human flesh in the garb of labor," a story in the union magazine lamented after the Delagua disaster, "is the cheapest commodity in the world."[11] Dead workers, as the WFM and Colorado's coal miners recognized, constituted a kind of industrial by-product.

The capricious nature of Colorado's collieries resulted, in part, from the particularities of Colorado's geology and climate. Methane and other highly explosive substances permeated many of the state's fuel deposits. To make matters worse, Colorado's coal seams typically contained little water. The famously arid Rocky Mountain atmosphere further exacerbated the risk of igniting the region's chemically dry coal; it also made it more difficult and costly for mine operators to mitigate dust hazards by sprinkling underground workings. Natural conditions alone, however, fail to explain the string of disasters that afflicted Las Animas County in 1910. Mine operators aggravated the dangers mine workers faced not only by skimping on sprinkling but also by hiring inexperienced workers who only dimly understood the hazards of coal mining; by co-opting local political, judicial, and law enforcement systems; and by adroitly lobbying lawmakers to stymy stricter safety regulations.[12]

JANUARY 31, 1910: PRIMERO

Just moments after Primero exploded, relatives of those caught underground and off-duty mine workers rushed frantically toward the mine opening. Upon reaching the tunnel, though, they discovered that "smoke and flame poured forth" from the pit "as if it were the crater of a volcano." An hour passed before "the air became clear," at which time "volunteer rescuers dashed into the air shaft next to the main tunnel. [. . .] Quickly half-crazed wives followed," only to find "portions of human bodies in the debris." A macabre newspaper account published the next day described how the desperate women rushed forward "in an endeavor to ascertain" if these body parts "belonged to their missing loved ones."[13]

CF&I managers and law enforcement officers worried that the sight of such carnage might provoke disorder or even violence outside the mine. To keep the peace, two National Guard officers, sheriffs from Las Animas County and neighboring Huerfano County, and several deputies hastened to Primero by train.[14] Meanwhile, superintendents and bosses from other CF&I properties joined forty-five of southern Colorado's most experienced coal miners just outside the wrecked mine.[15] "As gently as possible," the *Denver Post* explained, "the women and children were pressed back, ropes were stretched and a large space thus cleared to permit the men to work without interruption." Rescue crews donned breathing helmets supplied by CF&I's rescue car and plunged into Primero's shattered entries.[16]

These so-called helmet crews pushed through immense piles of debris, extinguished "a smoldering fire" in one of the mine's entries, and placed fans to bring fresh air underground. As they worked, they found the listless body of Leonardo Virgen, "a young Mexican, who came here recently from the sister republic," buried beneath "a heap of a dozen dead men and half as many dead mules."[17] When rescuers trained their flashlights on Virgen's face, his "eyes suddenly opened." The Mexican "sat up" and, in "an uncanny performance for one" seemingly entombed by a pile of corpses "blinked his eyes and quavered: 'Please, boss, can I go home now?'"[18] The rescuers "set to work to get Virgen" out of the mine workings; "in a few moments he was landed safely outside the chamber of the dead"—the sole survivor of the Primero blast.[19]

An incomplete tally of those killed in the "indescribabl[y] damage[d] mine" reflected the tremendous diversity of southern Colorado's coal mine workforce, including 34 Austrians, 9 Koreans, 6 Anglos, 6 Croatians, 5 African Americans, 4 Hungarians, 4 "Mexicans" (including Virgen's partner), 3 Italians, a Welshmen, and a German.[20] The *Post* expounded upon the explosion's human cost: "Cold in death forty-three men lie in sheds and warehouses near the main tunnel of the Primero coal mine, victims of their calling."[21]

CF&I placed these makeshift morgues "under guard, and no one [wa]s allowed in there except the physicians. As fast as the bodies were taken from the mine they were covered with blankets."[22] A day after the tragedy, Burney Sipe, deputy county coroner and proprietor of the Trinidad Undertaking Company, found that his firm "d[id] not have enough coffins to take care of the bodies," obliging him to order "eighty-five coffins to be sent down by express from Denver."[23] For its part, Trinidad's Catholic cemetery was forced to hire ten men willing to work around the clock, "digging graves for the unfortunate victims."[24]

By smashing many of the mine's passageways and demolishing its ventilation system, the blast had blocked escape routes while filling the mine with afterdamp.[25] Dauntless rescue crews nonetheless threw themselves into harm's way in hopes of saving any men who might have survived the explosion and taken refuge in pockets of good air. Three days after the mine blew up, however, most observers acknowledged that the heroic struggle to save the living had segued into a grim and dangerous campaign to recover the dead, as well as to account for the disaster and repair the mine workings so that extraction could resume.[26]

Viewed coldly and from a distance, the ensuing recovery efforts traced a curious but revealing arc. Exhuming corpses buried by the Primero explosion proved both dangerous and expensive. Workmen labored tirelessly to remove the rapidly decaying bodies of their comrades to the surface above, while state and company mine safety experts inspected each corpse for clues regarding the explosion's cause and course. After these examinations, undertakers placed the bodies in coffins, which were then hauled by horse-drawn wagons and coal-burning trains to family homes and houses of worship. Once the living had a chance to grieve over their loved ones' remains, grave diggers reburied the dead, thus completing a circuit that carried the lifeless bodies of the explosion's victims out of the profane earth of Primero's workings and into sacred cemetery ground.

Dead bodies traveled from pit to grave not simply because deep-seated traditions among Colorado's extraordinarily diverse mine workforce demanded proper burial, but also because the dead continued to make claims on the living for as long as their earthly remains lingered underground. Disasters temporarily transformed Primero's tunnels and chambers from CF&I's private property into quasi-public sites of mourning— "tombs," as newspaper accounts frequently characterized them. To resume production at Primero, CF&I had to relocate the bodies of the explosion's victims to more appropriate final resting places while also containing the risks that the mutilated and rotting corpses of the company's former employees posed to CF&I's precarious reputation among southern Colorado's miners and the West's coal-consuming public alike.

Indeed, the corporation went to great lengths to conceal Primero's horrors from public view. Two days after the explosion, the *Post* claimed that "the rescuers are only bringing to the surface bodies that while badly mutilated and burned, have no members missing." From that point onward, CF&I apparently elected to keep most of the corpses its crews found within the mine, removing these brutalized bodies long after nightfall to prevent the anguished crowds congregating just outside the mine-mouth from laying eyes on them. One account, for instance, related reports "from volunteers who had penetrated far into the workings that the entries and tunnels are strewn in places with limbs and trunks of the dead," with "some of the miners [. . .] blown into such small particles that it will be impossible to place them together, and thus ascertain whether these portions represent one or more bodies."[27] In identifying the dead, Deputy Coroner Sipe and his colleagues often had nothing more to rely upon than the numbered brass tabs known as checks, which miners used to claim credit for the cars of coal they loaded.[28] Some bodies, though, lacked even these commonplace markers; the *Post* reported that six bodies "remain[ed] unidentified and will probably be buried that way, as they are torn and mangled beyond recognition."[29]

The rapid decay of these "torn and mangled" bodies mandated swift burial for Primero's victims. "Funerals," the *Post* reported, "will continue as fast as the corpses can be shipped out."[30] The *Post* noted on February 2 that a return of cold weather "was welcomed as the bodies in the morgue, where the steam had been left on, had begun to putrify [sic] and the odor was almost unendurable."[31] Sixteen bodies recovered from the mine and sent to Trinidad on the evening of February 3 were decomposing so swiftly that they had to be interred the very next day.[32]

The outpouring of grief prompted by the hasty burial that followed temporarily blunted the outrage that many coalfield residents felt toward CF&I. One paper declared a mass funeral of disaster victims held on February 4 to be "the saddest [. . .] cortege the city of Trinidad has ever witnessed."[33] The eulogy delivered by Primero's priest struck a *Post* reporter as "the most powerful sermon I have ever heard." The journalist consoled himself that the disaster had at least yielded one glimmer of good: "Wonderful," he exclaimed, "the manner in which this commanding sorrow has made the whole country of Primero kin."[34]

And yet the solidarity extolled by reporter and priest alike was already wearing thin. The taint of death and disaster that loomed over Primero drove more than a few of its residents to flee. "Many of the miners," the *Post* reported two days after the explosion, "are rapidly deserting the camp as rats desert a sinking ship." The newspaper caricatured those most inclined to abandon the camp as "superstitious foreigners [who] are afraid to go back to work when the mine is repaired again, and so fast as they can gather up their effects they are pouring out."[35]

Another journalist's excursion into the mine itself hinted at the challenges that recovery workers faced as they struggled to reclaim the dead. "Yesterday I invaded this dark and hampered underground territory," a *Post* reporter related, "and for the first time realized in some measure the awfulness and extent of the explosion. There is only one description possible." Primero looked "as if hell itself had been purged with a tartar emetic and had spit out its spiteful vengeance into every nook and corner of this midnight microcosm."[36] The mass deathscape described in this passage constituted the mine workscape's doppelganger.

A few days earlier, the *Denver Republican* had morbidly explained that the bodies still remaining underground "are so torn and dismembered that the rescue work will consist mainly of picking up pieces of bodies. Rescuers, one after another," this account claimed, "have refused to go into the mine a second time, declaring that they could not bear to walk over the remains of human bodies."[37] The recovery effort slowed considerably because of the unwillingness of recovery workers to walk on the dead; their fear that "not another recognizable body" would be found; and the dawning realization that most or all of the corpses still underground lay "buried beneath tons of dirt, rock and timbers," strewn amid "fallen debris varying in depth from two to ten feet."[38] Although recovery crews had removed fifty bodies from the mine in the first few days after the explosion, the remaining dead took several more weeks to locate and extract.[39] Eight days after the blast, crews found seven additional bodies, most "in such a terrible condition from decomposition" that coal company officials were trying "to keep them out of sight until chemicals can be used upon them so that they will be in some sort of presentable shape for burial."[40]

Colorado's state coal mine inspector, who had been "almost constantly" at work since the rescue began, was joined on February 3 by the deputy state labor commissioner and George S. Rice, the U.S. Geological Survey's leading expert on coal dust's explosive potential. Together, these officials scoured the mine "for some sign, some suggestion, even the faintest clew [sic], which might enable them to eventually reach a conclusion" regarding the explosion's cause.[41] The state mine inspector needed several more weeks to complete his investigation. In the meantime, though, he confessed before a coroner's jury that he "could find nothing in all the hours he has spent in exploring [the mine's] cavernous recesses, since the morning following the explosion, which would in any way cast reflections upon the management."[42] The jury, presumably swayed by this expert testimony, absolved CF&I "from all blame."[43]

The Primero explosion produced bereaved widows as well as dead miners. One woman reportedly went "temporarily insane" after the explosion claimed her husband, son, and brother, while another widow "slipped quietly into the chapel of the undertaking parlors" during one of the mass funerals held for workmen "literally blown to atoms" and released "a moan like that of a wounded animal." No less important, the tragedy also propelled the federal government and CF&I to devote additional resources to mine safety.[44] Although the House of Representatives had already approved legislation to establish the U.S. Bureau of Mines prior to the blast, the Primero tragedy may have contributed to Senate approval of the measure.[45] CF&I, meanwhile, brought in a representative from Draeger, a leading manufacturer of self-contained breathing helmets, to instruct southern Colorado miners in rescue techniques and helmet use. This move undoubtedly reflected a sincere concern for work safety among some company officials. But it also suggested that the CF&I hierarchy saw mine explosions as an unavoidable part of the company's business.[46] Helmets, after all, served to mitigate the loss of life only *after* an explosion had already occurred.

Because CF&I had money to make and the people of the West needed coal to burn, the company resumed production as quickly as possible. Nine days after the blast hurtled through the Primero workings, the *Republican* reported that "work [wa]s going on as usual in the western slope of the mine, and the disaster is fast being forgotten in the maelstrom of the day's activities."[47] Nearly three months later, rescue crews finally removed the last corpse from the pit.[48]

OCTOBER 8, 1910: STARKVILLE

Eight months after tragedy struck Primero, the Starkville mine, another major CF&I producer lying two dozen miles to the east, exploded "without warning" on the evening of Saturday, October 8. Fifty-six members of the night shift were laboring underground at around 10 P.M. when the mine atmosphere ignited with such force that it "shook the earth" in Trinidad, five miles away, thrusting "huge rocks and boulders [. . .] hundreds of feet" into the air outside one of Starkville's openings.[49] What one reporter aptly called "the resistless force of the explosion" destroyed the mine entrance "as [if] it were made of fragile eggshell instead of being carved from the granite side of one of the eternal hills."[50] With the mine "wrecked" and its ventilation system decimated, the forty or so men who had survived the initial detonation were entombed "like rats in a trap"—"struggling," as a *Denver Post* writer imagined, "with maniac energy to win their way to freedom, or lie dead in ghastly confusion amid the tons of debris hurled upon them from the volcano-like outburst."[51]

Within minutes, "heartrending scenes" began to unfold outside the mine-mouth. "Women, clasping babies to their breasts, rushed frantically to the spot beseeching, demanding some news of their husbands." The *Miner's Magazine* described the sight of "the families of the murdered men gathered around the mine" as "the most pathetic that have ever been painted in the columns of a daily journal."[52] CF&I officials implored the wives of Starkville's night-shift workers to return home, but "not a woman would consent to leave."[53] As at Primero, the company dispatched "a special force of deputy sheriffs" to establish "strict guard" and "prevent disorder of any character."[54] First, though, deputies had to save Frank Greet, the father of a motorman caught in the blast, from his own desperation. Greet "hastened to the mine [. . .] and heedless of the danger, started to enter the slope to search for his son." A guard stopped Greet and "explained that he could not possibly come out alive if he entered the house of death." As the deputy held Greet fast, the stricken man cried: "Let me go. [. . .] I know where my son is, and I will bring him out."[55]

Crowds of people descended upon the small camp from near and far "as fast as steam could carry them."[56] In the first few days after the blast, "thousands of passengers from Trinidad and surrounding camps" stepped off of "railroads and interurban lines."[57] Many onlookers "attracted by idle curiosity" made haste to Starkville. So, too, did at least a hundred experienced mining men from the newly formed U.S. Bureau of Mines, CF&I's other properties, and mines owned by other operators. Rescuers organized themselves into crews, then "pushed their way forward" "with [the] feverish intensity of madmen," risking their own lives to save their comrades while "surmounting difficulties," as one reporter phrased it, "which would have made the heart of less strong men quall [sic]."[58] Each sally into the mine, however, brought more bad news about the blast's extent and severity. "Only the most persistent optimist," the *Post* noted, "can contemplate the awful devastation wrought by Saturday night's catastrophe and expect to see a single one of those ill-fated miners emerge from his prison alive."[59]

Crews surfaced from the disaster site and told reporters about the "unequal battle" they were waging against poisonous afterdamp.[60] "Even the helmet men, with their modern protective apparatus, including reservoirs of oxygen to supply the lungs with pure air for breathing," one account glumly related, "could do little, and once were themselves overcome when their oxygen became exhausted." Reporters tended to portray oxygen helmets as technological wonders that bestowed superhuman powers on any man who donned one. In this case, though, journalists noted that the stricken helmet men themselves would have perished if not for the intervention of miners whose bodies possessed the unusual "ability to withstand impure air without artificial assistance."[61] This precarious rescue reinforced the mounting fear among knowledgeable observers that every member of the Starkville night shift must already have perished in the pit. "Way down in our hearts, boys," one crew leader confessed to reporters two days after the blast, "we know our poor comrades are dead."[62] Even the anxious crowds waiting outside the mine-mouth lost faith: "As the hours dragged slowly on hope began to die and tears strengthened into conviction that none of the entombed men would be found alive."[63]

Indeed, every man toiling underground at the time of the explosion had already breathed his last breath. Rescue crews "worked savagely throughout the day" on October 11, eventually discovering a group of corpses two miles from the surface.[64] Crews exhumed the bodies "from the debris" and placed them along the main entry. They then waited until nightfall to bring the corpses out in a "funeral car" bearing a "pitiful heap of bodies," two of them "badly burned" and the rest "blackened but not mutilated."[65] In the face of mounting controversy, CF&I Fuel Department manager H. H. Weitzel justified the company's policy of removing burned or mutilated corpses at night on both moral and practical grounds. Weitzel declared to a reporter that he "shr[a]nk from the awful scenes that would occur" if "the hundreds of agonized women and children who are already worn almost to the breaking point by the terrific strain they have endured since they first heard of the great catastrophe" had to look upon the bodies of the dead "while the light of day remains to reveal in all its ghastliness the story of what those miners must have passed through."[66]

The discovery of a second group of explosion victims whose "blackened [. . .] bodies" were "roasted beyond all resemblance to human beings" further solidified Weitzel's resolve to continue his "successful plan of protecting relatives of the cruelly tortured miners from the anguish of having to look upon the mutilated remains of their loved ones."[67] As usual, public officials in southern Colorado adhered to CF&I policy. "So horrible is the appearance of the victims," the *Republican* reported, "that Coroner Guilfoil will not allow even the relatives of the dead to gaze upon the features of their loved ones."[68] As the *Republican* noted, the shocking condition of the initial two groups of bodies recovered underground spurred the efforts of H. H. Weitzel and John R. Guilfoil to hide them from the gaze of an excitable public. The *Post* painted a phantasmagoric portrait of one group of victims found deep within Starkville: "Not a single one [. . .] was identified by the lineaments of the face, for the flesh of neck and cheeks was swollen prodigiously, in most cases to such an extent that the head seemed elephantine in size. [. . .] Through the swollen lips protruded tongues distended to thrice the natural size, clenched between the teeth in that eternal grip that always marks death by violence and accompanied by terrible pain."[69]

Bodies marred by violent and painful death presented obvious threats to public health and order. Yet coal company officials felt even greater concern about a group of corpses discovered on October 13 that were not at all "burned or mutilated." These remains, which were found "in a recumbent posture with the hands held over the face," offered clear evidence that some miners had survived the initial blast only to perish "from the insidious, fatal gas known as 'after damp.'"[70] The state labor commissioner later reconstructed these workers' final moments. About a dozen survivors, he believed, had rendezvoused in the minutes following the blast. Whether motivated by hunger, resignation, or that common human tendency to hold fast to routine even in the face of catastrophe, the miners sat down, opened their dinner pails, and began to eat. Rescuers later found their bodies arrayed in a circle beside their empty pails, suggesting that they had joined together to consume a last supper just before carbon monoxide snuffed out their lives.[71]

The assertion that afterdamp intoxication had caused a majority of the deaths at Starkville sharpened and focused the anger that the explosion had already kindled among many mining folk.[72] "I shall always feel," one rescuer declared, "that my comrades, whom I saw yesterday dead without a mark, would be alive and well this minute if they had not been penned in like dumb brutes in a mine that was poorly equipped with means of escape as well as poorly ventilated."[73] Suspicion spread that CF&I's insistence on exhuming the explosion's victims under cover of night stemmed not so much from the company's consideration for the relatives of the dead as from scheming by Weitzel and his underlings to prevent it from "becom[ing] known that some of the men in the mine at the time of the explosion were not killed outright by the force of the shock," "but died later, possibly while trying to effect their escape."[74] The devil of CF&I's misdeeds lay in the details of the disaster scene. Consider an account in the *Denver Republican* of a dead miner found "lying face downward in the dust" with "his hand tightly clutched in the neckband of a companion's shirt. Death," the article's author concluded, "had overtaken both while one was making an heroic effort to drag his companion to safety."[75] The coal corporation's negligence, a growing chorus of detractors argued, had pushed brave men arm in arm toward their doom.

Starkville's corpses, like Primero's, were removed from the earth so they could be reinterred in more fitting burial places. Recovery workers carried the remains of explosion victims to a temporary morgue outside the mine, equipped with "washing tubs, burlap for wrapping the bodies, and rude benches."[76] Soon thereafter, the bodies traveled to the "the side entrance of a local undertaking establishment" in Trinidad where "a carload of coffins" had been shipped in from Denver.[77] At least a few corpses were freighted back to their homelands by rail, retracing the migration routes along which the disaster's victims had ventured to the southern Colorado coalfields in the months and years preceding the disaster.[78] Most of the dead, though, were interred at Trinidad cemeteries.

The burials proceeded day after day without incident before a funeral procession for five Poles slowed to cross some railroad tracks and found the lifeless body of electrician Fred Foster "cut in twain by a coke car [. . .] in full view of those in the carriages."[79] Foster, one of the "heroes" who had risked his life during the rescue, was reportedly "fatigued to the verge of exhaustion from 22 hours' incessant toil in an effort to reach the bodies of the entombed victims" and had "dropped asleep on a railroad track" while walking home. Moments after the cortege found Foster's corpse, the news came from the mine that three more bodies, all in a "shocking state" of decomposition, had been found underground. "In any place not already sated with horrors," a *Post* writer acerbically noted, "such a thrilling conjunction of tragedies would have created tremendous excitement. Yet here, where death has become a familiar sight, no one uttered a sound."[80] A little more than a week after the explosion, CF&I crews brought out the last of Starkville's dead, with "many of the victims' bodies" showing signs that these men had run "a considerable distance to escape the deadly fumes [. . .] and were either overtaken by the afterdamp or ran into it while attempting to make their escape."[81]

Company officials no doubt felt relieved that the recovery effort was drawing to a close. "Many" workmen were reportedly "at the point of refusing to work because of the small wage paid them for the heroic effort put forward in recovering the dead bodies." As for CF&I managers, they were intent on "clearing the mine of debris and preparing it for operation."[82]

By that point, the coal corporation was facing much closer scrutiny than it had after the Primero disaster. Detractors assailed CF&I's poor record of sprinkling its mines. "The charge of neglect," the *Denver Post* claimed, "is being freely made by men who worked in the mine and knew the conditions that existed" leading up to the explosion.[83] The *Post* related interviews with anonymous authorities who claimed the company had decided a few years earlier not to sink a new tunnel at Starkville.[84] "It is now established," the Denver paper's editor scolded, "that the Starkville disaster could have been averted had the CF&I expended $10,000 for an air and escape shaft."[85] Some observers even faulted the company's conduct of the rescue operations. The *Post* blasted the company's "aimless, poorly directed attempts to enter the mine," marveling that CF&I, "with two thousand volunteers to draw from, should spend nearly thirty-six hours without accomplishing more than placing a fan in a few hundred feet inside the opening of the mine."[86] The *Post* drove home its portrayal of corporate negligence by reporting that every time Starkville's electric mine locomotive emerged from the workings, "carrying messages and performing other duties for mine officials," the machine short-circuited the supply of power to the single huge fan responsible for ventilating the entire wrecked mine, forcing rescue workers to "scamper in all directions" in order to alert crews still working within the mine to beat a hasty retreat to the comparatively pure air outside the mine opening.[87]

Company managers bitterly denied any and all criticisms. Time and again, they argued that CF&I had always placed safety first. Mine explosions, CF&I manager Weitzel told the *Post*, saddled the firm's balance sheet with a "dead loss": "All our bookkeeping entries from now on for months will be set down in red ink."[88] "Do you think for a moment," Weitzel implored, that "those poor fellows knew what struck them? Do you think that if there had been forty inclined air shafts"—each of which would have provided alternate escape routes—"they could have made use of them?"[89] "The very fact that the explosion occurred," CF&I counsel Fred Herrington reasoned with breathless cynicism, "proves that the air in our mine was fresh. Fire, you know, feeds on oxygen. I want you to quote me on this."[90]

At least some folks in the coal camps interpreted the Starkville explosion in a manner consistent with the original meaning of *disaster* as "a mishap due to a baleful stellar aspect."[91] Rumors circulated in the days following the blast that an "aged Mexican woman" had supposedly issued a chilling warning weeks before the disaster. "Blood," she had predicted, "will run in Starkville inside of six months.'" The woman based this prophecy on the actions of Haley's comet, which she claimed had "rested" during its transit across the night sky earlier that year "for a considerable space of time over [. . .] the mine entrance," thus signifying "a direct warning from heaven that a disaster of some sort was going to occur."[92]

Those with longer experience underground, by contrast, generally traced the causes of the Starkville tragedy to earthly forces instead of celestial divination. "The dust theory for the explosion," the *Post* claimed two days after the disaster, "is clung to by many who have thought over all possible causes for the catastrophe, and experts say that the presence of this dangerous explosive must be accepted as fact because all other theories have no foundation of fact to be based upon."[93] Even some company officials, the *Post* continued, "declare that in their opinion the deadly explosion was caused by fine coal dust being ignited by spontaneous combustion or else by a spark carelessly dropped by some one [fiddling] with an open lamp."[94]

On October 13, less than five days after the explosion, state mine inspector John Jones suddenly left Starkville for Denver, leading "clearthinking miners familiar with the conditions in the mine" to conclude that he had "solved the problem in his own mind of the cause of the explosion."[95] That same day, someone in Trinidad allegedly placed "a stick of dynamite, cap and fuse" beneath a private railroad car used by CF&I officials "as headquarters" for "rescue work at the ill-fated Starkville mine," prompting the company to dispatch the head of its in-house detective force to the southern coalfields.[96] Before a coroner's inquest could meet or arrive at any verdict regarding the cause of the Starkville disaster, though, a third mine exploded.

NOVEMBER 8, 1910: DELAGUA

In the early afternoon of Election Day, "generally observed as a sort of holiday by mine workers," more than 150 men were working underground at Victor-American Coal and Coke Company's Delagua No. 3 Mine when a fire broke out.[97] At 1:57 P.M., flames from a burning door detonated coal dust thrust into the air by a mine locomotive as it sped toward the surface to summon reinforcements and retrieve firefighting equipment.[98] The resulting explosion "crushed and burned" the mine superintendent and several men who had joined him to investigate the fire.[99] As the blast reached the mine-mouth, it sent flames into the air outside while launching timbers and rocks into another group of workmen, killing three and injuring five. "It seemed like the lid had blown right off the bottomless pit," one miner recalled, "and all the fires of hell had broken into that mine."[100]

With "all those in immediate authority" struck dead by the explosion, confusion reigned.[101] It took half an hour to restart the mine's fan, giving carbon monoxide and other deadly combustion by-products a jump on the five hundred to six hundred rescuers who rushed to the mine in the hours after the blast, "firm in the belief that many of the men [were] alive" and could be rescued if only crews could reach them in time. By early evening, two teams of specially trained helmet men arrived on mine rescue cars outfitted earlier in 1910, one by CF&I and the other by the U.S. Bureau of Mines.[102]

After donning their special breathing devices, helmet men plunged underground, searching for survivors while attempting to repair the mine's ventilation systems. Around 7:00 P.M., CF&I rescuers found four men who had used canvas barriers to safeguard a pocket of breathable air against afterdamp. Willis Evans, a young Colorado School of Mines graduate who "ha[d] made a special study of mine rescue work," gave his helmet "to one of these men who was partly overcome."[103] Amidst the danger and darkness, no one noticed when Evans fell behind. "Overlooked in the confusion," he succumbed to carbon monoxide. Later found "in practically a state of coma," Evans died the next morning.[104] Eighty-eight other men, though, would survive the disaster, thanks to the tireless work of Evans and other rescuers, the "great rapidity" with which workmen restored the flow of clean air into the mine, and the "very mild" effects of the explosion on more distant sections of Delagua's immense underground workings.[105]

Moments after a rescue party surfaced with what turned out to be the last group of survivors, a second crew emerged. Hearts throughout Delagua sank as the request went out for carpenters to make the stretchers needed to carry thirty-five freshly discovered corpses out of the mine. Together with Willis Evans and several men killed at the mine-mouth, Delagua's death toll reached seventy-nine, breaking the record recently set at Primero. Afterdamp intoxication caused thirty-six of these deaths.[106] The rest of the dead were killed by the irrepressible heat and force of the explosion itself. Mine electrician Till Woodward, for instance, "was so riddled by the flying missiles and seared by the fierce heat of the exploding gas as to be scarcely recognizable."[107] The clothing of William J. Evans (no relation, it seems, to the unfortunate young rescuer Willis Evans), meanwhile, "was torn and rent as if a shower of bullets had passed through them," his face "burned [. . .] beyond identification save by those who had known him many years," and his neck "broken and twisted so that his head was pushed upward and back, giving him an uncanny appearance."[108] The bodies found closest to the source of the explosion, though, presented the most gruesome sight of all. One eyewitness described this "tangled mass" "of charred and disfigured bodies [. . .] with limbs torn and mangled" as forming a scene "tragical beyond all power of pen to describe."[109]

The unsettling speed with which decomposition afflicted such battered and burned corpses soon undermined the cooperative spirit that had prevailed between Victor-American and its employees since the mine went up. On the second day after the blast, the *Trinidad Chronicle-News*, which remained closely associated with southern Colorado's largest coal operators, claimed that harmony endured at Delagua. "Every employe [*sic*]," the newspaper approvingly noted, "displayed his loyalty to the company."[110] Two days later, though, even the *Chronicle-News* had to admit that in the wake of Las Animas County's third major mine disaster in less than ten months, "terror" was spreading "among the mine workers throughout the entire southern field."[111]

As the miners' resentment toward Victor-American roiled, the care previously shown for removing the dead and burying their earthly remains in consecrated ground faltered. Several days after the blast, the company encountered "the greatest difficulty" in convincing "foreign" mine workers to return to the same tunnels and chambers where some of them had nearly perished just days earlier. Having failed to persuade immigrant employees to venture into underground spaces filled with debris, deadly afterdamp, and rotting corpses, Victor-American compelled African American mine workers to undertake this ghastly work. After confronting the horrific sights and smells permeating the pit, however, these black recovery workers immediately walked off the job. On November 11, "a strike of the laborers pressed into service to take the bodies out of the mine" broke out. African Americans notified the company that "they would no longer work for $2.95 a day" in dangerous places where "the stench from dead mules and from the bodies themselves was overpowering."[112] Victor-American tried to abate the noxious odors underground by injecting the bodies of dead mules with "large quantities of formaldehyde [. . .] to arrest decomposition until they can be moved." At the same time, the company steadfastly refused to meet a demand from black recovery workers for a pay increase.[113] After Mexican mine workers proved similarly "averse to working until after the bodies are buried," Victor-American found that it could muster only "a small force of intrepid men" to "participate in the work" of reclaiming the corpses of Delagua's dead.[114]

For many days after the mine exploded, the masses assembled outside the pit beheld "rickety wagons driv[ing] up with terrible frequency and regularity" to a temporary morgue in Delagua's mine machine shop, where they "discharge[d] their ghastly burdens."[115] The *Post* explicitly contrasted Coroner Guilfoil's emotional reaction to the sight of Delagua's dead with the inscrutable reserve allegedly maintained by the friends and family of the fallen workmen. The coroner, though no stranger to death, "furtively wiped away many a drop of moisture from his eyes as he gazed on the pitiful array of wrecked human beings that passed in review before him with such nerve racking monotony."[116] By contrast, "the crowd of watchers across the way," among whom Americans of color and immigrants from Europe, Asia, and Mexico almost certainly figured prominently, were "penned in a rope corral like a flock of sheep," reacting to "the tragic panorama with the same dull lackluster stare that real sheep might have given it."[117] In the eyes of the Anglo-American reporter who penned these lines, the crowd's failure to express grief, sadness, anger, or, indeed, any sentiment at all marked them as brutish, uncivilized, and un-American.[118]

The prejudices revealed and exacerbated by the Delagua explosion also influenced the hard line Victor-American adopted toward the bodies of the many Mexican nationals killed in the explosion. Though CF&I had deliberately brought out mangled or mutilated corpses from

Starkville after nightfall, that company had nonetheless refrained from interfering with funerals or burials. Victor-American, by contrast, took pains to ensure that the more than two dozen "Mexicans" who perished in the bowels of Delagua would be laid to rest "silent and unattended, without funeral note or the peal of bells." The furtiveness of these burials, conducted "under cover of darkness [. . .] with only a cloud of dust as an escort" and "a few brief words hurriedly spoken by the priest in charge" for a eulogy, contrasted strikingly with the public mourning that followed the Primero explosion. Back in February, Trinidad's stores had closed early, flags throughout the town had fluttered at half-mast, and "great crowds" with "tear-dampened eye lids" had attended funerals, requiem masses, and even a show at the Dreamland Theater in which a local shutterbug displayed footage of rescue and recovery efforts in the wrecked mine.[119] Nine months later at Delagua, by contrast, Victor-American "compelled" the friends and family of the Mexican miners who perished in the blast "to bid [the dead] the last farewell at the camp, without even looking upon their faces," and denied the bereaved the opportunity "to be present even at what brief services were held."[120] Burying the men in secret, the company reasoned, was wiser than interring them in public, lest the discontent stirred up by 1910's third major coal mine explosion set off an open revolt.[121]

The company endeavored to reduce its liability in other ways, too. Following the Primero and Starkville disasters, CF&I had quickly sought to broker settlements "with any person having any claim whatever" against the corporation. Company officials reasoned that "the kindly way in which" their firm dealt with its employees "in case of accident [would] enable us to retain their goodwill." Swift financial recompense would also prevent "shyster lawyers" from talking next of kin into filing suit against the company—a worrisome proposition since Colorado law allowed judgments to reach as high as $5,000 for each person killed.[122] Victor-American sought to limit its potential legal liability of nearly $400,000 by pursuing a similar strategy, as documented by a detailed ledger that recorded the settlements company officials negotiated with the explosion victims' legal representatives.[123] Payouts to next of kin averaged about $750, somewhat more than most southern Colorado mine workers typically earned in a year.[124]

More revealing than this average, however, were the patterns of variation in the settlements Victor-American paid. Settlements ranged from a low of $200, the sum paid to the surviving relatives of several of the Mexicans buried so unceremoniously by Victor-American lackeys, to a high of $2,000, received by survivors of mine superintendent William Lewis. Although no written accounts of the settlement negotiations remain, the disparities in indemnity payments clearly reflected the some-times surprising interplay between ideas of race, nationality, occupa-tion, and masculinity, as well as the varying skill and persistence with which diplomats from several nations conducted negotiations on behalf of the citizens they represented. Indemnities for workers categorized as "Americans" started at $1,000 and reached as high as $2,000. Not surprisingly, the mine superintendent and other mine officials were val-ued more highly than rank-and-file workers. The widow of José "Plac-ita" Valdez received $1,000 for her loss, substantially more than any other survivor of a worker described on the ledger as "Mexican," sug-gesting that Valdez was an Hispano and an American citizen. Survivors of Jerry Davis, L. Smith, and James Sampson—all categorized as "Americans" yet further described in the ledger's "Remarks" section as "half breed Mexican," "Colored," and "Colored," respectively—were also paid $1,000, an all-too-rare instance of common nationality trump-ing racial distinctions in early-twentieth-century America.[125]

MEAN INDEMNITY PAYMENTS BY NATIONALITY

"American"	$1,192
"Austrian"	$1,075
"Italian"	$1,058
"Montenegrin"	$500
"Japanese"	$367
"Mexican"	$304

Indemnity payments for immigrant workers, by contrast, varied more widely. Race and diplomacy together explain why. Survivors of workers hailing from Italy and from the Austro-Hungarian Empire received an average payment exceeding $1,000. Victor-American paid substantially smaller settlements, by contrast, to the next of kin of men listed on the ledger as "Montenegrin," "Mexican," and "Japanese." Both Italians and Austrians had been fixtures of Colorado coal mine payrolls since the 1890s; more importantly, Italian and Austrian consular officials vigor-ously defended the interests of their national subjects in the wake of all three of Colorado's 1910 mine explosions.[126] Mexican and Montenegrin diplomats also parlayed with Victor-American, but with less success. The ability of Mexican representatives to drive a hard bargain with the coal company may have been compromised by the recent outbreak of revolution in that country.[127] The small and vulnerable kingdom of Montenegro probably wielded even less heft in talks with coal company lawyers, while indemnity negotiations on behalf of the Japanese dead were conducted not by consular officials, but by a Japanese labor contractor who presumably had strong incentives to placate the coal company functionaries with whom he probably hoped to continue doing business.[128] Racial discrimination probably also contributed to the comparatively meager settlements for Mexican and Japanese workers killed in the disaster. The *Chronicle-News* assailed the Mexicans who testified at the Delagua inquest, for instance, as "illiterate miners from the jungles of Old Mexico [who] could not grasp even through an interpreter the meaning of the questions put to them," while vitriolic anti-Asian sentiment throughout Colorado almost certainly made it easier for Victor-American to skimp on the settlements it paid out to the relatives of Delagua's Japanese victims.[129]

Family wage ideologies further inflected indemnity payments, pegging settlement amounts for workers with many dependents above those with few or none. According to patriarchal assumptions shared alike by company managers and the coalfields' diverse mining populations, the value of each man killed by the Delagua explosion usually varied according to the number of mouths his labor supposedly fed.[130] Those listed on the ledger as Italians illustrate the resulting disparities most clearly, with Victor-American paying just $500 to the next of kin of single miners, but $1,100 for those who were married with one child and $1,400 for miners who were married with four children. The male breadwinner ideal also accounts for the relatively small disparities in settlements for Japanese and Mexican victims. The coal company paid just $200 to $210 to protect itself from wrongful death suits involving unmarried Mexican nationals, and $300 in instances involving unmarried Japanese men. The ledger recorded disbursements of $350 to the next of kin of married Mexican miners, whether they had no children or several, and $500 in the case of S. Asaida, a Japanese miner who was married with one child.[131]

These variations in settlement payments testified to a hard truth: the residents of southern Colorado's coal-mining communities were simultaneously drawn together by shared travails of work and disaster, and divided against one another by the complicated entanglements of skill, nationality, international politics, and gendered ideals.

AFTERMATH: THE COALFIELD WAR OF 1913–14

In late November, several weeks after tragedy descended on Delagua, Coroner Guilfoil finally held an inquest over Starkville's dead. The ensuing proceedings brought long-simmering struggles over knowledge and power in the southern Colorado coalfields to the boiling point. Before the inquest, CF&I had convened a special meeting in which expert miners and long-time company officials struggled to account for the Starkville tragedy. A newspaper account of the resulting discussions led with a surprising declaration: "Thirty of more grizzled veterans of the coal fields [. . .] who until lately contended that dust would not explode" publicly declared their conviction that October's disaster was caused by a coal dust explosion.[132] CF&I consul Fred Herrington put his own spin on the concurrence of these "grizzled veterans": "The explosion we believe has established a hitherto unrecorded fact in mining science, that under certain conditions dust may explode without the contributing agencies of gas or fire."[133]

Far from an "unrecorded fact," however, the volatility of coal dust had long been recognized. Although some mining experts continued to maintain that coal dust could go off only when detonated by another explosive substance, such as methane, a growing number argued that coal dust could ignite in its own right.[134] By the turn of the century, British mine safety researchers had "elevated" coal dust "to the rank of principal agent" in coal mine explosions.[135] Blue-ribbon investigations on both sides of the Atlantic had gone on to blame coal dust for several infamous mine disasters, including the 1906 Courrières explosion, which killed 1,099 workers in northern France, and the 1907 Monongah explosion, the deadliest industrial accident in U.S. history, which killed at least 362 in West Virginia.[136] In a widely reprinted report published just two months after the Starkville inquest, George Rice of the U.S. Geological Survey proclaimed: "It is now exceptional to find a mining man who does not accept the evidence of the explosibility of coal dust. The question of the day no longer is 'Will coal dust explode?' but 'What is the best method of preventing coal-dust explosions?'"[137]

Jurors at the Starkville inquest had no opportunity to consult Rice's forthcoming report, but they did hear more than enough evidence to refute Herrington's brazen claim that the Starkville disaster had revealed a new truth. "All of the witnesses" called by Coroner Guilfoil on the inquest's second day "declared themselves convinced that a dust explosion [could] occur without the contributing agencies of gas and fire." Among those who spoke were CF&I's own mine inspector, who labeled "coal dust a menace to a mine with or without the presence of gas," and Colorado's newly appointed coal mine inspector, James Dalrymple, who declared that "for twenty years [he . . .] had believed coal dust explosive."[138]

The glaring discrepancy between such testimony and Herrington's spin called into question the coal companies' long-standing control over what passed for official knowledge in and around Las Animas County's coal mine workscapes. For years, CF&I, Victor-American, and their counterparts had lorded over the southern coalfields like absolutist monarchs. "Even in Russia," one union miner complained, there was "more liberty than in Southern Colorado."[139] King Coal's henchmen had grown accustomed to deciding what counted as "fact." But the 1910 mine disasters demonstrated the power of nonhuman factors to expose some of the inconvenient truths that the companies had long succeeded in suppressing.[140] And so the Starkville coroner's jury delivered a stunning rebuke, finding CF&I "guilty of gross negligence." Had the mine "been properly sprinkled," the jury continued, the disaster "would not, and could not have occurred."[141]

Although another inquest held a few weeks later exonerated Victor-American for the Delagua disaster, most coalfield residents had already begun to join forces with progressives and radicals elsewhere in Colorado and the United States to mount an overlapping pair of challenges to hazardous mine workscapes and coal company tyranny: mine safety reform and unionization. Most coal companies willingly embraced at least some new safety measures following the 1910 disasters. Victor-American, for instance, collaborated with Rice and the new Bureau of Mines to become "the first American producer to introduce rock dusting," a dust-mitigation technique that worked better than sprinkling in Colorado's arid climate. CF&I, meanwhile, pumped additional resources into equipping and training rescue crews.[142]

The coal companies' efforts to address the perils that mine workers faced, however, failed to thwart calls for stricter state regulation. A *Denver Post* editorial used the Delagua disaster's coincidence with Election Day to push for stronger mine safety laws. Incoming lawmakers, *Post* editors declared, "were selected to put statutes on the books that would protect the people of the state—all the people—the people who use the result of the miner's toil, the man who owns the mine, the endangered digger in that mine. Let those newly elected legislators act!" After three explosions in less than a year, the *Post* could confidently assert that "the state wants protective statutes—not philosophizing."[143] Even

the *Republican* jumped on the bandwagon: "Coal production is necessary," an editorial in its columns acknowledged, "but that the waste of
human lives that attends it in this country is not necessary is demonstrated by the fact that proportionately from three to four times as
many lives are sacrificed every year in American coal mines than in
those of Europe."[144] Governor John Shafroth responded to the resulting
outcry by appointing a special commission to investigate the 1910 disasters and draft new mine safety legislation.[145]

By early 1911, though, southern Colorado's mine disasters were
already becoming old news. Mine operators capitalized on public apa-
thy to block even the relatively modest reforms proposed by the governor's commission. CF&I and its counterparts then moved to draft a
milder safety bill through negotiations with state mine inspector John
Dalrymple and John Lawson, district organizer for the United Mine
Workers of America (UMWA), the main coal miners' union in the
United States.

Mine workers in the southern coalfields later accused the companies
of refusing to obey even this watered-down law, which Colorado legislators enacted in 1913. The UMWA, as Lawson's involvement in drawing up Colorado's new mine safety regulations indicated, actively supported legislative reform as part of its larger campaign to improve
working conditions in the Colorado coalfields. When southern Colorado's mine workers launched a massive strike in southern Colorado in
September 1913, their strike demands thus encompassed not simply the
recognition of the UMWA as collective bargaining agent for Colorado
coal mine workers, but also the enforcement of the new safety law
enacted in response to the deaths of 211 men at Primero, Starkville, and
Delagua three years earlier.

The 1910 mine explosions helped the union to leverage long-standing
unrest into a dramatic resurgence of organization and mobilization.
Las Animas County miners had been fighting to unionize for decades.
For the people of the coalfields, the Underground West combined peril
and promise—the ever-present possibility of death with the promise of a
better life. Since the coal industry's inception, the mines themselves
served as the vital crucibles in which workers from around the world
had come together to forge a powerful oppositional culture. Death
underground reinforced the everyday experience of subterranean labor
and solidified a workforce fragmented by race, nationality, ethnicity,
skill level, and many other factors into a cohesive and potentially militant movement.[146]

As UMWA organizers sought to build support for their cause—a campaign hindered by the coal companies' near-total victory in the southern coalfields' last major strike, which culminated in 1904—they sometimes leveraged the dust explosions of 1910 into scathing critiques of King Coal's reign and forceful calls for union recognition as the key to rectifying Colorado's abysmal mine-safety record. The toil and terror that had long suffused Colorado's coal mines thus animated solidarity and social mobilization, as well.[147] For striking miners and their allies, memories of Primero, Starkville, and Delagua epitomized the failure of mine operators to fulfill their legal and moral obligations to the workers on whose labor the profitability of these enterprises hinged. The trio of mine disasters triggered by a concatenation of material and social processes thus set the stage for the tumultuous Colorado coalfield war of 1913–14, a highly publicized, unusually violent labor struggle that exposed the dirty secrets of Colorado's southern coalfields to probing, nationwide scrutiny.

Back in 1910, Robert Uhlich, a tireless UMWA stalwart, had prophetically warned: "There may be bloodshed on[e] day in Southern Colorado." Because of "accidents" large and small, the miners were "aroused against this System which exist[s] here." Uhlich assured the state labor commissioner that he still believed that "we"—he and his fellow union leaders—"could prevent a class war but on[e] Day, we will lose control over the miners, and when this [sic] unorganized go on Strike, it will be a terrible lession [sic]."[148]

As Uhlich sensed, the 1910 disasters would go on to shape not only *why* miners went on strike but also *how* they would wage war against mine operators and Colorado National Guardsmen. This trio of dust explosions—along with the seemingly incessant falls of rock and coal that injured and killed dozens of miners a year, the beatings company guards meted out to union organizers, a regional culture of violence that made Las Animas County's homicide rate one of the highest in the nation, and the profligate miscarriages of justice that prevailed under coal company dominance—exemplified the shocking cheapness of human life in the southern coalfields. Is it any wonder, then, that after the Ludlow massacre claimed the lives of seventeen strikers in April 1914, battalions of armed mine workers showed their opponents no quarter? In ten days of fierce guerrilla fighting, strikers would kill more than thirty guards, strikebreakers, and militiamen. Rebellious workers also dynamited the entrances to several mines, temporarily closing off the subterranean workscapes in which their relatives, countrymen, and comrades had labored amid unnecessarily severe hazards—suffering and all too frequently losing their lives in the years leading up to the great coalfield war.[149]

NOTES

The author would like to acknowledge the editors of this volume for valuable and constructive criticism, as well as David Fouser and Alessandra La Rocca Link for research assistance. Preliminary versions were presented as the Livingston Lecture at the University of Denver, at the Huntington Library's "Under the West" Conference, and at the U.S. Mine Safety and Health Administration's Denver offices.

1. *Trinidad Chronicle-News,* Jan. 24, 1910; R.L. Herrick, "The Primero Disaster," *Mines and Minerals* 30 (March 1910): 463.

2. *Denver Post,* Feb. 1, 1910.

3. On workscape, see Thomas G. Andrews, *Killing for Coal: America's Deadliest Labor War* (Cambridge, MA: Harvard University Press, 2008), ch. 4.

4. I consider these the most reliable statistics, though many sources claim that 75 men perished in Primero, instead of 76.

5. Helpful works on coal mine disasters include Paul Anderson, "'There Is Something Wrong Down Here': The Smith Mine Disaster, Bearcreek, Montana, 1943," *Montana: The Magazine of Western History* 38 (1988): 2–13; David Jay Bercuson, "Tragedy at Bellevue: Anatomy of a Mine Disaster," *Labour/Le Travail* 3 (1978): 221–31; Anthony Fleege, "The 1947 Centralia Mine Disaster," *Journal of the Illinois State Historical Society* 102 (2009): 163–76; J. Davit McAteer, *Monongah: The Tragic Story of the Worst Industrial Accident in U.S. History* (Morgantown: West Virginia University Press, 2007); Robert G. Neville, "The Courrières Mine Disaster, 1906," *Journal of Contemporary History* 13 (1978): 33–52; Donald Reid, "The Role of Mine Safety in the Development of Working-Class Consciousness and Organization: The Case of the Aubin Coal Basin, 1867–1914," *French Historical Studies* 12 (1981): 98–119; Steve Stout, "Tragedy in November: The Cherry Mine Disaster," *Journal of the Illinois State Historical Society* 72 (1979): 57–69; Anthony F.C. Wallace, *St. Clair: A Nineteenth-Century Coal Town's Experience with a Disaster-Prone Industry* (Ithaca, NY: Cornell University Press, 1988), ch. 5; and Liping Zhu, "Claiming the Bloodiest Shaft: The 1913 Tragedy of the Stag Cañon Mine, Dawson, New Mexico," *Journal of the West* 35 (1996): 58–64.

6. George S. McGovern and Leonard F. Guttridge, *The Great Coalfield War* (Boston: Houghton-Mifflin, 1972); Andrews, *Killing for Coal.*

7. John Fabian Witt, *The Accidental Republic: Crippled Workingmen, Destitute Widows, and the Remaking of American Law* (Cambridge, MA: Harvard University Press, 2004), 2. Witt's analysis focuses on one era of U.S. history, and it seems likely that he is overstating the uniqueness of the American case; possible competitors for the title of world's most accident-strewn phase include the coal and iron boom in Song Dynasty China and the Soviet Five-Year Plans of the 1920s and '30s.

8. Quoted in Mark Aldrich, *Safety First: Technology, Labor, and Business in the Building of American Work Safety, 1870–1939* (Baltimore: Johns Hopkins University Press, 1997), 63. Statistics for 1906–10 from Mark Aldrich, "'The Needless Peril of the Coal Mine': The Bureau of Mines and the Campaign against Coal Mine Explosions, 1910–1940," *Technology and Culture* 36 (1995): 488.

9. Colorado's coal mine fatality rate of 6.81 deaths per thousand over the 1884 to 1912 period was exceeded by Utah's, highest among all major coal-mining states. James Whiteside, *Regulating Danger: The Struggle for Mine Safety in the Rocky Mountain Coal Industry* (Lincoln: University of Nebraska Press, 1990), 74–75. Over the longer run, "the coal mining fatality rate, as measured per employee, ranged between two and three times as high in the United States as in Great Britain between 1880 and 1930." Witt, *Accidental Republic*, 26.

10. *Denver Post*, Feb. 1, 1910; see also Andrews, *Killing for Coal*, especially ch. 4.

11. *Miner's Magazine* 11 (Nov. 17, 1910): 3.

12. These themes are explored more fully in Whiteside, *Regulating Danger*, and Andrews, *Killing for Coal*.

13. In time, "the rescuers discovered that the portions of humanity were once five human beings," so "cruelly mutilated" that "not one [. . .] could be identified." *Denver Post*, Feb. 1, 1910.

14. The *Post* described the sheriffs and their deputies as "armed [and] ready to preserve order." Ibid. Sixteen doctors from CF&I's Pueblo hospital, Colorado's state mine inspector, two mining engineers from the U.S. Geological Survey, and the Denver-based Italian and Austrian consuls also ventured by rail to the disaster site.

15. For details on the mine officials involved and the "forty-five sturdy miners from Cokedale, Starkville and Sopris" who constituted the first wave of rescuers, see ibid.

16. Ibid. (quoted) and *Denver Post*, Feb. 2, 1910. The *Denver Republican* also mentioned ropes and deputy sheriffs; Feb. 2, 1910.

17. *Denver Post*, Feb. 1, 1910. Virgen was found not far from ten corpses. Ibid. Reflecting broader confusion over the racial categorization of Mexican Americans and Mexicans, one paper referred to Virgen as "colored." *Denver Republican*, Feb. 3, 1910.

18. One report spelled his name Domacio Vergan. *Denver Post*, Feb. 1, 1910.

19. Ibid.

20. Virgen told much the same story to the coroner's jury. *Trinidad-Chronicle News*, Feb. 3, 1910. See also Herrick, "Primero Disaster," 464. Throughout this chapter, the placement of "Mexican" in quotation marks indicates that I am drawing this attribution from primary sources, which often used "Mexican" to refer to U.S. citizens of Mexican or Hispanic descent.

21. *Denver Post*, Feb. 1, 1910.

22. *Denver Republican*, Feb. 1, 1910.

23. *Denver Post*, Feb. 1, 1910.

24. Ibid.

25. Even in its initial reports, for instance, the *Denver Republican* (Feb. 1, 1910) claimed, "There is little hope that any of the men in the mine are alive."

26. *Denver Post,* Feb. 2, 1910.

27. Ibid.

28. Ibid.; see also *Denver Republican,* Feb. 1, 1910.

29. *Denver Post,* Feb. 4, 1910. That some of the men killed had only recently arrived in Primero and were not well known in Las Animas County made them even harder to identify. A final problem Coroner Guilfoil encountered resulted from the common Anglo-American perception of Asians as an undifferentiated mass (for Colorado context in this regard, see William Wei, *Asians in Colorado: A History of Persecution and Perseverance in the Centennial State* [Seattle: University of Washington Press, 2016]). "There are eight Japanese corpses," one newspaper claimed of men who were almost certainly ethnic Koreans, "which seem impossible of identification." *Denver Republican,* Feb. 2, 1910. A list of the dead published by the *Denver Post* reflected this broader confusion: six dead mine workers are listed as "Japanese," despite bearing common Korean names (Kim, Cho, and Chun, in addition to Clioy), while four others (two named Lee, plus Chim and Yar) were identified (correctly, one presumes) as Koreans. *Denver Post,* Feb. 2, 1910. Apparently, Guilfoil soon succeeded in identifying at least some of these men, as a few days later, another paper placed the number of unidentified corpses at just four. *Denver Republican,* Feb. 4, 1910.

30. *Denver Post,* Feb. 2, 1910.

31. Ibid.

32. *Denver Post,* Feb. 4, 1910.

33. *Denver Post,* Feb. 2, 1910 (quoted); see also *Denver Republican,* Feb. 3 and 4, 1910. It's not clear how many funerals there were; the *Republican* claimed on February 4 that "twelve funerals were held today," but this wording still admits of the possibility that just one funeral was held for twelve mine workers.

34. *Denver Post,* Feb. 2, 1910.

35. Ibid. By February 3, Primero seemed "indeed a deserted village," but appearances could deceive. "Most of the relatives and friends" of the several dozen dead found during the early days of the recovery had "gone to Trinidad to attend the funerals," closing up their houses and drawing their blinds behind them. *Denver Post,* Feb. 3, 1910.

36. *Denver Post,* Feb. 4, 1910.

37. *Denver Republican,* Feb. 2, 1910. Surprisingly, perhaps, not a single account of the rescue effort mentions the health dangers that decomposing corpses might have presented. This apparent lack of concern with hygiene and sanitation seems especially puzzling when compared to contemporaneous events in the American West and well beyond; e.g., Neville, "Courrières Mine Disaster," 37; Nayan Shah, *Contagious Divides: Epidemics and Race in San Francisco's Chinatown* (Berkeley: University of California Press, 2001); and William Deverell, *Whitewashed Adobe: The Rise of Los Angeles and the Remaking of Its Mexican Past* (Berkeley: University of California Press, 2004).

38. *Denver Republican,* Feb. 2, 1910.

39. *Denver Republican,* Feb. 3, 1910; *Denver Post,* Feb. 3, 1910. The two full trainloads of rock that workmen had already removed from the mine, it seems, represented but a drop in the proverbial bucket. *Denver Republican,*

Feb. 4, 1910. "An occasional corpse may be taken out," the *Post* (Feb. 4, 1910) declared, "but most of the missing men may be buried under the falls of thousands upon thousands of tons of rock, slate and coal."

40. *Trinidad Chronicle-News*, Feb. 7 and 8, 1910; *Denver Republican*, Feb. 9, 1910.

41. *Denver Post*, Feb. 3, 1910 (first quote), and Feb. 4, 1910 (second). See also *Denver Post*, Feb. 2, 1910, and *Denver Republican*, Feb. 4, 1910. For more on Rice's prominent role in dust explosion research, see Aldrich, "'Needless Peril of the Coal Mine,'" 490–503.

42. *Denver Post*, Feb. 3 and 4, 1910. Curiously, the *Republican* had reported (Feb. 3, 1910) that the jury's proceedings would last for weeks, but Guilfoil called testimony to a halt after just two days.

43. *Denver Post*, Feb. 4, 1910.

44. *Trinidad Chronicle-News*, Feb. 2, 1910 (both quotes). See also *Trinidad Chronicle-News*, Feb. 3, 7, and 8, 1910. Longtime Primero miner Samuel P. Lyon reportedly went insane in the weeks after the Primero disaster. Having lost many friends in the explosion, Lyon became "violent and is said to have gone about threatening to blow up everything in sight 'like it did at Primero.'" *Trinidad Chronicle-News*, Feb. 17, 1910.

45. "A Bureau of Mines," editorial, *Denver Republican*, Feb. 3, 1910; *Denver Post*, Feb. 7, 1910. Colorado had established the office of state inspector of coal mines following the Jokerville coal mine explosion of 1882.

46. *Denver Post*, Feb. 19, 1910.

47. *Denver Republican*, Feb. 9, 1910.

48. *Denver Post*, Oct. 10, 1910; *Trinidad Advertiser*, Oct. 10, 1910.

49. *Trinidad Advertiser*, Oct. 10, 1910; *Denver Post*, Oct. 9 and 10, 1910.

50. *Denver Post*, Oct. 9, 1910.

51. *Trinidad Advertiser*, Oct. 10, 1910; *Denver Post*, Oct. 10, 1910. The *Denver Post* echoed this language on Oct. 9, 1910.

52. "Another Explosion," *Miner's Magazine* 11 (Oct. 20, 1910): 6.

53. *Denver Post*, Oct. 9 and 10, 1910.

54. Ibid.

55. *Denver Republican*, Oct. 10, 1910.

56. *Denver Post*, Oct. 9, 1910. Elsewhere in the same day's edition of the *Post*, a story identified their main correspondent at Starkville as A.R. Brown, city editor of the *Trinidad Chronicle-News*.

57. *Denver Republican*, Oct. 10, 1910.

58. *Trinidad Advertiser*, Oct. 10 and 13, 1910; *Denver Post*, Oct. 9, 1910. The *Trinidad Advertiser* (Oct. 10, 1910) noted that the USBM's helmets had "never before [been] used." See also *Denver Republican*, Oct. 10, 1910.

59. *Denver Post*, Oct. 10, 1910. Curiously, state mine inspector Jones held out hope. "Without any stretch of the imagination," he told reporters, "these men could be alive and perfectly safe." Ibid.

60. *Denver Republican*, Oct. 10, 1910.

61. *Denver Post*, Oct. 10, 1910. Compare this version of fallible technologies and exceptionally tolerant human bodies with a more typical celebration of expertise and apparatus in the *Denver Republican*, Oct. 10, 1910. For recent

research on variations in how CO affects different individuals and groups, see J. A. Raub, M. Mathieu-Nolf, N. B. Hampson, and S. R. Thom, "Carbon Monoxide Poisoning: A Public Health Perspective," *Toxicology* 145 (April 2000): 1–14.

62. *Denver Post,* Oct. 10, 1910. "The grim fact that the men buried in Starkville mine are all dead," the *Post* conceded elsewhere in the same edition, "may as well be faced now as later."

63. *Denver Republican,* Oct. 10, 1910.

64. *Denver Republican,* Oct. 10 and 11, 1910.

65. Quotes from *Denver Post,* Oct. 10 and 11, 1910.

66. *Denver Post,* Oct. 10, 1910. Weitzel's experience overseeing the company's response to the Starkville disaster was already haunting him. "When I become exhausted and drop down on the ground for a moment's rest," Weitzel complained, "my mind at once begins to conjure up harrowing pictures of the events that passed in that awful place Saturday night, and it seems as if I shall go raving crazy. It is a fearful experience, and I pray that I may never be called upon to go through another like it." Ibid.

67. *Denver Post,* Oct. 12, 1910.

68. *Denver Republican,* Oct. 12, 1910.

69. *Denver Post,* Oct. 12, 1910. The account continued by describing these corpses as "seared by fire in a worse manner even than the victims of the Primero horror, still fresh in the minds of all citizens of Colorado."

70. *Denver Post,* Oct. 13, 1910.

71. *Twelfth Biennial Report of the Bureau of Labor Statistics of the State of Colorado, 1909–1910* (Denver: Smith-Brooks, 1911), 35. Intriguing in this regard is Jerome M. Chertkoff and Russell H. Kushigian, *Don't Panic: The Psychology of Emergency Egress and Ingress* (Westport, CT: Praeger, 1999).

72. *Trinidad Advertiser,* Oct. 16, 1910.

73. *Denver Post,* Oct. 13, 1910.

74. *Denver Republican,* Oct. 14, 1910; *Denver Post,* Oct. 13, 1910.

75. *Denver Republican,* Oct. 16, 1910.

76. *Denver Republican,* Oct. 12, 1910.

77. Ibid.

78. Relatives of Frank London, an African American mine worker, received his body in Walsenburg; another miner was shipped south to Raton, New Mexico; and the corpse of a third traveled by train to Oakland, California. *Denver Republican,* Oct. 13, 1910; *Denver Post,* Oct. 12, 1910.

79. *Denver Republican,* Oct. 13, 1910; *Denver Post,* Oct. 13, 1910.

80. *Denver Post,* Oct. 13, 1910.

81. *Denver Republican,* Oct. 16, 1910.

82. *Trinidad Advertiser,* Oct. 13, 1910 (first two quotes); *Denver Republican,* Oct. 17, 1910 (last quote). "Men were frequently overcome by the awful stench of the bodies in carrying them out," another source reported, "and were further weakened by the intense heat which existed in the lower workings." *Twelfth Biennial Report of the Bureau of Labor Statistics,* 37.

83. *Denver Post,* Oct. 10, 1910.

84. Ibid.

85. *Denver Post,* Oct. 11, 1910.

86. *Denver Post,* Oct. 10, 1910. The next day, the *Post* praised the company's progress while reinforcing the sense that the early rescue effort had been deeply flawed: "That yesterday's rescue parties got 5,000 feet further into the mine than any group had been able to reach before was a striking demonstration of the truth of the statement made in this paper yesterday that the company's forces were practically without organization or capable direction on Sunday."

87. *Denver Post,* Oct. 11, 1910. As was common, portions of this story appeared in almost the exact same form in other papers; e.g., *Denver Republican,* Oct. 11, 1910.

88. *Denver Post,* Oct. 11, 1910.

89. Ibid.

90. Ibid.

91. *Oxford English Dictionary Online* (2008), s.v. "disaster" (quoting Whitney, *Life Lang.* vi. (1875), 99).

92. *Trinidad Advertiser,* Oct. 11, 1910.

93. *Denver Post,* Oct. 10, 1910.

94. Ibid.

95. *Denver Republican,* Oct. 14, 1910.

96. *Denver Post,* Oct. 14, 1910; *Denver Republican,* Oct. 15, 1910.

97. *Denver Post,* Nov. 9, 1910 (quoted). One investigator placed the labor force at "135 miners and 22 company hands." At the time of the explosion, three other men were also in and around the mine—"a visitor at the drift mouth" and "two who went inside out of curiosity." George F. Duck, "The Delagua, Colo. Explosion," *Mines and Minerals* 31 (Jan. 1911): 378.

98. Duck, "The Delagua, Colo. Explosion," 376.

99. Ibid., 374–78.

100. *Trinidad Chronicle-News,* Nov. 12, 1910. As Duck explained, "While there were undoubtedly a series of explosions, the time interval between them was but a minute fraction of a second, so that to observers the action appeared instantaneous." Duck, "The Delagua, Colo. Explosion," 379.

101. Duck, "The Delagua, Colo. Explosion," 380.

102. *Denver Republican,* Nov. 9, 1910.

103. *Denver Post,* Nov. 9, 1910.

104. Duck, "Delagua, Colo. Explosion," 380.

105. Ibid. ("great rapidity"); John D. Jones to Shafroth, Nov. 9 and 10, Folder 5, Box 26732, Governor John Shafroth Papers, Colorado State Archives, Denver (other quotes).

106. The *Republican* (Nov. 11, 1910), currying favor with Colorado's conservative business interests, blamed these victims for their own demise. "Not one man in the crowd," the paper scoffed, "kn[ew] how to brattice, or shut the miners off from after damp, with nearby canvas."

107. *Denver Post,* Nov. 11, 1910.

108. Ibid.

109. *Trinidad Chronicle-News,* Nov. 11, 1910.

110. *Trinidad Chronicle-News,* Nov. 10, 1910.

111. *Trinidad Chronicle-News,* Nov. 12, 1910.

112. *Denver Post,* Nov. 12, 1910.

113. *Denver Republican,* Nov. 12, 1910.

114. Ibid. (first quote); *Pueblo Chieftain,* Nov. 12–14, 1910 (other quotes).

115. *Denver Post,* Nov. 10, 1910.

116. Ibid.

117. Ibid.

118. As with what James J. Farrell called "the spatial segregation of the living and the dead," portrayals of the coalfield working classes as unfeeling "paralleled other types of spatial differentiation in American middle-class life." Farrell, *Inventing the American Way of Death, 1830–1920* (Philadelphia: Temple University Press, 1980), 110. Significantly, southern Colorado's Hispanos, the descendants of New Mexican agriculturists who had first colonized Las Animas County a half century before, presented a revealing exception. "The only outward show of emotion in the town of Delagua," a *Republican* reporter claimed, occurred during a Hispano miner's funeral; a woman widowed by the tragedy allegedly "did not weep, but her eyes wandered over the throng of bystanders with a peculiar look, as if her sorrow was too deep for tears." *Denver Republican,* Nov. 12, 1910.

119. *Trinidad Monitor,* Oct. 14, 1910.

120. *Trinidad Chronicle-News,* Feb. 2 and 3 and Nov. 11, 1910. A deputy inspector from the Colorado Bureau of Labor Statistics had criticized CF&I for "permit[ting] a photographer to enter the temporary morgue for the purpose of taking a photograph of the bodies in order to put the same on the moving picture circuit," while barring family members of the dead from seeing the earthly remains of their loved ones. *Twelfth Annual Report of the Bureau of Labor Statistics,* 28.

121. In fact, it was the haste with which the Mexicans were buried that inspired the Serbian representative to implore Guilfoil to delay the burial of Serbs and Montenegrins. *Denver Republican,* Nov. 13, 1910.

122. Bowers to Murphy, Nov. 8, 1910, Folder 84, Box 27, Lamont Montgomery Bowers Papers, Special Collections, University of Binghamton, Binghamton, NY; on indemnity claims, *Denver Post,* Feb. 8, 1910.

123. Although the next of kin of some dead would also have received payouts from mutual benefit societies and insurance companies, most had little reason to hope for additional remuneration for their loss. Colorado had failed to enact workers' compensation laws despite repeated campaigns by unions and progressives. Local and state courts, meanwhile, generally decided civil suits in favor of the companies. Whiteside, *Regulating Danger,* 85–90.

124. The ledger is in Folder 34, Box 1, Victor American Fuel Co. Papers, History Colorado, Denver. Unless otherwise cited, all information in the paragraphs that follow is drawn from the ledger. The ledger contains seventy-six entries; it thus includes all of the men killed underground, but none of those killed outside the mine. It is unclear whether the next of kin for these victims received settlements or not.

125. Some of the men listed on the ledger as "American" were probably not U.S. citizens, yet their next of kin received a minimum of $1,000.

126. See, for instance, *Denver Post,* Oct. 11 and Nov. 13, 1910; *Denver Republican,* Nov. 13, 1910. After the Primero blast, the Austrian and Italian consuls secured indemnities from CF&I amounting to "$1,200 for each widow," with "fathers, mothers and children of the victims receiv[ing] $600 to $700 each." *Denver Post,* Oct. 12, 1910.

127. Montenegro was evidently chomping at the bit to exercise its sovereignty; it had become a kingdom only in late August 1910, but already its officials were busily attempting to safeguard the rights of its citizens abroad. The variations in the payments made according to race, ethnicity, and nationality at Delagua stand in contrast to the payouts made by a Phelps-Dodge subsidiary following the 1913 Stag Cañon explosion in Dawson, New Mexico. Liping Zhu claims that "in the end, each widow received a base of $1,200 and $100 for each child"; the company also "offered transportation to any part of the world" for the survivors of explosion victims. The difference between the indemnities paid at Delagua and Stag Cañon is all the more interesting since the latter mine was owned by a subsidiary of the Phelps-Dodge copper concern, which invariably employed racialized wage differentials at its copper mines and smelters in Arizona and New Mexico. Zhu, "Claiming the Bloodiest Shaft," 63. For more on Phelps-Dodge and race, see Katherine Benton-Cohen, *Borderline Americans: Racial Division and Labor War in the Arizona Borderlands* (Cambridge, MA: Harvard University Press, 2009).

128. All four entries for the Japanese read as follows under "Remarks": "Paid H.S. Okumura c/o W J Murray; Paid Shibata, PO Box 150 Delagua, Colo." It would stand to reason that "Shibata" was in charge of the "little colony" of Japanese in Delagua; he may or may not have forwarded the payment to family members of the deceased back in the homeland. On labor contractors in the American West during this era, see Gunther Peck, *Reinventing Free Labor: Padrones and Immigrant Workers in the North American West, 1880–1930* (New York: Cambridge University Press, 2000).

129. *Trinidad Chronicle-News,* Dec. 13, 1910.

130. "This model of the family," Witt explains of family wage thinking, "played an influential role in the development of the American law of accidents." Witt, *Accidental Republic,* 20. See also Maurine Weiner Greenwald, "Working-Class Feminism and the Family Wage Ideal: The Seattle Debate on Married Women's Right to Work, 1914–1920," *Journal of American History* 76 (1989): 118–49; Ron Rothbart, "'Homes Are What Any Strike Is About': Immigrant Labor and the Family Wage," *Journal of Social History* 23 (1989): 267–84; Linda Gordon, "Social Insurance and Public Assistance: The Influence of Gender on Welfare Thought in the United States," *American Historical Review* 97 (1992): 19–54; and Lawrence Glickman, "Inventing the 'American Standard of Living': Gender, Race, and Working-Class Identity, 1880–1925," *Labor History* 34 (1993): 221–35.

131. On Christmas Day, weeks after a coroner's jury had exonerated Victor-American of all blame, company couriers delivered settlement checks to the families whose "husbands and fathers ha[d] been laid away in the cold ground." The *Chronicle-News* reassured local readers that "none of the widows and orphans which this disaster made will become charges of the community," portraying the

indemnity payments as generous enough that "many of the families will thus be enabled to return to the old country from which they came years ago looking toward America as the great land of promise." *Trinidad Chronicle-News*, Dec. 26, 1910.

132. *Trinidad Chronicle-News*, Nov. 21, 1910.

133. The *Trinidad Chronicle-News* (Nov. 23, 1910) seconded Herrington, calling the explosion "historical in the annals of mining industry."

134. On dissenting nineteenth-century views, which held that coal dust was an explosion hazard in its own right, see Jacqueline Karnell Korn, "'Dark as a Dungeon': Environment and Coal Miners' Health and Safety in Nineteenth-Century America," *Environmental Review* 7 (1983): 261.

135. Neville, "Courrières Mine Disaster," 47.

136. Ibid., 45–48; Aldrich, "'The Needless Peril of the Coal Mine,'" 489–90.

137. George S. Rice, *The Explosibility of Coal Dust*, United States Geological Survey Bulletin 425 (Washington, DC: U.S. Government Printing Office, 1911); a shorter version of Rice's work also appeared in "Explosibility of Coal Dust," *Mines and Minerals* 31 (Jan. 1911): 369. Excerpts from the report appeared in several trade publications, particularly those focusing on mining in general and coal mining in particular. See also "How Uncle Sam Protects His Coal Miners," *Current Literature* 52 (May 1912): 536; and Aldrich, *Safety First*, 41. On the ongoing reluctance of many U.S. mine experts, especially practical miners, to accept the so-called "coal-dust theory," see Aldrich, "'Needless Peril of the Coal Mine,'" 497–501.

138. *Trinidad Chronicle-News*, Nov. 25, 26, and 30, 1910.

139. Robert Uhlich to Edwin V. Brake, July 17, 1910, Folder 4, Box 26732, John F. Shafroth Papers, Colorado State Archives, Denver.

140. The company's power was widely recognized—and widely condemned. See, for instance, *Twelfth Annual Report of the Bureau of Labor Statistics*, 26.

141. *Trinidad Chronicle-News*, Dec. 1, 1910. Many observers insisted that the Primero explosion, too, had resulted from dust. Although John D. Jones, Dalrymple's predecessor as state mine inspector, could not locate the source or proximate cause of the blast, his remarkably thorough investigations alleged that "had it not been for the presence of the dust in the mine, [. . .] the explosion would have been merely local and would have been confined, possibly, to the room where it started." The coroner's jury in that inquest nonetheless attributed the disaster to "causes unknown"—a predictable conclusion, perhaps, since five out of the six jurors were employed by Colorado Fuel & Iron. *Twelfth Annual Report of the Bureau of Labor Statistics*, 27, 32; *Trinidad Chronicle-News*, March 1, 1910. On state mine inspector Dalrymple's efforts to hold CF&I to account for excessive dust accumulation at Starkville, see ibid., Nov. 30, 1910.

142. As usual, corporations preferred to undertake their own reforms. CF&I called a conference of leading mining men in Las Animas County following the Delagua explosion and resolved to construct "a thorough system of piping" to facilitate the sprinkling down of coal dust, as well as steam jets to moisten the arid western air drawn into the mines by massive fans. CF&I also began to give cash prizes for safety. *Trinidad Chronicle-News*, Nov. 15, 16, and 21, 1910; Aldrich, "'Needless Peril of the Coal Mine,'" 503.

143. Editorial, *Denver Post*, Nov. 9, 1910.

144. Editorial, *Denver Republican*, Nov. 10, 1910.

145. One authority attributed the governor's change of heart to the death of the mine's superintendent alongside his men. *Denver Post*, Nov. 12, 1910.

146. This, at least, is the interpretation I advance in *Killing for Coal*.

147. Even as the Delagua rescue was unfolding, Slavic miners anxiously awaited delivery of "an immense marble shaft which is to stand in the Catholic cemetery as a permanent memorial to the Slavish miners who were victims of the Primero explosion." Paid for by Slavic mutual aid societies, the monument was made of blue Vermont marble and carried "the names of each one of the victims [. . .] as well as the names of the societies, the date and place of death and the emblems of the various societies"—a practice one newspaper declared "an old country custom." *Trinidad Chronicle-News*, Nov. 12 and Dec. 10, 1910.

148. Uhlich to Brake, July 17, 1910.

149. Andrews, *Killing for Coal*, introduction, ch. 7, and epilogue; Scott Martelle, *Blood Passion: The Ludlow Massacre and Class War in the American West* (New Brunswick, NJ: Rutgers University Press, 2007).

CHAPTER 6

Copper and Longhorns

Material and Human Power in Montana's
Smelter Smoke War, 1860–1910

TIMOTHY JAMES LECAIN

A deposit of copper ore possesses a certain type of power. We humans often use this power to generate other forms of physical and social power. Instinctually narcissistic, we just as often then ignore or deny the material origins of our own power. History books reserve their active verbs for the humans who "discovered" and "mined" mineral deposits; the copper ore itself is rarely granted any real agency. Copper is, to use the curtly dismissive English phrase, a mere "natural resource." Recently, however, developments in a variety of scholarly disciplines have begun to converge in ways that suggest we must rethink the agency and power of a mineral deposit. After several decades of intellectual neglect, matter—in the sense of that nonhuman material world somewhere "out there"—has begun to find its way back into our understanding of the past and present.

In this essay, I draw on and extend some of this new materialist thinking to analyze one of the worst cases of mining-related pollution in the history of twentieth-century North America. In 1902, the American copper corporation Anaconda erected an immense smelter in the Deer Lodge Valley of southwestern Montana. Because of the nature of the copper ores and the technologies used, the smelting process generated large amounts of sulfur dioxide and arsenic that harmed the valley's wild grasses, crops, and livestock. Ranchers and farmers fought to force the Anaconda to sharply reduce or eliminate the smelter pollution, in what was called the Montana Smoke War. An earlier generation of historians typically framed this and similar environmental battles primarily in political and economic terms, focusing on legislative and judicial arenas where the human power of individuals, states, or the corporation are most evident. More recently, historians of environment and technology have also revealed the complex ecological story of mining pollutants moving through various organisms and systems. Here, I attempt to bring the two together to argue that the ultimate source of human social power was a multifaceted material force that inhered in "things" such as cattle, copper, arsenic, and sulfur. Human power struggles like those at Anaconda, I argue, were the product of more fundamental collisions between the original material sources of power in the environment. Social history thus must be understood as intimately and inextricably connected with environmental history. Further, I suggest that the material power of Longhorns, sulfide copper, arsenic, and other types of matter drove the history of the Deer Lodge Valley as much as, if not more than, humans did acting on their own. My goal is thus to understand not only what humans make of matter but also what matter makes of us.

To make this argument, I begin with a brief overview of some recent materialist thinking. Then I turn to a more empirical analysis of cattle ranching and copper mining and smelting to suggest some of the ways these forms of matter created social power for humans. Finally, I illustrate how the material power generated by copper mining and smelting conflicted with that of cattle ranching and farming, a contest with disastrous results for the latter.

MATTER AND POWER

Materialist thinking is hardly new. Most famously, Karl Marx and Frederick Engels turned Georg Wilhelm Hegel's *Geist*-haunted idealism on its head to argue that social infrastructure was a product of material substructures. But while still useful, Marx and Engels's conception of matter is much narrower and more anthropocentric than that embraced by most new materialist thinkers.[1] Fernand Braudel and the so-called Annales School also opened up a once-popular vein of materialist thinking, arguing that in the *longue durée* material forces of geography and climate shaped the broad outlines of human history.[2] Despite the school's prominence in the historiography, however, relatively few historians followed in Braudel's steps.[3] Likewise, as with Marx and Engels, the *Annalistes'* concept of nature as a largely static and separate background influence on

168 I Industrial Catalysts

humanity differs sharply from newer materialist ideas of nature as a dynamic force deeply intermixed with the human and the cultural. More recently, a wing within World System Theory has also given considerable weight to the material as a driver of history, particularly in its emphasis on the role of peripheral states in providing raw materials to fuel the power of the core. Immanuel Wallerstein has noted the importance of considering the material biophysical world as a central aspect of world systems. Other World System theorists, such as Jason Moore and Alf Hornborg, have begun to develop promising new ways of integrating environmental influences into the theory.[4]

Many scholars in the history of technology and environmental history have also had close, if not always cordial, relationships with matter. Both fields began with an unusual focus on the material world, whether that world was predominantly a natural or a technological one.[5] Many environmental historians have long embraced a stubborn materialism, insisting that some form of nonhuman nature is an active force in history.[6] One of the most important spurs to the current new materialist thinking, however, came from scholars interested primarily in human creations. In the 1980s, John Law, Michael Callon, Bruno Latour, and other sociologists of science and technology developed their influential Actor Network Theory (ANT). In particular, as articulated in the imaginative work of Latour, ANT included all manner of potential nonhuman "actants" in its complex webs of networks. Agency was distributive and emerged from the interactions of actors, opening up an entirely new space for a type of materialist thinking.[7] These ideas also found confirmation in the influential work of the French philosophers Gilles Deleuze and Félix Guattari, whose ontological theories viewed the properties of all matter as continually emergent and fluid rather than static and fixed.[8]

As the geographers Owain Jones and Paul Cloke recently pointed out, however, ANT strikes many scholars of the environment as an inadequate means of dealing with the independent nonhuman agency of organic systems. Since agency in ANT emerges only from the interactions within a network, the theory tends to obscure what power might inhere in material things themselves, particularly organic things.[9] Further, the two geographers rightly note that many ANT scholars have been primarily interested in technological rather than organic systems, "a manoeuvre which somehow makes it easier to deny the specific non-human contribution to hybrid agency."[10]

Some scholars of a new materialist bent are now attempting to move beyond the limitations of ANT by suggesting that a type of discrete "thing power" inheres in all matter. The influential political theorist Jane Bennett argues that even seemingly inert and inorganic matter may possess such potential power. Under the banner of what she calls "vital materialism," Bennett strives to strip away both anthropocentrism and biocentrism in order to arrive at a concept of a matter that is much more than a merely passive or sometimes recalcitrant object of human action. Further, she argues that this dynamic concept of matter must change the way we conceive of human agency and power: "But the case for matter as active needs also to readjust the status of human actants: not by denying humanity's awesome, awful powers, but by presenting these powers as evidence of our own constitutions as vital materiality. In other words, human power itself is a kind of thing-power."[11] Drawing on contemporary scientific and technological insights into the nature of organic and inorganic matter, Bennett makes several attempts to illustrate the relationships between human and thing power. She even tries to imagine a "life of metal" using copper as her protagonist. However, as a political theorist, Bennett is primarily interested in exploring questions of ontology and ethics, not in developing a materialist methodology for analyzing the past.

The same can be said of what is thus far the most coherent assertion of a new materialist approach, the 2010 collection of essays *The New Materialisms*. In their introduction, volume editors Diana Coole and Samantha Frost assert that human beings "inhabit an ineluctably material world," but that this essential materiality has been marginalized in recent decades by "the dominant constructivist orientation to social analysis."[12] While observing that this "new materialism" need not be antithetical to constructivist methods, Coole and Frost call for recognizing a more vibrant role for matter in its interaction with humans and their social systems. Like that of Deleuze and Guattari, theirs is a matter that is "active, self-creative, productive, unpredictable," a matter that "becomes" rather than simply "is."[13] As promising as Coole and Frost's introduction is, however, the actual essays in the volume often fall short of really engaging the material world. Ironically, matter and "thing power" make far fewer appearances here than one might expect, and human ideas about these issues matter far more.

Fortunately, a number of environmental historians have provided just the sort of empirical micro studies of "thing power" in action needed to inject more materialism into the new materialism. Indeed, several important insights have emerged in the field of mining history, a subject that should inherently lend itself to materialist thinking. Thomas Andrews has convincingly argued that the material nature of coal—the demands it exacted from the men who attempted to mine it—was a key force in fostering solidarity and power among coal workers. Likewise, Andrews illustrates how the countervailing power of coal companies, and even the United States itself, was a product of the solar energy stored in hydrocarbon molecules.[14] Similarly, Timothy Mitchell persuasively argues that the material nature of coal deposits forces states to rely on large numbers of workers to extract it. As a result, state power based on coal extraction was more conducive to democratic control than oil, which could be obtained with far fewer workers and thus encouraged more centralized authoritarian regimes.[15] The importance of matter's energy content has also recently found a more general theoretical and methodological statement. Edmund Russell and several colleagues argue that material and social power are intimately linked and that energy power in particular is the first source of the ability of some humans to dominate others. "All power, social as well as physical," the authors conclude, "derives from energy."[16] Put in Bennett's new materialist terms, the work of Andrews, Mitchell, and Russell helps us see human power as a type of thing power, which in turn is sometimes a type of energy power.

The work of the environmental historian Andrew Isenberg suggests another form of systemic material power, one that inheres in the ability of environments to absorb destructive forces. In *Mining California*, Isenberg demonstrates how some humans extracted social power from the material world in the hydraulic mining of gold and silver deposits in California. By this Isenberg means not only the obvious point that the wealth derived from gold and silver could translate into human social power. Rather, his work suggests that gaining wealth and power from mining also came from exploiting the power of the environment to absorb the accompanying pollution and destruction.[17] Thus social power may be a product not only of harnessing material energy—as Russell argues—but also of harnessing or managing the material destruction and disorder generated from using that energy. As I argue below, this form of material power can be usefully thought of in terms of entropy.

While this linking of physical energy, entropy, and human social power is a critical step toward a useful new materialist methodology, it is also important to bear in mind that the energetic content of matter is only one of the many ways that matter may possess power. The power of copper, for example, has less to do with its energetic content than with its unique molecular ability to transmit heat and electricity. Unlike the energy content of coal or wheat, this type of material power cannot be easily quantified, yet it can play an equally important role in creating human social power. Likewise, the power of some forms of matter may inhere in both energetic and non-energetic forms. Such was the case with Longhorn cattle in the American West.

THE POWER OF LONGHORNS

Traditional anthropocentric accounts of the Euro-American settlement of the Deer Lodge Valley inevitably stress the story of ranchers and farmers overcoming the challenges of a wild land to create civilization. Yet even these frontier stories often implicitly recognize that the valley itself played some role in the process. As the pioneer rancher Conrad Kohrs later recalled, it was "one of the best, if not the best valleys in Montana, because the bunch grass was long and very nutritious."[18]

Indeed, the first Longhorn cattle to arrive in the valley found a ruminant grazer's paradise. Today, the area is classified as part of the Montana Valley and Foothill Grasslands ecoregion, a narrow patch of river valleys and rolling hills threading through the rugged mountains of southwestern Montana and up into southern Alberta. Compared to the extreme cold and heat of the northern plains of eastern Montana, the Deer Lodge Valley is relatively temperate. Annual rainfall is sparse, rarely much above 12 inches, and significantly less during cyclic droughts. The surrounding high mountain peaks capture much of the winter snow, helping to keep the valley grasses exposed for grazing. Prior to the invasion of cattle, the valley was home to several of their ruminant cousins, such as bison, elk, and deer. Their constant grazing had helped to create the open range that Longhorns and ranchers would later enjoy.[19]

The cattle arrived in the 1860s, led there by Euro-Americans such as Johnny Grant and Conrad Kohrs who were betting that the same grass that supported bison and deer could support Texas Longhorns. Several different types of hardy fescues and bunchgrass dominated the range. One of the best for cattle was bluebunch wheatgrass (*Pseudoroegneria spicata*). As the name suggests, bunchgrasses grow in tall scattered clumps rather than forming a dense contiguous mat like Kentucky blue or other lawn grasses. The plants can also develop extraordinarily deep root systems, up to 6 feet or more, that helped them to tap groundwater during the long Montana dry seasons.

To transform the lush bunchgrass into something humans could consume, the Deer Lodge Valley ranchers relied on one of the earliest and most successful animal-human partnerships in history. Domesticated since the early Neolithic, cattle have been critical to the survival and expansion of humans around the globe. Their ability to digest complex fibrous plant material like the cellulose of bunchgrass and turn it into muscle and milk provided humans with a new source of highly concentrated caloric energy. However, Kohrs, Grant, and other ranchers were able to *efficiently* harness the energy of the Deer Lodge Valley grass only because their Longhorn—and later, Shorthorn and Hereford—cattle were willing to work with them in doing so. The material power of the cattle thus resided not only in the caloric content of their meat but also in their genetic ability to cooperate with humans, which had resulted from domestication.

Most of the cattle that first arrived in Montana's Deer Lodge Valley were descendants of Texas Longhorns. The Longhorns were relatively lean and long-legged animals whose great sharp horns could stretch as much as 7 feet from tip to tip. Longhorns had descended from Spanish breeds shipped to the New World by the conquistadors of the sixteenth century. The animals thrived in the scrubby woodlands of southeast Texas, where many ran wild. Intelligent, fast breeding, and well armed, the Longhorns could fend off many predators and survive without being fed by humans. In this they resembled the American bison, which were as yet still the most populous large herbivore in North America.

Even these semiferal Longhorns, however, still carried the genetic markers of their earlier coevolutionary history with humans.[20] When horseback-mounted men took a renewed interest in them in the second half of the nineteenth century, the Longhorns were not so skittish or aggressive that they saw the men as mortal threats to be attacked or resisted at all costs. By manipulating the Longhorns' social instincts, cowboys could usually (though not always) get the animals to cooperate and move as an orderly herd. Some were eventually herded up into sheltered northern areas like the Deer Lodge Valley. Here Kohrs and other ranchers continued to depend on a delicate balance between the Longhorns' hardy independence and their willingness to tolerate human direction. In Montana, a mother cow had to be aggressive enough to protect her calf from danger—primarily wolves and coyotes—but not so aggressive that she attacked any human who came near.

In the late nineteenth century, American ranchers also began looking to fatter and gentler cattle breeds to stock their ranges. British Shorthorns and Scottish Herefords were the most popular breeds, as well as a Longhorn-Shorthorn hybrid that was larger and meatier than the original Texas Longhorn yet better able to survive disease and harsh conditions than a purebred Shorthorn.[21] Conrad Kohrs was among the first to introduce Shorthorns to Montana in 1872, and two decades later, he also imported the first registered Herefords. Careful breeding helped the animals to adapt to their new environment even more effectively, maximizing their ability to make grass into meat without destroying their ability to survive with minimal human care.[22]

Just how much social power the Longhorns and other cattle breeds in the Deer Lodge Valley created for humans is difficult to precisely delineate. Nick Bielenberg, Conrad Kohrs's half-brother and sometime partner, estimated that by the early twentieth century he had raised more than 100,000 head of cattle.[23] Kohrs himself reported he had raised some 200,000 head since he arrived in the valley in 1864, and he eventually came to own nearly a million acres of land scattered around four states and two Canadian provinces. Both men became very wealthy. Bielenberg preferred to remain in the Deer Lodge Valley actively running his ranching operations, but Kohrs invested his cattle wealth in other businesses and used it to pursue a successful political career.[24] Kohrs was among the original "Cattle Kings" who dominated the early politics of many western plains states. He became a territorial and later a state senator, was a delegate to the Montana constitutional convention, and served as president of the Montana Stockgrowers Association.[25] However, in contrast to Wyoming, where the Cheyenne ranching interests dominated the state, in Montana, Kohrs and other ranchers had to compete for political power with timber, farming, and especially mining interests.[26]

Obviously, the social power of Kohrs and other successful ranchers was a product of the wealth generated from their cattle. When Kohrs built a fine new mansion in Helena, the Montana state capital, it might have seemed irrelevant whether the dollars that paid for it came from cattle raising, mining, or even betting on horse races. One of the key characteristics of capitalism is its ability to reduce everything to the same unit of measurement, an abstract price that obscures the original source of value. A materialist energy flow analysis should remind us that Kohrs's mansion was, in part, a reformulation of the energy first captured in the bunchgrass of the Deer Lodge Valley and subsequently concentrated into the muscles of Longhorn cattle.

The energy content of Longhorn meat, however, constituted only part of the material power of cattle. Equally important was their ability to survive largely on their own, yet also cooperate with humans when asked. This type of material power can be thought of, in part, in terms of another closely related concept borrowed from physics: entropy. In physics, entropy is typically understood as a measure of the disorder in a moving or energy-using system. In a steam engine, for example, fuel is burned to generate heat that creates high-energy pressurized steam that is initially concentrated in a compressed cylinder, resulting in a relatively high-order or low-entropy state. When the steam expands and thus does useful work in moving the cylinder, the system loses order and entropy goes up as the useful organized energy of the steam is dissipated through a loss of pressure and heat. Thus as entropy or disorder of a system increases, the system contains less useful energy available for doing work.

According to the second law of thermodynamics, all closed systems will either remain stable or increase in entropy; like the steam engine, once all the coal or other energy source in the system is used up, it simply stops working. However, the input of new energy from outside the system (as when we add coal to the steam engine, or the sun adds energy to the earth) can again increase order and decrease entropy. Likewise, humans and other animals can use their bodily energy and cognitive abilities to increase the order of a system. For example, humans might gather coal dispersed over a large mine into one concentrated area where its chemical energy content can be used to continually fuel a steam engine.

Viewed in this sense, the behavioral abilities of Longhorn cattle helped ranchers like Kohrs to decrease the entropy of the Deer Lodge Valley. First, by grazing widely using their own intelligence and skills, the cattle concentrated the solar energy dispersed over a wide area into their relatively compact and portable bodies. Second, thanks to their long coevolutionary history with humans, the Longhorns had brains and nervous systems that allowed ranchers and their cowboys to gather them efficiently in one spot during a roundup. Entropy thus decreased as dispersed cattle concentrated in one area, such as a corral or feedlot, simultaneously increasing their potential to do further useful work for humans as a ready source of the caloric energy we call food.[27]

In sum, the social power of Deer Lodge Valley ranchers such as Kohrs derived from a surprisingly complex system for energy conversion and concentration that depended on the unique physical and behavioral powers of the Longhorn cattle themselves. In this sense, what is typically seen as solely the human intelligence behind successful open-range cattle ranching is better understood as a type of distributed intelligence, one in which human and animal abilities merge almost seamlessly.[28] Order emerged from relative disorder, and usable energy replaced entropy. Ultimately, the energy that had been so efficiently concentrated in the muscles of the cattle would feed thousands of humans, most of them well beyond the boundaries of the Deer Lodge Valley.

The human intersection with another form of matter in the region, however, functioned very differently. For mining and processing the copper ores buried under the neighboring city of Butte, humans had no efficient and cooperative organisms that could consume the copper for them, concentrate it in their own bodies, and efficiently carry it to a central location. To the contrary, in mining, the energetic and entropic patterns were largely reversed from those in ranching.

THE POWER OF COPPER

In contrast to bunchgrass or cattle, the material power of copper has relatively little to do with its chemical energy content, but instead lies in its unique atomic structure. Like gold and silver, with which its shares a column on the Periodic Table of Elements, copper's outer orbit is occupied by only one electron, which is only weakly bonded to the nucleus. As a result, when electrical current or heat is applied to a wire made up of copper atoms, this single outer electron is easily stripped away and can efficiently conduct electricity or heat. This is why copper is the human metal of choice for many electrical applications as well as for use in pans, radiators, air conditioners, and other heat-transfer technologies.[29]

The way in which the atoms of copper pack together at room temperature is also critical to its material power. When properly treated, a piece of pure copper takes on an internal crystalline structure known as a face-centered cube in which a copper molecule occupies every corner of a cube as well as the center of each face. Since each unit is made up entirely of triangles, this crystalline lattice of atoms *should* be extremely strong and resistant to deformation—far too strong for humans to easily bend or hammer it into new shapes. However, copper's seemingly uniform crystalline structure is actually riddled with imperfections called "dislocations"—areas where the lattice pattern of copper crystals is not precisely aligned and connected. The presence of millions of these small dislocations allow what would otherwise be perfectly rigid cubes to slide past each other, making it possible to easily shape the metal without having to heat it, as is generally necessary with harder and more brittle metals like iron. Usefully, this ductility also rapidly decreases as the copper metal continues to be bent or hammered and these dislocations begin to pile up, rather like a log jam. As a result, soft copper metal can be substantially hardened simply by pounding it with a rock or a hammer, a property that many humans have found useful when making copper knives or arrowheads.[30]

The copper mined in Butte, Montana, occurred mainly in two forms: chalcocite, an ore consisting of two atoms of copper bonded with one atom of sulfur (Cu_2S), and enargite, which had three atoms of copper, four of sulfur, and one of arsenic (Cu_3AsS_4). Thus the same solitary outer electron that made the copper atom so ductile and useful in conducting heat and electricity also means that it is often accompanied by atoms of sulfur and arsenic, elemental substances that have their own types of material power. Humans who wanted Butte's copper would also have to deal with its less desirable colleagues.

An Irish-born mining entrepreneur named Marcus Daly discovered the rich copper deposits under the town of Butte in 1882 while mining for silver in the Anaconda shaft. Daly was able to convince his old friend George Hearst and several others to invest heavily in developing the large-scale copper mining and smelting business that became the Anaconda Copper Mining Company. Mining eventually revealed an immense area of copper mineralization between 300 and 1,000 feet beneath the earth, one of the largest such deposits in the world. Daly and his partners invested in the latest mining technologies to remove the valuable ore. Deep underground mining was an energy-hungry enterprise, so the Anaconda operated its own coal mines elsewhere in the region to fuel the steam engines that drove the ore lifts, compressors, rock drills, and other mining machinery.[31]

In 1883, Daly and the Anaconda erected a smelter some twenty-six miles to the west of Butte in the Deer Lodge Valley. The smelter was nestled in the narrow ravine of Warm Springs Creek, which limited the spread of the resulting smoke to the wider valley. The company transported the ore from Butte via a dedicated steam (and later electric) rail line. By 1900, however, ore production from the mines had increased rapidly, and Daly determined the company needed an even bigger smelter. The new smelter, called the Washoe, was constructed on a slope that faced directly out onto the wide Deer Lodge Valley. When the workers smelted the first charge of copper ore in early 1902, the clouds of smoke and mist could now easily sweep down onto the valley's ranches and farms.

Even the richest ore in Butte was only 4 or 5 percent copper. The remaining quartz, iron, silica, and other less valuable material had to be removed by crushing, concentrating, and smelting. In essence, the Washoe smelter operated by reversing some of the geochemical processes that had created the Butte copper ore deposits in the first place, though the source of heat was now wood and coal instead of geothermal energy. Relatively low-temperature roasting of the concentrated copper ore provided enough energy to drive much of the sulfur off and into the atmosphere. This roasted ore was subsequently superheated to around 2,700 degrees Fahrenheit, which pushed more sulfur and arsenic into the air and permitted the now–relatively pure copper to be separated from any remaining iron, silica, and other substances, which were poured off as waste slag.[32]

Once released into the sky, the liberated atoms of sulfur and arsenic became powerful historical agents in their own right. The sulfur immediately bonded with atmospheric oxygen to form sulfur dioxide, which in turn could interact with water in the atmosphere to form a highly corrosive sulfuric acid. The arsenic also bonded with oxygen to form arsenic di- and trioxides, both of which can be highly toxic to animal life. By the autumn of 1902, the smoke began to cause devastating crop and livestock losses for big ranchers like Kohrs and Bielenberg, as well as many smaller ranchers and farmers. Bielenberg alone lost more than a thousand head of cattle, eight hundred sheep, and twenty horses in the course of just a few weeks.[33]

Ranching in the valley had relied on the ability of cattle to intelligently concentrate solar energy and reduce the entropic disorder of the valley. In contrast, copper smelting depended on the large-scale *application* of energy, mostly from hydrocarbons, which had the effect of increasing the overall disorder and entropy of the system. The Anaconda first used energy to break up the highly stable geological structure of the underground copper deposit, to transport it for smelting, to crush and concentrate the ore, and finally to smelt it. At every stage, the industrial system used large amounts of energy to concentrate and purify the copper. But this decrease in entropy gained from isolating the small amount of copper from the massive amounts of waste simultaneously resulted in a large increase in entropic disorder for the system as whole. One result was that the molecules of sulfur and arsenic that had previously been concentrated in a small area of stable subsurface rock were now broken up and randomly dispersed into the atmosphere. Indeed, had the Anaconda been unable to disperse the smoke pollution from the immediate area around the smelter, the levels of arsenic and sulfur would have quickly become deadly to humans.

The Anaconda's success and its attendant social power thus derived from at least three material sources. First, the copper itself had unique material properties that permitted humans to develop electrical and heat transfer technologies. Second, the stored solar energy extracted from the company coal mines and lumber operations. And third, the ability of the surrounding environment to absorb the entropy created by the application of this energy in mining and smelting—particularly the random dispersal of arsenic and sulfur compounds that would otherwise have been immediately toxic to human life.

The extent of the resulting social power was apparent. By 1909, the Anaconda had produced 590 million pounds of copper, which supplied about 10 percent of the entire world demand and as much as 20 percent of the U.S. demand.[34] The Anaconda paid about 30 percent of the total tax revenues of the city of Butte. Over just the previous seven years, it had also spent more than $7 million for labor, $4 million for coal, $4 million for coke, and more than $1 million for machinery.[35] Indeed, the Anaconda exercised so much economic might that in 1899 the Rockefellers' infamous Standard Oil trust purchased it in a failed attempt to establish monopoly control over the world copper industry.[36] Although Daly and most of the other executives who ran the company did not personally seek political office, there was little need for them to do so anyway, as the Anaconda kept a tight "copper collar" on the state of Montana. The company was by far the largest single employer and economic presence, and it also controlled most of the major newspapers in the state, giving it considerable power to pick cooperative candidates for state and federal offices.[37]

As with the power of Kohrs, Bielenberg, and other ranchers in the Deer Lodge Valley, the power of the Anaconda can all too easily be reduced to the abstraction of money or capital. Yet we must make a conscious effort to recognize that the company's economic influence and social power also ultimately derived from the material power of copper, the energy of coal, and the ability of the environment to absorb entropic disorder. Most important, when the entropy-generating system of mining and smelting collided with the entropy-reducing system of ranching, both human power and material power clashed.

LONGHORNS VERSUS COPPER

In 1905, a front-page article in the *Butte Inter Mountain* celebrated Anaconda's three-year-old Washoe smelter with the headline, "It Is the Largest in the World."[38] Just a month later, the *Anaconda Standard* reported on the "Monster Beef Cattle" from a neighboring valley that were on their way to Chicago via the Anaconda Stockyards: "There are 450 head in the herd and some of the largest will weigh more than 1,700 pounds, while there is one monster that is estimated at close to 2,000 pounds."[39] These two articles suggest the material and political nature of the escalating conflict in the Deer Lodge Valley. Both the Anaconda and the ranchers were using energy and other forms of material power to maximize the size of their output, but they were doing so in fundamentally incompatible ways.

By 1905, the Washoe smelter was generating about 2 billion cubic feet of smoke every day that carried some 48,100 pounds of arsenic and immense volumes of sulfur gas.[40] Nick Bielenberg estimated that the poisonous smoke subsequently spread over an area approximately 30 miles long and 12 to 14 miles wide.[41] Residents described it as a "white mist" or having a "a bluish color."[42] Some days, the smoke settled in low-lying areas and shifted with the surface winds. As the rancher Angus D. Smith noted, some mornings the cloud of smoke would be so thick on his property that he could not see more than two or three hundred feet.[43] But the ranchers and farmers noticed that the behavior of the smoke varied depending on atmospheric conditions. William T. Stephens reported, "Sometimes it comes in a stream across, and other times it settles and spreads out more, and sometimes it goes clear over head."[44] Regardless of their personal experience, many of the ranchers and farmers agreed that "the smoke from the stack is charged with large quantities of sulphur dioxide, arsenic, antimony, copper and other noxious and poisonous substances, which are deposited upon the farms of the valley, burning and dwarfing the crops, poisoning the soil and causing large numbers of horses, sheep, cattle and other livestock to sicken and die."[45]

The sulfur and arsenic in the smoke stream attacked the material basis of the ranchers' power in at least three fundamental ways, each with its own dynamics. First, the sulfur dioxide and sulfuric acid undermined the energetic basis of ranching by killing or limiting the growth of the valley grasses—both the wild bunchgrasses and the cultivated grasses such as hay, oats, alfalfa, and other feed crops. Sulfur compounds were once a significant part of the earth's atmosphere, and volcanic activity can still occasionally discharge large amounts of sulfur into the air. But like copper, sulfur is highly reactive and easily bonds with many other elements, including iron and copper. Over millions of years, much of the previous atmospheric load of sulfur was bound up in rocks in the lithosphere or absorbed by the oceans.[46] During the ages when most of the chlorophyll-based plant life of today evolved, atmospheric levels of sulfur compounds were low in most areas. When the

Anaconda smelted the Butte sulfide ores, it reversed this primordial biogeochemical cycle and shifted lithospheric sulfur back into the atmosphere at levels that harm most modern plants.

Sulfur dioxide damages grasses and other plants by directly interfering first with their ability to generate and then to store and use energy from sunshine. The sulfur dioxide dispersed by the Washoe smelter entered plants through the small holes, or stomata, that penetrate the protective waxy cuticle of leaves and can open and close in response to environmental conditions.[47] Once sulfur dioxide gas enters the stomata, it spreads through the intercellular spaces in the leaf, where most of it dissolves in water to form sulfuric acid and sulfite ions. These sulfur compounds attack the plant's chlorophyll-filled chloroplasts, destroying their ability to transform solar energy into the sugars that can be consumed for energy. The sulfite also interferes with the plant's mitochondria, the cells that subsequently consume these sugars and generate adenosine triphosphate (ATP), the molecule that provides usable energy to cells in both plant and animal life.[48]

High concentrations of sulfur dioxide can cause almost immediate death of plant leaves, while lower concentrations slow growth and reduce yields.[49] While any terrestrial plant can be harmed or killed by sulfur dioxide gas, sensitivity varies. Effects within a region are also highly variable depending on topography, soil conditions, wind directions, and other factors. For example, trees and other plants on mountainsides might be damaged at a greater distance from the pollution source than flat areas closer to the source.[50]

The farmers and ranchers observed all these effects. Several noticed that the smoke was particularly harmful to the quaking aspen trees that grew wild in moist areas of the valley.[51] One farmer reported that the smoke only sporadically affected plants in his kitchen garden: "Some things will stand a pretty good siege, while other garden truck will not recover from its effects." Bielenberg recalled that the smoke "once cut a path right through my grain crop, leaving about 200 yards [on each side] it did not touch."[52] The very complexity of the interactions between the sulfur and the plants made it extremely difficult to prove definitively that the Washoe smelter was responsible. The company-owned *Anaconda Standard* argued that since some ranches seemed to be untouched by smoke damage, ranchers like Bielenberg who complained must have been "unthrifty" farmers who failed to properly care for their land.[53]

By killing or reducing the nutrient value of wild grasses and cultivated crops, the sulfur smoke reduced the supply of energy available to the valley's ranching industry. Prior to the Washoe, hay yields were typically around a ton per acre.[54] Chronic low-level smoke exposure appears to have cut yields by at least two-thirds.[55] Acute exposure could kill an entire crop in just the course of a few days or even hours.[56] What hay and other feed crops did survive might still be "smoked," a consequence of arsenic and other poisons deposited on their surface. Even if hay from some areas of the valley may have actually been arsenic free, many potential buyers believed it to be poisonous, driving down prices or making it impossible to sell at all.[57]

The second way the smoke attacked the material basis of the ranchers' power occurred when arsenical compounds undermined the ability of the Longhorns and other cattle to efficiently transform plant energy into meat, milk, and baby Longhorns. Released by the Washoe as a gas or mist, the arsenic trioxide (As_2O_3) molecules, when cooled, formed a fine white powder that settled unevenly over the valley. Ranchers and farmers quickly recognized that the white powder was toxic to humans. Many reported that it caused blisters on the mouth and nose. One woman noted, "I would become so dizzy that I could not walk across the room without staggering. My daughter was worse than myself, very much worse."[58] The rancher Nicholas A. Liffring recalled being poisoned: "Why there was a white dust on the straw, and, while baling it it made me sick; I broke out in boils around the hat band and sores upon the body."[59]

To some degree, ranchers and farmers could limit their internal exposure to arsenic dust through careful cleaning of garden crops and staying away from the smoke stream as much as possible. But avoiding the dust was more difficult for cattle and other stock animals in the valley that consumed the wild and cultivated grasses. Longhorns possessed formidable defensive skills, but were ill prepared to deal with a danger like arsenic. The Anaconda's power emerged in part from the ability of the surrounding environment to dilute and absorb the Washoe arsenic, but the cattle's energetic and entropic basis drove them to do just the opposite. Because wild grasses are relatively low in caloric content, cattle must spend most of their waking hours grazing over a wide area just to consume enough to stay alive. Typically, a steer or cow must eat every day about 2 percent of its body weight in grass (excluding the water content), which for a large animal might be as much as 70 pounds of actual forage. What had previously been the valuable ability of cattle to concentrate the dispersed solar energy of the valley's grass into muscle now had the damaging effect of reconcentrating the dispersed arsenic to poisonous levels. Further, since the sulfur dioxide was simultaneously reducing the size and caloric content of the valley grass, the cattle had to eat more grass over an even wider area, further increasing their uptake of poisonous arsenic.

When consumed in relatively small doses, arsenic trioxide can actually be a mild stimulant. One sheepherder reported that his animals deliberately ate one another's arsenic-contaminated wool because they were addicted to its effects.[60] But in higher amounts, arsenic trioxide is devastating to the basic biological functions of most animals. The biochemical toxicity of arsenic oxides comes in part from their molecular structures, which are very similar to that of phosphate (a phosphorous atom bonded with four oxygen atoms, PO_4), an essential cellular building block of all currently known organic life. Indeed, arsenic is immediately beneath phosphorous on the Periodic Table of Elements, and it can easily bond with oxygen to form arsenate (AsO_4), a molecule structurally almost identical to phosphate. In a normally functioning cell, phosphate is used in the mitochondria, the cellular "power plants" where the caloric energy of sugars from food is broken down to make the ATP (adenosine triphosphate) to power cellular metabolism. Because of arsenate's structural (though not functional) similarity to phosphate, some of the cell's mitochondria bind with it instead, destroying their ability to generate ATP. Literally starved of energy, the cell begins to die, which in turn causes internal lesions and bleeding, organ failure, coma, and death. Cattle, horses, sheep, and humans all shared this biochemical vulnerability to the energy-robbing effects of arsenic.[61]

Although they could not have known the complex biochemical causes, ranchers clearly observed the acute and chronic effects of arsenic in their cattle and other animals. Morgan Evans, a seventy-two-year-old rancher whose property was about two and a half miles from the Washoe stack, noted that the smoke "turned my place into a graveyard. I lost from 75 to 80 head of cattle in 90 days." As mentioned earlier, big ranchers like Kohrs and Bielenberg lost thousands of head of cattle, horses, and sheep. Postmortem autopsies revealed the internal signs of arsenic poisoning: "lesions affecting the stomach, intestines, liver, kidneys, spleen, heart, respiratory organs, and membranes of the brain."[62] The death rate was so high that the Anaconda dug mass graves to quietly bury the thousands of dead animals that, in an implicit recognition of their responsibility, managers initially agreed to buy from ranchers.[63]

However, the effect of chronic arsenic exposure on the overall vitality and energetic capacity of stock animals could be harder to prove, though the anecdotal evidence seemed clear. Morgan Evans, for example, reported, "The horses have sore noses now and seem to be weak. They cannot stand the work they formerly stood." Another small rancher, George Parrott, said, "[The horses] are not nearly so good as they used to be, and they are soft, and sweat easy if you go to driving them, and they don't seem to have strength and cannot stand work like they used to."[64] Since the biological muscle power of these draft horses was as critical to successful ranching as the chemical power of coal was to mining and smelting, the Anaconda's arsenic was sharply reducing the energy available to ranchers for transporting supplies, hay, and other materials.

Even more critical, the arsenic interfered with the beef cattle's key task: to transform low-energy forage into high-energy meat and milk. Jerry Ryan, a small-holder rancher, noted, "I have 40 head of cattle and all they do is stand up and eat without seeming to derive any benefit from it. They are sickly and weak." George Parrott observed that his cattle "seem to eat hearty enough and a good deal of it, but it did not seem to do them any good."[65] Although Ryan and Parrott did not know that the arsenic was interfering with ATP production in their cattle's mitochondria, they did recognize that the poison was somehow robbing the animals of the essential life-giving energy that made profitable meat. What meat cattle did manage to put on often had little fat, and the slaughtered animals looked suspicious to local butchers. Ryan noted that he tried to conceal the effects of the smoke on one beef cow: "Took him up and sold him to Mr. Wegner; and had to take some fat off his stomach and put it over his kidney to make it look respectable."[66]

The arsenic also affected the valley's dairy cows, reducing or eliminating their ability to give milk. Acute arsenic poisoning in the dairy cows typically began with loose bowels followed by constipation. The cows then stopped eating, and as dairy farmer Angus D. Smith testified, "the hair turned on them and they would not lick themselves, and their noses got dry, which is unnatural for a healthy cow [. . . and] it would take two or three months and sometimes weeks and then they would die."[67] As with smoked hay, former customers in the valley were suspicious of milk that came from smoked cows. Indeed, valley residents were wise to be concerned, as scientists now know that dairy cows can concentrate arsenic in their milk.[68] Some dairy farmers reported that the milk from smoked cows had a strong "garlicky" smell, which is characteristic of arsenic content.[69] As Kenneth Smith noted, the drop in consumer demand for milk was matched by a drop in supply, as his dairy cows soon stopped giving milk and he was forced to abandon his milk wagon business.[70]

Much of the material power of ranching derived from cattle's (particularly the independent Longhorn's) ability to reproduce without assistance, thus transforming the energy of plants not only into more meat but also into more cattle. But as rancher Eli Dehourdi noted, the smoke pollution also interfered with calving. Dehourdi had previously found that 75 to 80 percent of his heifers would bear a live calf each year. After the Washoe opened, the rate dropped to 50 percent, either because the cows were unable to conceive or because they miscarried their fetuses.[71] Others found that as many as 40 percent of mares miscarried their colts.[72] George Parrott had previously kept a purebred (probably Shorthorn) bull to breed with his cows, but he had since castrated the animal. "In the winter time," Parrott noted, "the cows were throwing [miscarrying] so many dead calves that I saw it was no use in trying to breed."[73] William T. Stephens testified that even if calves did survive to term, they "are weak and puny when they come [and] some of them die, and some of them never do well afterwards, a great many of them."[74] Just as it could poison children, the arsenic-contaminated milk could also poison calves, since the poisonous dose of arsenic is much lower for both small cows and small humans.[75]

The third and final way in which the smoke damaged the material basis of the ranchers' power came from the way the sulfur and arsenic undermined the unique coevolutionary bonds between humans and cattle that had made the previously lucrative "open range" possible.[76] Prior to the opening of the Washoe smelter, many big ranchers in the valley such as Bielenberg and Kohrs still provided little or no hay or other feed crops to many of their cattle, even in winter. They depended instead on the ability of their Longhorns and hybrid Longhorn-Shorthorn breeds to graze widely over vast areas of range largely on their own.[77] But the pollutants from the Washoe made such free-ranging behavior nearly impossible. In 1905, the Montana state veterinarian advised the Deer Lodge Valley ranchers to stop allowing their cattle or horses to range on open pastures and to instead corral them in smaller pens where they would need to be fed uncontaminated hay or other fodder daily.[78] William Stephens, who had previously let his cattle run in the pastures, was now "keeping them shut up in corrals and barns and feeding some of them bran and hay and oats."[79]

The Anaconda sulfur and arsenic thus attacked the practice of open-range grazing in the valley, and hence affected the cowboys and roundups that had already become the mythic stuff of the American popular imagination. By forcing ranchers to confine and feed even their hardy Longhorns and hybrid Shorthorns, the arsenic undermined the material power that ranchers had previously derived from the ancient evolutionary bonds between humans and cattle. The Longhorns' ability to survive on their own, find the best forage, and fend off predators could no longer contribute to ranchers' human power. What the cattle had once done for themselves the Deer Lodge Valley ranchers were now forced to do for them.

In the years following the Washoe's opening, the ranchers and farmers of the valley fought a long battle to limit or eliminate the smoke pollution. In 1905, many joined together to sue the Anaconda, asking for more than a million dollars in damages and the closing of the smelter. The resulting trial stretched on for many months, and both sides presented convincing expert testimony supporting contradictory positions. When the judge finally issued his ruling in 1909, however, he declined to comment in much detail on whether or how the smoke was harming the valley's animals and crops. Instead, he decided to rule for the company, primarily on the grounds that the copper produced was economically essential to the valley, state, and nation. Eventually, the threat of further litigation from the federal government pushed the company to develop or adopt more powerful pollution-control technologies. However, increased production rates worked to minimize the effectiveness of even these measures, and many ranchers and farmers finally had little choice but to abandon ranching and sell their properties to the company.[80] In the end, the material power of copper and its miners defeated the material power of Longhorns and the ranchers who depended on them.

CONCLUSION

Can a copper deposit make history? Not on its own, perhaps. But the same might well be said of the humans who have developed so many novel ways of *using* the material world to survive and create their variegated cultures. To emphasize the centrality of nonhuman matter in the history of the Deer Lodge Valley Smoke War is not to assert some sort of primitive environmental determinism. Rather, it is an attempt to resist the powerful human tendency to see the world solely as a reflection of ourselves, to suggest instead that we do not use matter so much as cooperate with it in ways that form and define us. Copper, Longhorns, sulfur, and arsenic did not dictate the course of events in the Deer Lodge Valley, but neither did humans, precisely because these "things" were much more than mere natural resources that could be bent freely to human will. The matter contained a type of power, a material power that was a fundamental—though not the only—basis for human social power. This material power took many forms, some of them identifiable with physical concepts like energy and entropy, others more subtle, like the distributed intelligence created by the interactions between Longhorns and ranchers. But when the material power of copper collided with that of cattle, the humans whose social power derived from them also became entwined in the conflict. The humans involved could, of course, have handled their own social power conflicts in any number of fascinating ways, all of them well worth the attention of historians. Yet, in our inevitable fascination with ourselves, we should also take care to remember that we are not so far removed from the world of matter as we like to think. Matter makes us as much as we make it.

NOTES

I wish to thank the Rachel Carson Center for Environment and Society in Munich, Germany, for a generous fellowship year during which the conceptualization for and writing of this essay was completed. Also, much of the archival research was done with the support of a National Science Foundation grant (Award No. 0646644) in collaboration with my colleague Brett Walker. Robert Gardner and Constance Staudohar provided invaluable research assistance.

1. Diana H. Coole and Samantha Frost, eds., *New Materialisms: Ontology, Agency, and Politics* (Durham, NC: Duke University Press, 2010), 29; J.R. McNeill, José Augusto Pádua, and Mahesh Rangarajan, eds., *Environmental History: As If Nature Existed* (New Delhi: Oxford University Press, 2010), 4.

2. The seminal text being Fernand Braudel, *La Méditerranée et le monde méditerranéen à l'époque de Philippe II* (Paris: Colin, 1949).

3. See McNeill, Pádua, and Rangarajan, *Environmental History*, 5.

4. Immanuel Wallerstein, "What Are We Bounding, and Whom, When We Bound Social Research?" *Social Research* 62 (1995): 839–56. Jason Moore, for example, notes the roots of historical capitalism in exploiting and undermining complex socio-ecological webs in early Peruvian silver mining. Jason Moore, "'This lofty mountain of silver could conquer the whole world': Potosí and the Political Ecology of Underdevelopment, 1545–1800," *Journal of Philosophical Economics* 4:1 (2010). See also Alf Hornborg and Carole L. Crumley, *The World System and the Earth System: Global Socioenvironmental Change and Sustainability since the Neolithic* (Walnut Creek, CA: Left Coast Press, 2007); Alf Hornborg, J.R. McNeill, and Juan Martínez Alier, *Rethinking Environmental History: World-System History and Global Environmental Change* (Lanham, MD: AltaMira Press, 2007); and Jason Moore, "*The Modern World System* as Environmental History? Ecology and the Rise of Capitalism," *Theory and Society* 32:3 (2003).

5. Indeed, when the two are fused into the hybrid called enviro-tech, they gain entirely new analytical purchase. See the insightful introduction in Sara B.

Pritchard, *Confluence: The Nature of Technology and the Remaking of the Rhône* (Cambridge, MA: Harvard University Press, 2011).

6. For a good review, see Richard C. Foltz, "Does Nature Have Historical Agency? World History, Environmental History, and How Historians Can Help to Save the Planet," *History Teacher* 37 (2003): 9–28.

7. John Law and John Hassard, *Actor Network Theory and After*, Sociological Review Monographs (Oxford, UK; Malden, MA: Blackwell/Sociological Review, 1999); Bruno Latour, *Reassembling the Social: An Introduction to Actor-Network-Theory*, Clarendon Lectures in Management Studies (Oxford: Oxford University Press, 2007).

8. Gilles Deleuze and Félix Guattari, *A Thousand Plateaus: Capitalism and Schizophrenia* (Minneapolis: University of Minnesota Press, 1987).

9. Owain Jones and Paul Cloke, "Non-human Agencies: Trees in Place and Time," in *Material Agency: Towards a Non-anthropocentric Approach*, ed. Carl Knappett and Lambros Malafouris (New York: Springer, 2008), 79–96, 80.

10. Ibid., 81. See also M. Fitzsimmons and D. Goodman, "Incorporating Nature: Environmental Narratives and the Reproduction of Food," in *Remaking Reality: Nature at the Millennium*, ed. Bruce Braun and Noel Castree (London: Routledge, 1998), 194.

11. Jane Bennett, *Vibrant Matter: A Political Ecology of Things* (Durham, NC: Duke University Press, 2010), 10.

12. Coole and Frost, introduction to *New Materialisms*, 6.

13. Ibid., 9.

14. Thomas G. Andrews, *Killing for Coal: America's Deadliest Labor War* (Cambridge, MA: Harvard University Press, 2008).

15. Timothy Mitchell, *Carbon Democracy: Political Power in the Age of Oil* (London: Verso, 2011).

16. Edmund Russell, James Allison, Thomas Finger, John K. Brown, Brian Balogh, and W. Bernard Carlson, "The Nature of Power: Synthesizing the History of Technology and Environmental History," *Technology and Culture* 52 (2011): 248. David Nye has also suggested the ways Americans used conspicuous consumption of power as a means of creating and demonstrating social power, though he puts much less emphasis on the importance of the physical energy of material power sources. David E. Nye, *Consuming Power: A Social History of American Energies* (Cambridge, MA: MIT Press, 1998).

17. Andrew C. Isenberg, *Mining California: An Ecological History* (New York: Hill and Wang, 2005).

18. "Washoe and Anaconda Are Good Customers," *Anaconda Standard* (18 February 1906).

19. Taylor H. Ricketts et al., *Terrestrial Ecoregions of North America: A Conservation Assessment* (Washington, DC: Island Press, 1999), 285–87; "Intermountain/Foothill Grassland Ecotype," in Montana Fish, Wildlife, and Parks, *Comprehensive Fish and Wildlife Conservation Strategy* (Helena, MT, 2012), 37–42.

20. On coevolution and history, see Edmund P. Russell, *Evolutionary History: Uniting History and Biology to Understand Life on Earth* (Cambridge: Cambridge University Press, 2011).

21. Ibid., 39–40.

22. Paul McGrew, "Conrad Kohrs—Pioneer Cattleman," *Pacific Northwesterner* 31 (1987): 8.

23. "Mr. Bielenberg Discusses Smoke," *Anaconda Standard* (3 March 1906).

24. "Washoe and Anaconda Are Good Customers."

25. *Progressive Men of the State of Montana* (Chicago: A. W. Bowen & Co., 1903); McGrew, "Conrad Kohrs," 1–10; Conrad Kohrs, *Conrad Kohrs: An Autobiography* (Helena: C. K. Warren, 1977, 1913); Lewis Atherton, *The Cattle Kings* (Lincoln: University of Nebraska Press, 1972), 183.

26. Atherton, *The Cattle Kings,* 67.

27. Of course, the further extraction of that caloric energy from the cattle involved other systems with their own unique energetic and entropic dynamics. Initially, cattle were further concentrated at meatpacking plants, but subsequently, humans used energy to distribute the meat to widely dispersed markets and homes.

28. On the powerful concept of extensive or distributed intelligence, see Andy Clark, "Where Brain, Body and World Collide," in Knappett and Malafouris, *Material Agency*, 1–18.

29. Timothy James LeCain, *Mass Destruction: The Men and Giant Mines That Wired America and Scarred the Planet* (New Brunswick, NJ: Rutgers University Press, 2009), 32–31.

30. C. R. Hammond, "The Elements," in David R. Lide, ed. in chief, *The Handbook of Chemistry and Physics,* 81st ed. (Boca Raton, FL: CRC Press, 2000); Stephen L. Sass, *The Substance of Civilization: Materials and Human History from the Stone Age to the Age of Silicon* (New York: Arcade Publishing, 1998), 44–45.

31. LeCain, *Mass Destruction,* 41–42.

32. James E. Fells, *Ores to Metals: The Rocky Mountain Smelting Industry* (Lincoln: University of Nebraska Press, 1979), 27–30, 273–74; Donald M. Levy, *Modern Copper Smelting* (London: Charles Griffin, 1912).

33. Donald MacMillan, "A History of the Struggle to Abate Air Pollution from Copper Smelters of the Far West, 1885–1933" (Ph.D. diss., University of Montana, 1973), 111.

34. Montana Historical Society Archives, "Anaconda Copper Mining Company Records," Collection 169 (hereafter cited as MHS), Box 21, *Bliss v. Washoe,* "Brief of Appellees," 333, 345.

35. MHS, Box 22, Folder 1, *Bliss v. Washoe,* "Opinion," 25 January 1909, 212–13.

36. Charles K. Hyde, *Copper for America* (Tucson: University of Arizona Press, 1998), 94–100.

37. K. Ross Toole, *Montana: An Uncommon Land* (Norman: University of Oklahoma Press, 1984, 1959); John McNay, "Breaking the Copper Collar: Press Freedom, Professionalization and the History of Montana Journalism," *American Journalism* 25:1 (2008): 99–123.

38. "It Is the Largest in the World," *Butte Inter Mountain* (19 March 1905).

39. "Monster Beef Cattle," *Anaconda Standard* (19 April 1905).

40. MHS, *Bliss v. Washoe,* vol. IV, 1218–19.

41. MHS, *Bliss v. Washoe,* vol. I, 114.

42. MHS, *Bliss v. Washoe,* vol. IX, 3232.

43. MHS, *Bliss v. Washoe,* vol. II, 646.

44. MHS, *Bliss v. Washoe,* vol. IX, 3232.

45. "Farmers Bring Action," *Anaconda Standard* (21 May 1905).

46. M. Pham, J.-F. Muller, G. P. Brasseur, C. Granier, and G. Megie, "A 3D Study of the Global Sulfur Cycle: Contributions of Anthropogenic and Biogenic Sources," *Atmospheric Environment* 30 (1996): 1815–22.

47. "Effects of Sulfur Dioxide on Vegetation: Critical Levels," in *WHO Air Quality Guidelines* (Copenhagen: World Health Organization, 2000), 6.

48. Wilhelm Knabe, "Effects of Sulfur Dioxide on Terrestrial Vegetation," *Ambio* 5–6 (1976): 213–18, 213.

49. "Effects of Sulfur Dioxide on Vegetation," 1.

50. Knabe, "Effects of Sulfur Dioxide on Terrestrial Vegetation," 215.

51. MHS, *Bliss v. Washoe,* vol. IX, 3223.

52. "Nick Bielenberg and His Alfalfa," *Anaconda Standard* (6 March 1906).

53. "Green Fields and Fat Herds Down Deer Lodge Valley," *Anaconda Standard* (4 June 1905).

54. MHS, *Bliss v. Washoe,* vol. I, 53.

55. MHS, *Bliss v. Washoe,* vol. II, 741.

56. MHS, *Bliss v. Washoe,* vol. IX, 3295.

57. MHS, *Bliss v. Washoe,* vol. I, 202–4.

58. MHS, *Bliss v. Washoe,* vol. IV, 1196–97.

59. MHS, *Bliss v. Washoe,* vol. IX, 3290.

60. "Abstract of Testimony of Lay Witnesses," MHS, 84.

61. Brett Walker, *The Toxic Archipelago: A History of Industrial Disease in Japan* (Seattle: University of Washington Press, 2010), 96.

62. MHS, *Bliss v. Washoe,* "Opinion," 202.

63. MHS, *Bliss v. Washoe,* vol. IX, 3403.

64. MHS, *Bliss v. Washoe,* vol. IX, 3179.

65. MHS, *Bliss v. Washoe,* vol. IX, 3172.

66. MHS, *Bliss v. Washoe,* vol. I, 85.

67. MHS, *Bliss v. Washoe,* vol. II, 648.

68. B. K. Datta et al., "Chronic Arsenicosis in Cattle with Special Reference to Its Metabolism in Arsenic Endemic Village of Nadia District West Bengal India," *Science of the Total Environment,* 409 (2010): 284–88.

69. "Jerry Ryan of Smoke Association Goes into Details of the Case," *Anaconda Standard* (28 February 1906).

70. MHS, *Bliss v. Washoe,* vol. II, 774.

71. MHS, *Bliss v. Washoe,* vol. I, 44–46.

72. MHS, *Bliss v. Washoe,* vol. I, 114.

73. MHS, *Bliss v. Washoe,* vol. IX, 3170.

74. MHS, *Bliss v. Washoe,* vol. IX, 3222.

75. Datta, "Chronic Arsenicosis," 284–8.

76. David Igler, *Industrial Cowboys: Miller & Lux and the Transformation of the American West, 1850–1920* (Berkeley: University of California Press, 2005), nicely illustrates the way a large California rancher used space and

topography to maximize cattle growth, though it is less successful in recognizing the cattle's own role in the process.

77. MHS, *Bliss v. Washoe,* vol. I, 114.
78. MHS, *Bliss v. Washoe,* vol. VI, 2126.
79. MHS, *Bliss v. Washoe,* vol. IX, 3228.
80. LeCain, *Mass Destruction,* 72–73.

CHAPTER 7

Efficiency, Economics, and Environmentalism

Low-Grade Iron Ore Mining in the Lake Superior District, 1913–2010

JEFFREY T. MANUEL

On April 20, 1974, United States District Court judge Miles Lord ruled that the Reserve Mining Company's enormous iron ore–processing plant perched on the edge of Lake Superior—the largest of the Great Lakes—must immediately stop dumping tailings into the lake. The tailings were an enormous stream of waste generated as Reserve Mining crushed boulders of taconite with some 25 percent magnetite iron content and separated the valuable iron from the worthless siliceous tailings. Since opening in 1955, the mammoth Reserve Mining plant had dumped approximately 67,000 tons of tailings into Lake Superior each day. This enormous volume of tailings soon created a spreading delta of finely powdered rock. The smallest tailings particles did not settle in the delta, but instead dispersed throughout the lake, creating billowing green clouds and infiltrating the water supplies of coastal towns. In 1969, the U.S. government filed a lawsuit against Reserve Mining alleging that the company was polluting Lake Superior. The trial, *Reserve Mining Company v. United States of America*, which lasted until 1980, was one of the longest and most expensive environmental lawsuits in U.S. history. As a result of the trial, Reserve Mining was forced to stop dumping tailings into the lake and to deposit them on land instead.[1]

Considered alone, *Reserve Mining Company v. United States of America* fits an often-told narrative about mining's relationship with the environment and with the modern environmental movement. According to this narrative, the American mining industry either ignored or was hostile to environmental concerns until the modern environmental movement of the 1960s and 1970s. Mining historian Duane Smith, for example, argues that an "environmental whirlwind" swept through the mining industry in the 1960s. Although the mining industry avoided environmental regulation for most of the nineteenth and twentieth centuries, by the late 1950s and early 1960s, the American mining industry "reeled under the barrage of condemnation" from environmental groups. The mining industry realized it could no longer resist growing pressure for regulation and slowly began to ameliorate the worst aspects of earlier pollution.[2]

This narrative about the conflict between mining and the environment in the 1960s and 1970s builds on the argument that the environmental movement emerged from consumer sensibilities that were external to industrial production in the post–World War II era. In his classic account of the rise of the modern environmental movement, Samuel P. Hays argues that in the decades after World War II, Americans began seeing the environment as a fundamental source of their quality of life rather than merely a resource to be exploited.[3] Industrial mining was especially problematic from the environmental viewpoint since it devastated large swaths of land, in some cases destroying areas that would otherwise have been valuable recreational spaces. Countering the environmental perspective, the mining industry argued that it provided minerals that Americans demanded in ever-increasing quantities. Yet this argument fell on deaf ears among an American public eager for nature's products but unwilling to consider consumer culture's roots in the natural world. This narrative of an American mining industry that polluted freely until the environmental movement forced miners to change their ways certainly captures the drama and tension of the era, but it misses a longer and more complex history of the relationship between mining technology, economic growth, and ecological realities.

Recent literature has painted a more complex picture of the long relationship between mining and the environment in the United States. Environmental historians have drawn attention to the enormous scale of global mining in the twentieth century. A global perspective reveals a worldwide mining industry moving more and more of the earth to satisfy growing populations and increasing consumer demand. In his environmental history of the twentieth century, J.R. McNeill argues, "Humankind moved mountains in the twentieth century and for the first time became a significant geological agent." Humans' enormously expanded capacity to move (and remove) mountains was largely the result of the technological and energy intensification of large-scale mining operations.[4] A global perspective also complicates Smith's idea that mining companies cleaned up their polluting practices because of U.S. environmentalism. Much as manufacturing in the United States and Western Europe was outsourced to low-wage nations, mining has expanded in countries where pollution is less tightly regulated.

Recent environmental histories of U.S. mining before the mid-twentieth century have also complicated binary understandings of mining versus environmentalism. Hydraulic mining in California provides an excellent example of nineteenth-century mining practices involving a hybrid of human ingenuity, hydrology, and geology. As Andrew Isenberg has noted, hydraulic mining for gold in California was a form of mass-destruction mining that preceded the open-pit mines later developed for coal, copper, and iron ore mining elsewhere in the United States. Hydraulic mining also illustrates how industrial mining's environmental consequences led to popular agitation and legal action long before the modern environmental movement of the 1960s. Following widespread complaints from farmers and downstream interests, hydraulic mining in California was stopped in 1884.[5] The history of low-grade ore mining also replaces binary histories of mining and the environment with more complicated understandings. Timothy LeCain describes the dramatic expansion of low-grade copper mining to meet the needs of the electrical age, creating a system of "mass destruction" open-pit copper mines. Using railroads, steam shovels, and massive ore separators, copper mines in the western United States profitably mined deposits containing as little as 2 percent copper. Low-grade ore mines turned the age-old logic of mining on its head. Rather than finding valuable minerals and removing them from the ground, the modern mass-destruction mine removed everything and used industrial technology to separate valuable ore from worthless tailings. The results included both profitable new mines and enormously expanded ecological consequences.[6] Overall, recent environmental histories suggest a complex, hybrid history of mining and the environment that has spread across the globe and has deep historical roots.

Low-grade iron ore lacks glamour. It has received surprisingly little attention from historians, especially considering that it revolutionized the global steel industry in the second half of the twentieth century. Many of those observers who have commented on low-grade iron ore, especially low-grade iron ore from the Lake Superior mining district, have emphasized that the industry was a techno-scientific fix for depleted natural resources and an ailing economy in the mining region. Journalists described low-grade iron ore as a "Cinderella story" to indicate how the fairy godmother of mineral engineering transformed the rock from something worthless to something cherished. "If ever there was a Cinderella in the world of nature," one journalist wrote, "it is taconite."[7] Although this enthusiastic endorsement captures the modernist belief that science and engineering overcame natural limits, it obscured a more complex reality. Low-grade iron ore was not a triumph of science and engineering over natural limits, and it did not unequivocally save the economy of the mining region. Rather, it is more accurate to say that low-grade iron ore rearranged ecologies, economies, and technologies into new alignments, with significant, often unintended consequences for humans and nonhumans in the region.

In the long view, it is useful to see the Reserve Mining Company's iron ore–processing plant as a link between two distinct eras in the history of American iron ore mining. In the first era, which lasted until the mid-twentieth century, Americans treated iron ore strictly as a mineral resource. This first era was dominated by fears of resource depletion and a frantic search for new ore fields. The second era, which continues into the present, relies on low-grade iron ore deposits that can, through extensive industrial processing, be manufactured into high-iron pellets that the iron and steel industry can use in blast furnaces (i.e., pellets containing 50 percent or more iron). This procedure resolved worries about depletion, but the result has been a constant tension between the development of mining districts and concerns about the ecological consequences of low-grade iron ore mining on a mass scale.

Pollution from low-grade iron ore mining in the late twentieth century was a direct result of changes to iron ore mining made in response to resource depletion, a modernist quest for technological efficiency, and attempts to promote economic development in the mining region. To understand how these interconnected histories played out in the Lake Superior mining region, this chapter describes a series of crises and enviro-technological fixes, beginning with depletion of high-grade iron ore in the early twentieth century, followed by the shift to low-grade pelletized iron ore as an efficient technological fix to the problem, and the ultimate environmental consequences that resulted from low-grade iron ore production.

The history of low-grade iron ore mining in the Lake Superior district deserves environmental historians' attention for several reasons. First, it offers a case study of how economic development in declining mining regions emphasized new technologies and expanded mining with little regard for possible environmental consequences. Throughout the Lake Superior mining district, miners, politicians, and engineers actively promoted low-grade iron ore mining as an economic development project to preserve the region's mining economy in the face of declining reserves of high-grade ore. The environmental problems that twentieth-century mining posed often developed in response to calls for economic development in depressed mining regions. This builds on arguments that the twentieth century's focus on sustained economic growth—the "growth fetish"—is crucial to understanding the ecological consequences of modern mass-removal mining practices.[8] It is impossible to understand the history of mass-removal mining in the United States—including mountaintop-removal mining in Appalachia, open-pit iron ore mining in the Lake Superior region, or western copper mining—without considering how economic development of depressed regions was a guiding principle in the twentieth century.

Second, the history of low-grade iron ore mining offers a case study of how one form of mining (low-grade pelletized iron ore) replaced another (high-grade, direct-shipping iron ore) and the environmental consequences that resulted. At the core of iron ore pelletizing and modern mass-removal mining is a quest for efficiency. The costly processing involved in low-grade iron ore mining meant that it was profitable only if conducted on a massive scale. The Lake Superior iron-mining region thus illustrates how the quest for efficiency in modern industrial mining led to industrial pollution on a scale that was correspondingly efficient. As the scale of production increased, so too did the scale of tailings waste.

THE LAKE SUPERIOR MINING DISTRICT

The Lake Superior mining district encompasses several deposits, or "ranges," of iron ore around Lake Superior. The district's iron deposits occur in six major ranges near the western and southern shores of the lake in Michigan, Wisconsin, and Minnesota. The largest and most significant of these deposits is the Mesabi Range in northeast Minnesota. Whereas several of the district's deposits lie deep underground, the Mesabi Range's iron ore is near the surface, allowing for open-pit mining operations. After a century of iron ore mining, the Mesabi Range has developed some of the world's largest open-pit mines. For example, the Hull-Rust-Mahoning mine near Hibbing, Minnesota, covers five square miles and is colloquially referred to as the "man-made Grand Canyon of the North." Mining historian Richard Francaviglia describes "the blood-red tablelike escarpments of Minnesota's Iron Ranges" as "some of the grandest manmade topography on earth."[9]

The region's history is inseparable from the history of mining. Euro-American prospectors first arrived in the area in the 1860s following reports of gold. These prospectors found large deposits of iron ore but no gold, and they soon left the region. A handful of miners remained, however, hoping to exploit the iron ore deposits. By the 1870s, an expanding U.S. steel industry needed larger quantities of iron ore, making the area's deposits increasingly valuable. Wealthy financiers soon joined local prospectors to develop the iron ore mines for commercial production. By 1884, seven mines were operating on the northern Vermillion Range. Soon after, prospectors discovered the far richer iron ore deposits of the Mesabi Range to the south. The Mesabi Range contained pockets of extremely high-grade iron ore, often 60 to 70 percent iron. And its gravel-like consistency meant it could easily be removed after scraping off a thin layer of overburden. The Mesabi Range developed the enormous open-pit mines that made it internationally famous by the early twentieth century. From the late 1890s through the first decades of the twentieth century, open-pit mining in the region grew quickly. The mines were vertically integrated into the largest American steel companies, including Carnegie Steel and U.S. Steel. The availability of huge quantities of high-grade iron ore from the region's open-pit mines was essential to the growth and development of the U.S. steel industry.[10]

To understand the region's history, it is necessary to differentiate between high-grade iron ore and low-grade ore. High-grade iron ore typically contains 50 percent or more iron content. High-grade ores on the Mesabi Range generally take the form of hematite, a rusty-red, gravel-like ore. The Lake Superior district's high-grade ores revolutionized the U.S. iron ore industry because they could be scooped from the ground and shipped directly to the steel mill with little further processing. Thus, the steel industry often called them direct-shipping ores. With their high iron content, they could be added directly to the blast furnace. Low-grade iron ore generally contains less than 50 percent iron as natural ore. Several types of low-grade iron ore exist in the Lake Superior mining district, with the most common being taconite (in Minnesota) and jasper (in Michigan). Both ores contain approximately 20 to 30 percent iron. Through industrial processing, low-grade ore can be concentrated (or beneficiated) to remove the noniron material and create a manufactured product containing enough iron to be used in steel mills. Rudimentary forms of concentration involved sifting and washing iron ores to eliminate noniron material. But the more complicated industrial processing of the mid-twentieth century involved extensive crushing and grinding, followed by separation using magnets or flotation, with the iron-bearing particles finally concentrated and bound together into high-iron chunks or pellets.

DEPLETION AND DEVELOPMENT, 1910–1919

By the first decades of the twentieth century, the fundamental problem facing the Lake Superior mining district was depletion of high-grade iron ore. Despite the district's abundance of rich ore, mechanized open-pit mining removed the ore at a fantastic rate. As early as the 1910s, Iron Range residents worried about the future of their communities. What would happen to their towns when the ore ran out?

Iron mining on the Mesabi Range removed so much ore so quickly because it was among the first instances of fully mechanized open-pit mining in the world. Unlike the slow, labor-intensive process required to dig ore from underground mines, gravel-like Mesabi Range hematite could be easily scooped from the ground using the era's primitive steam shovels. Transportation of the ore was fully mechanized as well. Railroads ran directly into the pits and carried the ore out to docks on Lake Superior, where it was loaded onto boats for water transport to steel mills in Pittsburgh, Cleveland, and Chicago. As Peter Temin described the process, "By the turn of the [twentieth] century the transport of Lake ores had become an intricate ballet of large and complex machines."[11] Mechanization quickly displaced human labor in the mines. From 1910 to 1920, the amount of ore produced annually per worker rose from 1,522 tons to 4,257 tons.[12]

During the late nineteenth century, mining engineers and operators were concerned that wasteful mining practices were costing millions of dollars each year. The Mesabi Range, in contrast, was upheld as a model of modern conservation practices, which mining engineers at the time defined strictly as preventing costly waste. The combination of open pits and full mechanization meant that the ore body could be completely and efficiently removed. The Mesabi Range mines wasted little ore—certainly far less than the 10 percent typically lost in underground mining.[13]

Mechanization shifted the mines' energy source from human and animal power to hydrocarbon-fueled shovels and locomotives. As Timothy LeCain notes, "The steam shovel was nothing more than a device [. . .]

to channel large amounts of concentrated hydrocarbon or hydropower energy into the previously slow and labor-intensive process of mining."[14] Replacing humans with hydrocarbon-powered machines allowed miners on the Mesabi Range to extract ore at rates unimaginable decades earlier. Mesabi Range mines extracted over 41 million tons of hematite iron ore in 1917 alone, five times more than in 1900.[15] The enormous volume of iron ore extraction from the Lake Superior district was a source of both awe and dread. Observers were clearly impressed by the size of the giant open-pit mines. Many mines set up crude observation stands so that the curious could watch the growing man-made canyons. Such vast mines offered a vision of the industrial sublime, David Nye's term for "a man-made landscape with the dynamism of moving machinery and powerful forces" that "evoked fear tinged with wonder" yet "threatened the individual with its sheer scale, its noise, its complexity, and the superhuman power of the forces at work."[16] Indeed, the observers' wonder battled with alarm that removing so much ore meant that the district's mines would soon be tapped out.

One indication of concern over iron ore depletion in the United States was the 1909 *Report of the National Conservation Commission*. Initiated by President Theodore Roosevelt, the report emphasized the key tenets of the Progressive era conservation movement, including concern that rampant industrialization was wasting finite natural resources; worry that nonrenewable resources, including minerals, would be depleted; and urgent calls for efficient use of resources to conserve them for future generations.[17] Regarding the nation's iron ore reserves, the report described the high-grade ores of the Lake Superior region as quickly diminishing even though the district was only several decades old. The report noted a "remarkable rate of increase" in American iron ore production between 1880 and 1909, with most of this increase in the Lake Superior district. Although an estimated 2.5 billion tons of high-grade iron ore remained in the district, this was far below the 6 billion tons of ore that the United States was predicted to need before 1940. If the Lake Superior district could not keep up with demand, the United States would be forced to import iron ore from foreign sources, a troubling prospect in an era that associated steel with military strength.[18]

Worse, the Lake Superior mining district's highest grades of ore were already showing signs of depletion. The percentage of iron in the ore mined on the Mesabi Range was falling steadily in the years before 1909. The report concluded with a pessimistic prediction for the region and the U.S. iron ore supply. "It is evident, therefore, that the present average rate of increase in production of high-grade ores can not continue even for the next thirty years, and that before 1940 the production must already have reached a maximum and begun to decline."[19] A 1914 textbook on iron ores confirmed the report's findings, noting a "growing scarcity of high-grade ores in the Lake Superior district" and describing various efforts, such as washing and roasting, used to improve the quality of remaining ores.[20] By the first decade of the twentieth century, news of the depletion of high-grade iron ore reserves on the mighty Mesabi Range had reached a national audience.

Fears of depletion only grew in the following decades. In a 1920 technical bulletin, Edward W. Davis, the engineer who later perfected low-grade pelletizing technology, predicted that the Mesabi Range's high-grade iron ores would soon be depleted. "Production has been at an enormous rate," Davis noted, "and the question naturally arises as to how much longer this district can continue to supply the demand."[21] Production slowed during the Great Depression, but World War II demand for ore rapidly depleted the Mesabi Range. During the war years, the United States consumed approximately 480 million tons of iron ore, with two-thirds of this supply coming from the Mesabi Range.[22] By 1950, the federal government was holding meetings to discuss the looming shortage of high-grade iron ore in the United States.[23]

At the same time that high-grade iron ore was waning, the towns of the Lake Superior mining district committed themselves to permanence through economic development and continued growth. Their emphasis on permanence and ongoing economic growth atop a foundation of finite (and quickly vanishing) natural resources set the stage for low-grade iron ore mining. The population of Iron Range towns boomed in the early twentieth century as thousands of workers flocked to the region to work in the iron ore mines. The population of the Mesabi Range increased from 15,800 residents in 1900 to 84,180 in 1920. The region's towns during these decades were, as geographer John Borchert notes, "the fastest-growing urban areas in the Upper Midwest."[24] City leaders made plans for the new towns' permanence. During the 1910s, Mesabi Range towns such as Hibbing began building extensive public facilities in anticipation of continued growth and community development into the future. By the 1920s, Hibbing was known as the "richest little village in the world" due to its high spending on public facilities such as lavish schools, parks, skating rinks, and day-care facilities.[25]

The desire to create permanent towns atop finite mineral resources ran against earlier precedent that imagined mining communities as transitory.

Mining camps sprouted next to the mines and were quickly abandoned when the mineral boom ended. Observers in the nineteenth-century United States saw hundreds of mining ghost towns littering the American West. John Muir, for example, described Nevada's mining region as "strewn with ruins that seem as gray and silent and time-worn as if the civilization to which they belonged had perished centuries ago."[26] In the Lake Superior district, some observers also believed that the district would turn into a string of ghost towns. A sociologist visiting the region in 1938 wrote: "It is obvious that the days of the Mesabi towns are numbered. The greatest iron range of all history will some day, perhaps during the next generation, be worthless and desolate."[27] Throughout the late nineteenth and early twentieth centuries, observers emphasized the inherently temporary nature of mining settlements.

While the economic and geologic reality of mass-removal mining pulled toward transitory communities and resulting abandonment, the era's prevailing ethos of development pushed toward permanence and stability. The towns of the Lake Superior mining district developed primarily during the late nineteenth and early twentieth centuries. Thus, town builders were immersed in western modernist thought that emphasized economic growth, development, and permanence. Throughout the region, civic boosters argued that their towns would thrive well into the future. "A high spirit of civic pride," historian Clarke Chambers writes, "came to infuse every activity on the Iron Range," reflecting the widespread belief that the Iron Range towns had a bright future. Iron Range residents devoted enormous energy to education and children's welfare programs, believing that their progeny had bright prospects on the Iron Range. "The faith was in the future," Chambers notes, "and the dynamic of hope was unleashed in concerted efforts in which the entire community joined to set loose the potential for good and for progress that the younger generation promised." Plans for permanent economic growth and future development in the region fit into a larger pattern of thought at the turn of the twentieth century that emphasized science and technology's ability to overcome natural limits.[28]

The tension between the mining economy and the modernist development impulse led to blind spots. The Lake Superior mining district was, in fact, littered with towns abandoned after their resources had been depleted. Among the earliest of these were the many logging camps abandoned in the nineteenth century after the region's white-pine forests had been cut. There were numerous mining ghost towns as well. The small town of Babbitt, for example, was abandoned in 1924 after the town's major mine shut down.[29] The largest example of how the region ignored the transience inherent in mining was the gradual relocation through the 1920s of Hibbing, Minnesota, one of the largest towns on the Mesabi Range, to make way for an expanding open-pit mine. Buildings were placed on enormous logs and rolled to new locations several miles away. The abandoned neighborhood, which still contains ghostly sidewalks and street signs, was turned into a city park and mine overlook. The desire to create permanent towns atop a foundation of finite natural resources forced residents to ignore the reality of change and depletion.

The Lake Superior iron-mining district faced a fundamental paradox by the first third of the twentieth century. On the one hand, the region's economy and society were built atop a finite natural resource that, due to mechanized open-pit mining, was vanishing rapidly. On the other hand, residents espoused a rhetoric of continued growth for their towns, the mining economy, and their way of life. To resolve the contradiction, the region turned to low-grade iron ore.

TACONITE AS TECHNOLOGICAL FIX

The Lake Superior region turned to low-grade iron ore as a technological fix that promised to address both the dangers of resource depletion and the lure of economic development through the miracles of modern engineering. Low-grade iron ore was an especially attractive solution because it seemingly allowed the mining region to postpone a reckoning with nature and to grow far into the future. Yet implementation of a low-grade iron ore industry in the district led to numerous unintended consequences, including widespread environmental pollution.[30]

For decades, iron miners had screened and separated ore to concentrate the higher-iron ores. In the 1880s, Thomas Edison designed a process that used magnets to concentrate crushed iron ore. Although Edison perfected the engineering process, his New Jersey plant was a commercial failure due to changing market conditions in the U.S. steel industry.[31] By the first decades of the twentieth century, several firms had tried and failed to make low-grade iron ore commercially viable on a large scale. Geologists knew that there were enormous deposits of low-grade iron–bearing rock throughout the Lake Superior mining district. Most of this ore occurred as taconite, a flintlike rock containing approximately 30 percent magnetite. For many decades, miners ignored this huge volume of low-grade ore.

But by the early twentieth century, the rapid depletion of high-grade ores led to increased interest in low-grade ores. The primary advocate for low-grade ores, especially Minnesota's taconite, was Edward W. Davis, an engineering professor at the University of Minnesota's Mines Experiment Station. Davis first began experimenting with the separation of iron from low-grade ore in 1915. By the 1920s, he had largely perfected the engineering process, but low-grade ore was not yet commercially viable in the existing market.

To promote his engineering project, Davis argued that low-grade iron ore offered a technological fix for the depletion of direct-shipping ores in the Lake Superior district. Further, he warned that if development of low-grade iron ore did not begin immediately, the entire district would be in jeopardy. In a 1920 technical bulletin, Davis painted an alarming picture of depletion and argued that low-grade iron ore was the obvious solution:

> It will not only be necessary to utilize these [low-grade taconite] ores in order to maintain the production of the district, but it will be necessary to begin the utilization of them in the very near future. If the furnace companies that have only a few years' supply of ore available are allowed to invest large amounts of capital in developing and bringing into production new mining districts, the Lake Superior region will immediately start on its decline. If, on the other hand, these furnace companies find that the low-grade ores can be utilized, the fact that the Lake Superior district is already in large production and is so well equipped to handle immense tonnages will cause them to contemplate seriously investing new capital in this district for the purpose of developing the low-grade iron ores. The development of such an industry on a large scale will extend the life of the district into the far distant future.[32]

By the 1940s, Davis's proposals attracted widespread attention from politicians and steel industry executives. The rapid depletion of high-grade iron ore during World War II—and the belief that more reserves were necessary for the dawning Cold War—made the iron ore supply crucial to U.S. national security. Davis argued during World War II that the federal government should support the development of low-grade iron ore mines, but he was rebuffed because of the need for immediately available high-grade iron ore during the war. After the war, however, arguments in favor of government support for low-grade iron ore mining picked up steam. Both trade economists and President Harry Truman's administration worried that the United States was running out of domestic mineral supplies and may become a "have-not" nation in the global race for strategic minerals. Presidents Truman and Eisenhower supported the development of the low-grade iron ore industry and other domestic mineral industries via the Strategic and Critical Materials Stockpiling Act of 1946 and the Defense Production Act of 1950.[33] Low-grade iron ore came to be seen as vital to national security. Steel companies could invest in the expensive processing technology with little financial risk thanks to government subsidies and tax incentives.

In 1955, the Reserve Mining Company, a joint subsidiary of Armco and Republic Steel, opened the first mill for processing low-grade iron ore in northern Minnesota. Reserve Mining Company mined taconite rock at its giant open-pit mine. After it was blasted out of the ground in piano-size chunks, the rock was transported by rail to the shore of Lake Superior where it entered the mammoth E. W. Davis Works. Here the rock was crushed and then ground to the consistency of flour. Using magnets, the high-iron magnetite was separated from the siliceous gangue. The magnetite was then mixed with bentonite for binding and rolled into marble-size round pellets. These pellets were baked hard and then shipped to blast furnaces. The tailings of noniron siliceous rock—which made up 60 to 80 percent of the original taconite—were mixed with water, and the resulting gray slurry was pumped into Lake Superior via two giant tubes.

The plant handled an enormous quantity of iron ore. Initially designed to produce 3.75 million tons per year, the Silver Bay plant was producing 6 million tons of taconite annually by 1961. Following successful tests of its ore in a blast furnace, the Reserve Mining plant increased its annual output to 9 million tons in the early 1960s. It was producing 10.7 million tons annually by 1970.[34] During the 1970s, Reserve Mining's Silver Bay taconite plant supplied 15 percent of the United States' iron ore.

The Reserve Mining operation was the first in a wave of pelletizing plants that transformed the global iron ore industry in the second half of the twentieth century. During its construction and early years of operation, Reserve Mining's mill "was praised by the mining engineering community [. . .] as a technological marvel."[35] Journalists breathlessly hailed the mill as the beginning of what they anticipated would be a billion dollars of corporate investment in iron ore–pelletizing plants across North America.[36] The demand for pelletized iron ore skyrocketed, rising by almost 300 percent from 1961 to 1966, and newly built plants were operating at full capacity from the moment they came online.[37] By 1970, thirteen low-grade iron ore pellet plants were operating in the vicinity, and the majority of the Lake Superior district's iron ore was concentrated before being used in a blast furnace.

The commercial and technical success of low-grade iron ore alleviated earlier concerns over the depletion of high-grade ore in the Lake Superior district. The combination of low-grade iron ore processing and new foreign sources served to "eliminate worry over adequate supplies of iron for centuries."[38] Davis, who had done more than anyone to promote low-grade iron ore, noted that high-grade ore had gone from shortage to surplus: "In a few short years the success of low-grade ore concentration and the demonstrated economy of carefully prepared blast furnace ore has brought about a complete re-evaluation of iron ore production and reserves. Instead of a shortage, a great surplus of high-grade natural ore now exists, and blast furnace operators who are willing to use direct ores are in a position to pick and choose among several sources."[39]

From one perspective, low-grade iron ore succeeded as a technological fix. But taconite was cheap only because the costs of low-grade iron ore mining and processing were shifted onto the environment. When the environmental costs of low-grade iron ore mining were later revealed, as described below, the fix appeared far less painless. Overall, low-grade iron ore offers a potent example of the high-modernist belief that the latest science and technology could obliterate all natural limits, including the finite deposits of high-grade iron ore, and the larger mindset that fundamentally separated nature from technology.[40]

Low-grade iron ore mining turned iron ore mines into high-throughput industrial factories. It was part of a more general change in twentieth-century mining systems, described by mining historians Logan Hovis and Jeremy Mouat as "the abandonment of low-volume, high-value, selective mining techniques and the adoption of higher-volume, nonselective methods that emphasized the quantity rather than the quality of the ore brought to the surface." The massive open-pit metal mines of the twentieth century increased the speed and size of the process so much that LeCain has called them "rural envirotechnical factories" able to create a system of "mass destruction."[41] Indeed, the scale and complexity of the low-grade iron ore mines baffled those familiar with an older type of mining that only removed the valuable ore and left everything else. In the Lake Superior district, one mine operator asked Reserve Mining executives how much gold was extracted during processing. The man "could not believe that such an elaborate project had been built merely to produce iron."[42]

The elaborate processing of low-grade iron ore was profitable only because of the mill's enormous volume. Low-grade iron ore mills made very little profit per ton.[43] High throughput was the key to a successful business. Since the overhead of factories and supplies was largely fixed, industries such as automobile manufacturing made money by pumping more products through their plants more quickly. Speed was also the essence of the open-pit mine. By moving so much earth so quickly—largely with the assistance of heavy machinery powered by hydrocarbon fuels—the large open-pit mines could economically extract and process very low grade ores.[44]

It is also useful to consider the changing temporal scales involved in modern mining. Low-grade mineral mining was little more than a drastic acceleration (aided by hydrocarbon-fueled technology) of the geological processes that created pockets of high-grade iron ore by leaching low-grade taconite ores over eons of geological time. Mining engineers soon conflated the language of geology and industrial processes. By the 1950s, engineers defined high-grade iron ore as ore in which geology, rather than machines, had done the work of concentration. Geology was imagined as a giant iron ore concentration machine, and one that worked very slowly at that.[45] This intensification of the temporal scale was common in industrialization. Historians Sara Pritchard and Thomas Zeller emphasize that industrialization, especially the use of hydrocarbons, "significantly altered the temporal scale of natural-resource extraction by intensely tapping deep geological time."[46] Low-grade iron ore mining complicates the temporal scales involved, since the mines certainly tapped into "deep geological time" by using hydrocarbon-fueled machines. But those same machines allowed miners to replicate the effects of geological time on the compressed timeline of industrial production.

ENVIRONMENTAL COSTS

Although low-grade iron ore offered a technological fix to the problem of depleted high-grade ore reserves and the threat of community abandonment, it was not a costless solution. Processing the low-grade ore required rivers of water and produced mountains of tailings. The combination of water and tailings, and the resulting environmental consequences, was vividly illustrated by a landmark water pollution lawsuit in the 1970s, *Reserve Mining Company v. United States of America*, that ultimately shut down the Reserve Mining mill.

The major environmental pollution stemming from low-grade iron ore mining was water pollution. Separating iron particles from the surrounding siliceous gangue required enormous quantities of water. The process that Davis designed required approximately 45 tons of water for every ton of iron ore concentrate produced. A typical iron ore–pelletizing plant used 100,000 gallons of water per minute, approximately the same amount of water as flowed every minute through the Mississippi River in northern Minnesota.[47] This water was ultimately filled with tailings slurry. The challenge for low-grade iron ore mining was how to dispose of those tailings.

Low-grade ores, no matter the mineral, generate enormous quantities of tailings. The tailings problem was less pronounced in low-grade iron ore mining than in other types of mining, such as copper sulfates, which often processed ores as low as 2 percent copper and left behind sulfuric acid and other highly poisonous chemicals. In contrast, low-grade iron ore created approximately 2 tons of tailings for every 3 tons of ore. And the tailings were a siliceous rock similar to that found on the bed of Lake Superior. The large volume of tailings still needed to be dumped somewhere, and the region's water system soon emerged as the logical dumping ground. From the moment that the earliest pilot plants for low-grade iron ore processing were developed in the Lake Superior mining region, the question of where to dump tailings was problematic. The first pilot plants dumped the wet tailings in nearby swamps or small lakes.[48]

The tailings issue became more important when the first industrial processing plants were built in the mid-twentieth century. While engineering the processes for low-grade iron ore mining, Davis immediately turned to the possibility of dumping the tailings into Lake Superior. From Davis's perspective, Lake Superior offered several advantages. The plant could use its enormous quantity of cold freshwater in processing operations. More important, if the tailings were dumped directly into the lake, they would seemingly disappear as the lake's currents pulled the tailings—so finely ground that they were suspended in the water—deep below the surface. Davis and Reserve Mining conducted several small laboratory tests and determined that depositing the taconite tailings in the lake would not lead to any serious problems. Davis wrote:

> As a result of these investigations, we [. . .] concluded that the best place for fine taconite tailings was in the deep valleys at the bottom of Lake Superior. There they would be out of sight forever and posterity would not have to cope with them. [. . .] It was our conclusion that the fine tailings from all the magnetic taconite on the Mesabi could be put into the deep water of Lake Superior and would have no harmful effect on its usefulness or beauty.[49]

Reality proved more complex, however. At the Mines Experiment Station, Davis and his staff tested how tailings might affect Lake Superior by putting taconite tailings into large water tanks where they "observed what they would do under controlled conditions—just how they would act and what would happen to the water and what pollution would occur, if any, and what discoloration would occur, if any." The University of Minnesota ultimately made a small-scale model of Lake Superior and dumped taconite tailings into the model in the same proportion as would the Reserve Mining plant.[50] Satisfied that the tailings did not pose a threat to Lake Superior's clarity, Davis went ahead with his plan to dump tailings into Lake Superior.

Reserve Mining obtained twelve permits before dumping tailings into Lake Superior. These were divided into three main categories: permits from Minnesota's Department of Conservation that allowed Reserve Mining to remove water from Lake Superior and use it in the mill, permits from the Minnesota Pollution Control Agency that allowed the company to discharge water and tailings back into Lake Superior, and permits from the U.S. Army Corps of Engineers granting the company permission to construct a dock and breakwater required for shipping.[51] These permits were quickly approved given the plant's promise of economic development and jobs in the mining region. The only significant opposition came from sportsmen's groups and the fishing industry, both of which were worried that tailings would affect fishing in Lake Superior. Ultimately, the quest for economic development overshadowed these concerns. In December 1947, Reserve Mining received the necessary permits to use 130,000 gallons of water per minute from Lake Superior.[52] Worries about possible pollution carried little weight against the promise of an expanded mining economy in the region. As Davis recalled, the head of the state's Department of Conservation "was very anxious to get something of that kind [a low-grade iron ore plant] started" because of the economic benefits it offered in a declining mining region. Overall, Davis noted, "everybody was very helpful and enthusiastic about getting our plant started [. . .] because it would employ three or four thousand men."[53] The permits also allowed Reserve Mining to dump tailings into Lake Superior. From 1955 to 1980, the Reserve Mining plant dumped an average of 67,000 tons of taconite tailings into Lake Superior each day. Historian Thomas Huffman notes that this amount was "more than two times the estimated solid waste garbage produced by New York City during the same period."[54]

Beginning in the late 1960s, government officials, along with citizen environmental groups, brought lawsuits against Reserve Mining to force the firm to stop dumping tailings into the lake. Although the details of the case were specific to low-grade iron ore mining and highly technical, the Reserve Mining lawsuit was just one event in what Samuel Hays identified as the "second phase" of the U.S. environmental movement, stretching from roughly 1965 to the early 1970s. During these years, Hays argues, environmental politics in the United States focused on "the reaction against the adverse effects of industrial growth."[55] Indeed, the lawsuit represented a fundamental shift in attitudes as the ecological costs of low-grade iron ore mining were accounted for alongside its economic benefits.

The case began in the late 1960s when the Reserve Mining Company attempted to renew permits to dump tailings into Lake Superior. Whereas its initial dumping permits had met little opposition in the 1940s, by the 1960s fishermen, scientists, and lakeside lodge owners were worried about how the tailings were affecting the lake. Residents complained that billowing clouds of green water were spreading across the lake's surface, which was typically a crystal-clear blue. Longtime fishermen described how water clarity had dropped drastically in just a few years.[56] Environmental scientists, often working closely with concerned citizens, started documenting the spread of fine tailings throughout the lake and noting how the tailings were affecting the lake's ecosystem. Water pollution experts found that the tailings were spreading hundreds of miles from the plant at Silver Bay and increasing turbidity throughout the lake. The tailings were also harming fish. According to an expert on the Great Lakes' sedimentary geology, the Reserve Mining mill was nothing less than "a major geological event in the history of the world's largest expanse of fresh water, comparable to the arrival of European civilization."[57] After several years of data collection, government officials believed they had enough evidence to show that Reserve Mining's tailings were harmfully polluting the lake. Using new water pollution regulations, the U.S. government and the state of Minnesota filed suit in 1972 to force Reserve Mining to quit dumping tailings into the lake.

The lawsuit initiated several years of conferences and administrative procedures that pitted the environmental costs of mining against its economic benefits. At the time, the case was framed as a classic example of the jobs-versus-the-environment debate. The citizens groups and government agencies suing to stop the dumping argued that tailings from low-grade iron ore mining were industrial pollution that threatened the health of Lake Superior and everyone living near its shores. Opposing them were Reserve Mining and many industrial workers in the region who argued that the economic benefits of low-grade iron ore mining outweighed any potential costs from pollution. Among those who supported Reserve Mining were many skeptics who questioned whether the tailings were even pollution, including several scientists who testified on behalf of Reserve Mining. Just as Edward W. Davis had done years earlier, Reserve Mining argued that the tailings were inert, pulverized rock that could not possibly harm the lake's ecosystem. According to Reserve, the fine tailings were "pure sand" that were "inert, inorganic, insoluble in Lake Superior, and biologically inactive."[58]

The case gained international attention in the early 1970s when Reserve's tailings were linked with asbestos. A researcher noticed that taconite tailings were similar in appearance to amphibole asbestos fibers, whose connection to cancer was being discovered at this time. There was almost no concrete scientific evidence of how asbestos-like fibers affected those who drank water containing them. Nonetheless, the revelation was a bombshell that turned the case from a pollution lawsuit into an enormous public health debate. Reacting to the news, Judge Miles Lord exclaimed that the pollution was "potentially [. . .] the number one ecological disaster of our time."[59] Soon thereafter, Judge Lord ruled that Reserve Mining was polluting the lake. The company was ordered to stop dumping tailings into the lake and begin depositing its tailings on land. Reserve Mining immediately appealed the decision and was allowed to continue dumping in the lake for several more years. In 1980, the company finally shifted to on-land tailings disposal near its Silver Bay plant.

The decision to stop Reserve's dumping proved to be an important precedent in U.S. environmental law. The most important legal decision to emerge from *Reserve Mining Company v. United States of America* has been cited extensively in American environmental case law. For example, the decision was a crucial precedent for *Ethyl Corporation v. EPA*, a 1976 decision that justified limits on leaded gasoline because of possible health risks. But the decision also had an economic impact on the Lake Superior iron-mining district. By the mid-1980s, Reserve Mining's Silver Bay plant was closed. Although Reserve Mining blamed the environmental lawsuit for the shutdown, it should be noted that several other low-grade iron ore plants shut down in the early 1980s amid an economic crisis in the U.S. steel industry.[60]

At the center of this story stands the engineer who designed the system for processing low-grade iron ore and dumping tailings into Lake Superior, Edward W. Davis. He was heavily criticized for Reserve Mining's pollution. Davis held views that were typical for mining engineers in the early twentieth century. Like many others in his field, Davis prioritized economic growth and the efficient use of natural resources over respect for complex ecologies. More important, he was certain that modern science and technology offered the tools to solve any potential problems from pollution, just as modern techno-science allowed him to turn worthless low-grade ore into valuable iron pellets.

Many of Davis's statements appear ridiculous in light of modern environmental science. Davis thought that so long as pollutants were not visible to the human eye, they were not a problem. For example, in 1968 he speculated that the enormous open-pit iron mines in the Lake Superior district could be filled with junk cars once mining ended. "At least you would be rid of the old automobiles," Davis speculated, adding that the cars could be blended with water and chemicals so that oxidation would turn the cars back into iron ore.[61] More alarming was Davis's suggestion that atomic bombs could be used as tools for open-pit mining. In 1964, he hypothesized that huge deposits of low-grade iron ore located deep underground could be easily exposed for open-pit mining by using atomic bombs to blast away overburden. "You have to strip off the top of the earth to get to the taconite now," Davis told a reporter, "but it may be done even easier for the next generation. The atomic energy commission [. . .] may come in some day and blow off the whole top."[62] This is not to say that Davis was entirely careless about nature and the environment. In his personal life, Davis was an avid outdoorsman. He summered at a remote cabin north of the Mesabi Range, and in unpublished personal journals he wrote loving descriptions of his time fishing and canoeing amid the clear, cool waters of northern Minnesota lakes.[63]

Instead of seeing Davis as either a hero of technological triumph over nature or a heartless destroyer of Lake Superior's ecology, it is more appropriate to view Davis's attitudes toward the natural world as a product of his time. As mining historians have emphasized, prior to the mid-twentieth century, miners and mining engineers often paid little attention to the ecological consequences of their trade. Reviewing the mining literature of the late nineteenth century, Duane Smith finds little mention of the environment among mining's proponents. When they did discuss the environment, it was only to emphasize its usefulness in mine operations. The few miners and engineers who spoke of the environment described the wilderness as a site for character building, but not preservation.[64] The point is not to apologize for Davis, but rather to emphasize that the mass-destruction mines of the early twentieth century and their resulting pollution derived from a worldview that separated nature from technology and insisted that modern techno-science could overcome any natural limits.

When the Reserve Mining controversy is viewed within the broader context of mass-destruction mining and its environmental consequences in the twentieth century, it is clear that the enormous scale and speed of low-grade iron ore mining, the very factors that made it profitable to mine and process ores containing only 25 percent iron, also created pollution on an enormous scale and with tremendous speed. As LeCain has noted for low-grade copper mining, the "insight that speed was the essence of mass production" also suggests "an equally profound insight into understanding modern environmental degradation."[65] In other words, the high-speed throughput of low-grade iron ore necessarily resulted in an equally high-speed throughput of tailings waste. The Reserve Mining plant produced 20 million tons of tailings waste per year. These tailings, a mixture of rock and water that poured into Lake Superior in a giant pipe-fed waterfall, contained, in Davis's words, "enough sand to cover 40 acres nearly 200 feet deep each year."[66] When dumped into Lake Superior, these tailings threatened the giant lake's ecology within a few years due to the sheer volume of pollution.

CONCLUSION

All histories of mining are environmental histories. Extracting minerals from the earth's crust is necessarily a hybrid activity combining nature, technology, and human labor. Amid the long history of mining, twentieth-century low-grade iron ore mining deserves attention for several reasons. First, low-grade iron ore mining in the Lake Superior district highlights how the need for continued economic development often fueled mining's expansion in the twentieth century. Rather than accept that high-grade iron ore supplies, and the future of mining communities dependent on them, were finite, the Lake Superior mining district was committed to the continuous growth of both mining operations and the towns that developed near the open pits. Globally, the push for sustained economic growth and development fueled a massive expansion of mining in the twentieth century. Attention to the demand for economic development in depressed mining regions also highlights causal factors in mining's development other than mining firms seeking to maximize profit. Low-grade iron ore mining in the Lake Superior district was not purely a response to capitalist pressures or technological change; it also emerged from a deep-seated desire for permanence and long-term growth built on a foundation of resource extraction industries.

Second, low-grade iron ore mining in the Lake Superior mining district highlights how changes in mining practices during the twentieth century often involved an intensification of the speed and scale of mining. Increased throughput and large-scale surface mining subsequently increased the speed and scale of environmental degradation. This increase in ecological consequences suggests a far more complex portrait of the changing patterns of pollution in the modern mining industry. It is undeniable that some negative ecological impacts from mining have been ameliorated. In the Lake Superior district, for example, large reclamation efforts have reforested tailings dumps and prevented widespread erosion. But these reclamation efforts take place alongside industrial processes dedicated to mining more intensely than ever and using such low-grade ores that huge volumes of tailings waste are a necessary by-product. Narratives that emphasize a constant progression away from the "bad old days" of mining pollution and narratives portraying a bleak descent into ever-more-destructive mining are both too simple to account for the largest iron ore operation in North American history.

NOTES

1. For an overview of the trial, see Robert V. Bartlett, *The Reserve Mining Controversy: Science, Technology, and Environmental Quality* (Bloomington: Indiana University Press, 1980); Thomas F. Bastow, *"This Vast Pollution . . .":* United States of America v. Reserve Mining Company (Washington, DC: Green Fields, 1986); and Frank Schaumburg, *Judgment Reserved: A Landmark Environmental Case* (Reston, VA: Reston, 1976).

2. Duane Smith, *Mining America: The Industry and the Environment, 1800–1980* (Lawrence: University of Kansas Press, 1987), 134–47.

3. Samuel P. Hays, *Beauty, Health, and Permanence: Environmental Politics in the United States, 1955–1985* (New York: Cambridge University Press, 1987). See also Samuel P. Hays, *A History of Environmental Politics since 1945* (Pittsburgh: University of Pittsburgh Press, 2000).

4. J. R. McNeill, *Something New Under the Sun: An Environmental History of the Twentieth Century World* (New York: W. W. Norton, 2000), 21, 31–32, 85–86.

5. Andrew Isenberg, *Mining California: An Ecological History* (New York: Hill and Wang, 2005).

6. Timothy J. LeCain, *Mass Destruction: The Men and Giant Mines That Wired America and Scarred the Planet* (New Brunswick, NJ: Rutgers University Press, 2009). See also Logan Hovis and Jeremy Mouat, "Miners, Engineers, and the Transformation of Work in the Western Mining Industry, 1880–1930,"

Technology and Culture 37:3 (1996): 429–56; Charles K. Hyde, *Copper for America: The United States Copper Industry from Colonial Times to the 1990s* (Tucson: University of Arizona Press, 1998); Larry Lankton, *Hollowed Ground: Copper Mining and Community Building on Lake Superior, 1840–1990s* (Detroit: Wayne State University Press, 2010); and Timothy J. LeCain, "When Everybody Wins Does the Environment Lose? The Environmental Techno-Fix in Twentieth Century American Mining," in *The Technological Fix: How People Use Technology to Create and Solve Problems,* ed. Lisa Rosner (New York: Routledge, 2004), 137–53.

7. Wendell Weed, "Minnesota Taconite: Nature's Cinderella," *Minneapolis Tribune Picture Magazine* (October 18, 1953), 21; "The State and Taconite," *Minneapolis Morning Tribune* (September 30, 1957). See also Terry S. Reynolds and Virginia P. Dawson, *Iron Will: Cleveland-Cliffs and the Mining of Iron Ore, 1847–2006* (Detroit: Wayne State University Press, 2011), 160–68. For a contrasting opinion, see Peter J. Kakela, "The Shift to Taconite Pellets: Necessary Evil or Lucky Break?" *Michigan History Magazine* (November–December 1994): 70–75.

8. McNeill, *Something New Under the Sun,* 335–36.

9. Richard V. Francaviglia, *Hard Places: Reading the Landscape of America's Historic Mining Districts* (Iowa City: University of Iowa Press, 1991), xviii. On the general history of Minnesota's iron ore ranges, see David A. Walker, *Iron Frontier: The Discovery and Early Development of Minnesota's Three Ranges* (St. Paul: Minnesota Historical Society Press, 1979).

10. Thomas Misa, *A Nation of Steel: The Making of Modern America, 1865–1925* (Baltimore: Johns Hopkins University Press, 1995), 158–60.

11. Peter Temin, *Iron and Steel in Nineteenth-Century America: An Economic Inquiry* (Cambridge, MA: MIT Press, 1964), 197.

12. Paul H. Landis, *Three Iron Mining Towns: A Study in Cultural Change* (Ann Arbor, MI: Edwards Bros., 1938; reprint, New York: Arno Press, 1970), 107; E.D. Gardner and McHenry Mosier, *Open-Cut Metal Mining* (Washington, DC: U.S. Government Printing Office, 1941), 70–71.

13. Smith, *Mining America,* 86, 103.

14. LeCain, *Mass Destruction,* 159.

15. Ernest F. Burchard, "Iron Ore, Pig Iron, and Steel," in *Mineral Resources of the United States, 1918,* ed. G.F. Loughlin (Washington, DC: U.S. Government Printing Office, 1921), 545.

16. David E. Nye, *American Technological Sublime* (Cambridge, MA: MIT Press, 1994), 126.

17. Henry Gannett, ed., *Report of the National Conservation Commission,* 3 vols. (Washington, DC: U.S. Government Printing Office, 1909). On the general context of Progressive era conservation, see Samuel P. Hays, *Conservation and the Gospel of Efficiency: The Progressive Conservation Movement, 1890–1920* (1959; reprint, Pittsburgh: University of Pittsburgh Press, 1999).

18. On the association of steel with military might in the Progressive era, see Misa, *A Nation of Steel,* 91–132.

19. C.W. Hayes, *Iron Ores of the United States* (Washington, DC: U.S. Government Printing Office, 1909), 520, 490.

20. Edwin C. Eckel, *Iron Ores: Their Occurrence, Valuation, and Control* (New York: McGraw Hill, 1914), 202.

21. E. W. Davis, "The Future of the Lake Superior District as an Iron-Ore Producer," *Bulletin of the University of Minnesota School of Mines Experiment Station* 7 (1920): 1.

22. William T. Hogan, *Economic History of the Iron and Steel Industry in the United States* (Lexington, MA: Lexington Books, 1971), 4:1481; George Eckel, "To Tap Taconite as a Source of Iron," *New York Times* (September 28, 1947).

23. "House Group Maps Inquiry on Steel," *New York Times* (April 15, 1950). It should be noted that claims of iron ore depletion were always contested. Iron is one of the earth's most plentiful elements and can be found throughout the globe. Some observers noted that the depletion of the Lake Superior region's high-grade ore reserves meant that the U.S. steel industry would have to turn to new sources.

24. Arnold R. Alanen, "Years of Change on the Iron Range," in *Minnesota in a Century of Change: The State and Its People since 1900*, ed. Clifford E. Clark Jr. (St. Paul: Minnesota Historical Society Press, 1989), 159; John R. Borchert, *America's Northern Heartland* (Minneapolis: University of Minnesota Press, 1987), 69.

25. Alanen, "Years of Change on the Iron Range," 164–65; Clarke A. Chambers, "Welfare on Minnesota's Iron Range," *Upper Midwest History* 3 (1983): 7–8.

26. John Muir, *Steep Trails* (Boston: Houghton Mifflin, 1918), 195; Smith, *Mining America*, 19–20.

27. Landis, *Three Iron Mining Towns*, 55.

28. Chambers, "Welfare on Minnesota's Iron Range," 11, 28; McNeill, *Something New Under the Sun*, xxii; James C. Scott, *Seeing like a State: How Certain Schemes to Improve the Human Condition Have Failed* (New Haven: Yale University Press, 1998), 4.

29. Edward W. Davis, *Pioneering with Taconite* (St. Paul: Minnesota Historical Society Press, 1964), 24.

30. The use of technological fixes in mining is discussed at length in LeCain, "When Everybody Wins Does the Environment Lose?"

31. On Edison's failed iron ore venture, see W. Bernard Carlson, "Edison in the Mountains: The Magnetic Ore Separation Venture, 1879–1900," *History of Technology* 8 (1983): 37–59.

32. Davis, "The Future of the Lake Superior District as an Iron-Ore Producer," 5.

33. Alfred E. Eckes Jr., *The United States and the Global Struggle for Minerals* (Austin: University of Texas Press, 1979), 121–73. This is not to say that iron ore mining did not proliferate outside the United States. American steel companies greatly expanded their iron ore operations around the world during the 1940s and 1950s, especially in Canada, Venezuela, Brazil, Chile, Liberia, Gabon, and Australia. Hogan, *Economic History of the Iron and Steel Industry*, 4:1481–86; Eckes, *The United States and the Global Struggle for Minerals*, 125–26.

34. Davis, *Pioneering with Taconite*, 181, 191; Hogan, *Economic History of the Iron and Steel Industry*, 4:1487.

35. Thomas R. Huffman, "Exploring the Legacy of Reserve Mining: What Does the Longest Environmental Trial in History Tell Us about the Meaning of American Environmentalism?" *Journal of Policy History* 12:3 (2000): 340, 345.

36. "The Taconite Rolls," *Newsweek* (October 24, 1955), 78. See also Gerald Manners, *The Changing World Market for Iron Ore, 1950–1980: An Economic Geography* (Baltimore: Johns Hopkins University Press, 1971), 159–72.

37. "Ore Pellet Plant Near Capacity," *New York Times* (September 1, 1967).

38. Thomas E. Mullaney, "Steel Industry Cheered in Quest for New Raw Materials Sources," *New York Times* (January 11, 1953).

39. Davis, *Pioneering with Taconite*, 194.

40. LeCain, *Mass Destruction*, 179.

41. Hovis and Mouat, "Miners, Engineers, and the Transformation of Work in the Western Mining Industry," 434–35; LeCain, *Mass Destruction*, 150.

42. Davis, *Pioneering with Taconite*, 187.

43. Strictly speaking, the plants were not all profit-maximizing entities. Most of the plants in the Lake Superior district were directly owned by steel companies, either individually, as in the case of U.S. Steel, or as joint ventures, as was the case for Reserve Mining, jointly owned by Armco and Republic Steel.

44. On the general history of high-throughput mass production, see Alfred D. Chandler, *The Visible Hand: The Managerial Revolution in American Business* (Cambridge, MA: Belknap Press of Harvard University Press, 1977), 240–82.

45. "The Taconite Story," unpublished manuscript, University of Minnesota Mines Experiment Station, July 25, 1952, University of Minnesota Archives, Minneapolis.

46. Sarah B. Pritchard and Thomas Zeller, "The Nature of Industrialization," in *The Illusory Boundary: Environment and Technology in History*, ed. Martin Reuss and Stephen H. Cutcliffe (Charlottesville: University of Virginia Press, 2010), 86, 91–92.

47. Davis, *Pioneering with Taconite*, 125; Hogan, *Economic History of the Iron and Steel Industry*, 4:1488.

48. Davis, *Pioneering with Taconite*, 43.

49. Ibid., 127–28.

50. E. W. Davis, "Pollution," unpublished manuscript, 1971–1972, pp. 7–8, Lake Studies Folder, Box 1, Edward W. Davis Papers, Minnesota Historical Society, St. Paul. Ecology students at the University of Minnesota later discovered a key flaw in these experiments. Davis had suspended the tailings in cold water and used room temperature water in the holding tank. The situation in Lake Superior was reversed. The lake water was very cold and the tailings were suspended in warm water. As a result, the tailings behaved very differently in the lake than they had in the test tank. Bastow, *This Vast Pollution*, 33.

51. Schaumburg, *Judgment Reserved*, 46.

52. Davis, *Pioneering with Taconite*, 135.

53. Davis, "Pollution."

54. Huffman, "Exploring the Legacy of Reserve Mining," 340–41; Bastow, *This Vast Pollution*, 8.

55. Hays, *Beauty, Health, and Permanence*, 55.

56. Wendy Adamson, *Saving Lake Superior: A Story of Environmental Action* (Minneapolis: Dillon, 1974), 1–2.

57. Bastow, *This Vast Pollution*, 82.

58. Ibid., 36–40. Environmental groups argued that the tailings were much finer than sand, more like fine silt or clay.

59. Ibid., 104–6.

60. Huffman, "Exploring the Legacy of Reserve Mining," 342–44. Reserve Mining's shutdown was also precipitated by consolidation within the steel industry. Joint owner Republic Steel was considering a merger with another steel company that owned a larger and more profitable low-grade iron ore plant. See Thomas McGinty, "Competitive Status of Reserve Mining in the Current Iron and Steel Environment," unpublished report, 1983, Box 30, Hogan Steel Archive, Fordham University Archives, New York, NY.

61. "Ore-able Fate? He'd Treat Old Cars Like Dirt," *St. Paul Pioneer Press* (January 20, 1968).

62. Carl Hennemann, "He Hopes Amendment Will Pass—Dr. Edward Davis: 'Mr. Taconite,'" *St. Paul Dispatch*, October 21, 1964; Davis, *Pioneering with Taconite*, 197. Speculation that atomic blasting could be used for mining was widespread in the late 1950s and early 1960s. See Frederick Reines, "The Peaceful Nuclear Explosion," *Bulletin of the Atomic Scientists* (March 1959), 121–22; and Scott Kaufman, *Project Plowshare: The Peaceful Use of Nuclear Explosives in Cold War America* (Ithaca, NY: Cornell University Press, 2013).

63. See, for example, E. W. Davis, "The Walleyes and Val," unpublished manuscript, September 28, 1949, Box 4, Edward W. Davis Papers, Minnesota Historical Society, St. Paul.

64. Smith, *Mining America*, 42–46.

65. LeCain, *Mass Destruction*, 148.

66. Davis, *Pioneering with Taconite*, 126.

PART THREE

Health and Environmental Justice

CHAPTER 8

Mining the Atom

*Uranium in the Twentieth-Century
American West*

ERIC MOGREN

In 1776, friars Francisco Atanasio Domínguez and Silvestre Vélez de Escalante led a small expedition to locate an overland route from Spain's northern frontier to its California missions. Winter, and poor planning, forced them to abandon their goal after a few weeks, and they barely survived their punishing retreat back to Santa Fe. Nevertheless, their journal of the ordeal was professional and detailed what royal and Catholic administrators valued—lands for settlement, Native Americans to baptize, landmarks and distances, and deposits of precious metals. In their failure, the Spaniards became the first Euro-Americans to explore and describe the intermountain region known as the Colorado Plateau.[1] Visitors today can travel the friars' route, but may conceptualize the landscape with two centuries of historical perspective. Eighteenth-century Spaniards saw opportunities for expanding imperial influence. Nineteenth-century Mormon pioneers honed their reclamation skills to green the deserts, and many of their communities survive where Escalante once imagined Catholic empire. Commercial development in the past century attracted millions more people to one of the harshest deserts in the world, and Native Americans still consider the plateau their physical and spiritual home. Today, bridges span the Colorado River, while Crossing of the Friars, the place near present-day Lee's Ferry, Arizona, where the freezing and starving priests forded the Colorado River, is the launching point for tourist excursions down one of the most feared rivers in Spanish north America.

In the twentieth century, the plateau region also became the most important domestic source for America's radioactive materials. Mining and milling mountains of uranium ore became a national priority that precluded romantic idylls rooted in agrarian aesthetics, Old World imperial dynamics, or religious ecstasy. Industry produced the uranium oxide "yellowcake" necessary to fabricate atomic weapons to defend the nation and to fuel a peaceful atomic-powered renaissance. Popular imagination transformed uranium miners and millers from industrial laborers into cultural icons of rugged individualism; uranium companies symbolized the triumph of American market enterprise over Communism. Uranium mining, and the atomic cultural assumptions that supported it, was also a celebration of another historical American virtue—our determination to force even the most hostile environments to yield to our will. The earlier, pastoral expectations of agricultural Colorado Plateau residents survived only when they supported, or at least did not interfere with, the new mining imperatives. Morality lay not in the farmers' desire to make the desert bloom and sustain communities, but in national security imperatives and profits to be gained from exploiting mineral resources on behalf of people who lived beyond the plateau, with little regard for the harmful consequences of their actions.[2]

This commodification of radioactive ores triggered environmental and social changes that reverberate today. Mines and mills left a scarred and contaminated landscape. Millions of tons of mill tailings sit in massive quarantine facilities whose man-made, inorganic architecture contrasts with the surrounding natural horizons. Roads and industrial infrastructure that supported radioactive mining scarred fragile desert ecosystems. The uranium industry attracted thousands of workers and their families, changing many of the region's remote settlements into bustling towns and cities. Today, ghosts of those times haunt communities touched by uranium mining's boom-and-bust cycles. Many residents also live with the risks of the industry's pollution, and aging workers remember friends and family members who died from diseases linked to their radioactive occupations. The uranium business, with all of its benefits and harms, was also exceptional in another way: at its height after World War II, it was a mining industry created, promoted, financed, and regulated by the federal government.

THE FIRST URANIUM BOOM

In 1789, about a decade after the priests survived their harrowing trek, Martin Klaproth announced to the Prussian Academy of Science that he had isolated an unknown metal and named it after the Greek sky god. His classification was new, but hundreds of years earlier, Roman glassmakers used uranium compounds to tint tesserae yellow-green, and Native Americans used its yellow-orange oxides for decoration. Klaproth's analysis triggered scientific curiosity about the metal, but it was industrial demand that gave uranium compounds their earliest commercial value.[3] Europeans rediscovered Roman uranium glassmaking techniques during the 1830s, and by the turn of the century, glassworks produced luxurious yellow-green "Bohemian" decorative glassware. The ceramics industry used uranium compounds in glazes and slip paints. Uranium shaded calico dyes, and photographers occasionally substituted it for silver to produce red-brown "uranotype" prints. Around the turn of the century, steel foundries experimented with ferro-uranium alloys. Yet, despite these diverse commercial uses for uranium, it remained a scarce and relatively expensive additive used in small quantities.[4]

In the nineteenth and early twentieth centuries, the most important source for uranium compounds was European pitchblende, an amorphous composition containing uranium in the form of uraninite. Pitchblende also occurred in the United States, but the costs of locating, extracting, sorting, and transporting domestic pitchblende to European refiners discouraged exploration and development of the rare deposits. The most significant early American source for rare metals was carnotite, a form of potassium vanadate. Carnotite deposits were more widespread than pitchblende in America, but it was even less marketable to European refiners because of its low uranium content. In addition, most known carnotite deposits were located in remote areas, especially deep within the American West where the high costs of extracting and transporting the low-grade ore rendered most of it unmarketable. A few U.S. mining companies attempted to make domestic carnotite profitable by building primitive reduction mills to concentrate the uranium and vanadium and thereby cut the exorbitant shipping costs. American Rare Metals, for example, the most successful American uranium company at the turn of the century, built a mill that ultimately produced about 15,000 tons of uranium concentrate before it closed in 1905. But despite miners' optimism about the possibility of exploiting the nation's vast reserves of low-grade domestic ore, American uranium production was insignificant compared with European output. By 1903 there was so little demand for domestically produced uranium that miners sold only 30 tons of uranium concentrates for $5,625; two years later total national uranium production was valued at a mere $375.[5]

It was medicine, not industry, that launched the first American radioactive mining boom. In 1896, Henri Becquerel proposed the idea that spontaneous decay of uranium atoms released energy and atomic particles. Two years later, Pierre and Marie Curie proved that radium, an exceptionally rare element that occurred in conjunction with uranium, accounted for much of the radioactivity of uranium compounds. Researchers also discovered, sometimes tragically, that radiation damaged living tissue, tending to destroy new and rapidly dividing cells more than mature ones. Because cancers arise from uncontrolled cell division, doctors found radium to be a nearly miraculous treatment for cancers that would otherwise require disfiguring surgery, if they could be treated at all. New radiotherapy applications generated overwhelming international demand for the few grams of radium that existed in the world.[6] Radium prices surged, and by 1913, and with radium selling for between $60,000 and $80,000 per gram—roughly $2.25 million per ounce—the once-lackluster climate for American radioactive ores containing radium improved. Miners imagined reaping vast profits from their mountains of low-grade ores. The United States Bureau of Mines and industrial experts reinforced that dream when they reported that domestic carnotite deposits contained as much as 900 grams of radium, at the time among the most extensive estimated radium deposits in the world.[7]

The economic realities of the radium market, however, quickly tempered miners' aspirations. Carnotite usually contained less than 2 percent uranium and yielded only about one part radium to three million parts uranium by weight. In other words, thousands of tons of carnotite yielded infinitesimal amounts of radium. Most carnotite was mined in remote areas, and the cost to extract and sort—often by hand—marketable grades of raw ore for the European refineries meant that miners sent only their finest output to the continent (sometimes as little as a few hundred pounds) and threw five times as much carnotite onto their tailings piles as they shipped overseas. Europeans paid up to $95 per ton for the best grades of American ore, but domestic extraction costs and transportation to the Continent were as much as $75. Those returns, combined with all the risks involved with shipping, made it barely worth the capital and effort. Europeans dominated the world radium market, produced the majority of the world's radium, dictated prices to the American uranium industry, sold their radium to American medical institutions at exorbitant prices, and returned little profit to Colorado Plateau miners.[8]

Progressive era reformers found this de facto European radium monopoly intolerable and in 1913 campaigned to nationalize the struggling domestic radium industry. Progressives proposed government-sponsored domestic competition for the European radium producers that would conserve the nation's radium reserves, produce radium economically from low-grade domestic ores in American refineries, and thereby lower its price to the nation's hospitals, research institutions, and patients. Private mining companies disagreed, opposed the policy as socialistic, and waged a successful countercampaign to keep radium development in the private sector. Frustrated Progressives settled for a cooperative agreement between the Bureau of Mines and the privately owned National Radium Institute to research and develop efficient methods to extract radium from domestic ore. Although by the time it closed in 1917, the National Radium Institute produced 8.5 grams of radium at $37,000 per gram from America ores, about a third of the market price of radium, it was never more than a demonstration and experimental facility and was incapable of challenging European refiners' market dominance. It was, however, an early federal experiment with public-private partnership to develop radioactive materials from domestic ores, a model that came to dominate the uranium industry a generation later.[9]

Yet, having derailed federal radium conservation, the private sector continued to struggle with the economic and technical challenges of developing its carnotite reserves. One approach was to concentrate the marketable elements of their ore near the mines, which in turn would cut transportation costs to European radium refiners and boost domestic profits. But even with stratospheric radium prices, the market landscape remained daunting. There were too few domestic reduction mills to process the most common grades of ore on the scale required to generate substantial radium profits, and there were too few American radium refineries for domestic producers to challenge the Europeans' radium market dominance. Extraction and transportation costs of the most common ores, even to regional mills, were high, and increasing domestic milling capacity in the remote carnotite fields was a significant financial undertaking. Despite the difficult commercial climate, however, producers remained bullish. With access to vast quantities of low-grade uranium ore, these companies continued to envision their future as the world's leading suppliers. A handful of American companies plunged forward, invested heavily to acquire and locate ore deposits, slowly expanded mill capacity, and steadily increased their output of marketable uranium concentrates.

The outbreak of World War I briefly disrupted international trade for radium and threatened domestic carnotite production, but surging wartime medical and military radium consumption, especially for radium-based luminescent paint, boosted demand for American ore and seemed to confirm the most optimistic predictions about radium's promising future. The warring nations also purchased vanadium and uranium to produce ferro-vanadium and ferro-uranium alloys. Domestic companies retrofitted their mill circuits to extract both uranium and vanadium, reprocessed old carnotite mine and mill tailings to extract vanadium, and bought or located new carnotite claims to meet the surging demand. Mining companies cooperated with local governments to improve public transportation infrastructure to facilitate increased production and the steadily growing population of mine and mill workers. With access to labor, capital, technology, vast quantities of raw material, and most important, a seemingly insatiable wartime demand for their product, Americans were finally poised to reap the carnotite profits that had eluded them for nearly two decades.[10]

Domestic producers invested heavily to modernize and expand their operations to provide radium for medical and industrial uses, and they anticipated long-term profitability, but their market dominance and wartime bounty proved fleeting. In 1922 Union Minière du Haute Katanga, a Belgian mining company, announced that it had found pitchblende reserves in Africa that were vastly superior to any in the world—as much as twenty-five times richer than Colorado Plateau ores—and built the refining capacity to process it. When the Belgians undercut radium prices by 30 percent, demand for American ore evaporated. A year later, new discoveries of vast pitchblende deposits in the Great Bear Lake region of northern Canada further undercut international radium prices. U.S. mining companies closed mines and mills. A handful of domestic firms continued to produce limited quantities of vanadium throughout the 1920s and the Depression years in order to meet European strategic military demand for the steel additive, but by end of the 1920s the nation's first rare metal boom was effectively over.[11]

THE SECOND URANIUM BOOM

In 1939, refugee scientists Leo Szilard and Albert Einstein wrote to President Franklin Roosevelt confirming that uranium could fuel a new class of atomic weapons of unimaginable destructive power, and in August 1942, the federal government launched a secret program to build them. A fundamental challenge confronting Manhattan Project scientists was the shortage of uranium from which to extract enough fissionable isotope to conduct experiments and ultimately build the weapon. African and Canadian pitchblende was the most important wartime source for fissionable uranium—ultimately 90 percent of the uranium used to develop the first atomic weapons came from abroad.[12] The government also worked with private industry to improve uranium prospecting and metallurgical research, to develop new methods to improve refining of uranium-bearing Colorado Plateau ores, and to encourage vanadium companies to reconfigure existing mills, and build new ones, to reprocess vanadium and radium mill tailings to extract uranium. National security drove a new strategic demand to locate, develop, and process domestic uranium reserves, and rare-metal producers again saw a promising future for their ores.

In August 1945, the United States launched the atomic era with blinding fireballs that destroyed two Japanese cities, and shortly afterward the Japanese surrendered. The national relief from the Allied victory, however, was tempered by lingering unease about the menacing mystery of the devastating weapon. The end of the war also occasioned, for the first time, political assessment of the secret technology. Civilian and military leaders reasoned that the destructiveness of atomic weapons—real and threatened—reduced the need for large, conventional armies and the likelihood of catastrophic future conflicts. Scientists, popular writers, and technocrats imagined war's alternative—a peaceful, prosperous future made possible by atomic power. Their extravagant predictions that fission reactors would produce seemingly limitless energy gave the atom a more reassuring postwar image than the mushroom clouds and smoldering Japanese cities that dominated the public's initial perceptions of the new technology. Utility companies and electrical equipment manufacturers, too, promoted visions of atomic-powered miracles—the potential of atomic energy, after all, promised their profits from both atomic technology and the electricity it generated. After a decade of depression followed by another world war, such utopian assurances resonated with Americans. In addition, there was growing consensus among civilian leaders that the federal government dominate the new technology to maintain the nation's atomic supremacy and foster peaceful applications in a manner consistent with American political and social values.

Congress ended the wartime military monopoly of atomic technology and reaffirmed civilian control of military affairs in 1946 with the Atomic Energy Act. It placed management and promotion of the nation's atomic program in the hands of the five-member civilian Atomic

Energy Commission, or AEC. Congress also created the bipartisan Joint Committee on Atomic Energy, or JCAE, to supervise the new agency and ease fears that legislators had ceded too much control of the nation's atomic enterprise to the executive branch. The AEC used its vast discretionary authority to implement its increasingly divergent mandates to both control the nation's atomic infrastructure and to encourage peaceful atomic applications. It had absolute authority to acquire fissionable materials, regulate the manufacturing of fuel for weapons and reactors, and oversee everyone who built, tested, and utilized atomic technology. It had the responsibility to ensure the safety of atomic industry employees at facilities regulated by the commission. The act, in short, created a government monopoly of the atom that gave the federal authorities historically unprecedented control of nearly every aspect of the new technology, including the uranium-mining industry.

Among the priorities for postwar strategists and civilian atomic power advocates was expanding domestic uranium production. Wartime consumption meant that little fissionable material was available after 1945 for further weapons fabrication and nonmilitary purposes. The perceived need to expand domestic uranium production compelled policy makers to embrace the traditional American faith in the marketplace as the framework in which to secure the nation's uranium independence. They believed that uranium resources would best be developed to meet future military and civilian applications by the private sector working for profit, but without relinquishing comprehensive federal oversight. It was a creative amalgam of government control of atomic technology and private enterprise idealism.

The earliest steps in stimulating postwar uranium production came in 1948, when the AEC announced its ore-buying program. The commission offered financial incentives to miners, including substantial cash bonuses for the discovery and production of uranium ore from newly discovered deposits, price guarantees, government operation of two reduction mills on the Colorado Plateau, and haulage allowances for transporting ore from remote mines to the government's buying stations. The commission also sponsored a uranium-prospecting program. Other federal agencies used their powers to promote private uranium development; the U.S. Geological Survey, for example, provided uranium prospectors with geological data. These inducements, however, attracted little interest from miners, because they generally rewarded the location and exploitation of high-grade uranium ores, rather than the predominately low grade ores commonly found in the American

West. Consequently, the AEC's initial commercial incentives did little to promote development of the most common domestic uranium ores, and even with price guarantees the miners were hard-pressed to compete with the bargain-basement prices of African and Canadian uranium.

As the Cold War intensified in 1949 and demand for domestic uranium became acute, the AEC expanded its incentive policies so that Colorado Plateau uranium ores could play the dominant role in the nation's atomic future. Several factors underscored the new approach. The amount of uranium required for a renewed Cold War weapons program on the scale envisioned by strategic planners was more than they wished to import, and fear that our national security rested upon foreign uranium amplified the calls for increased domestic production. Additionally, stimulating demand for uranium for civilian atomic reactors could help create economies of scale favorable for the production of uranium from low-grade American sources. Most important, the tentative postwar AEC purchasing program resulted in a significant reassessment of domestic uranium resources. Wartime estimates of American uranium reserves understated the amount of domestic low-grade uranium ore that might be exploited under favorable economic conditions, and for the first time, policy makers envisioned that domestic uranium supplies could satisfy both the nation's projected demand for atomic security and private atomic power. Consequently, in 1951 the AEC amended its purchase program for the most common grades of Colorado Plateau ores. It also provided extensive advice about prospecting, along with financial incentives for vanadium companies to reopen their mills and process local ores for their uranium content. The commission established a series of ore-purchasing stations to make it easier and more cost-effective for miners to deliver their output to buyers. It issued policy statements that reinforced the its preference for a government-controlled private uranium-mining and -processing industry designed to codify a predictable government purchasing strategy through the early 1960s, a move intended to reassure and stabilize a uranium-mining industry historically wracked by devastating market cycles. The government's escalating demand for fissionable materials finally made domestic low-grade uranium ore marketable, and the postwar uranium boom that reshaped the American West began.[13]

The nation's wartime uranium acquisition program was small and shrouded in secrecy, and the AEC's immediate postwar uranium procurement activities attracted little interest from miners, and even less from the public. The uranium boom of the 1950s and 1960s, however, was different: for an insecure Cold War nation steeped in romantic nostalgia and seeking comfort in cultural myths, it resonated with significance. Uranium became a symbol of power in an uncertain and dangerous world. Mining in the undeveloped West echoed themes of muscular patriotism, rugged individualism, and frontier independence. And, of course, uranium mining promised government-guaranteed wealth. Public enthusiasm for the AEC's purchasing program matched any of the mining booms in American history. Thousands of fortune hunters descended onto the harsh and sparsely populated uranium regions. Prospecting on the Colorado Plateau increased dramatically—a 20,000 ton uranium ore body made national headlines in 1950, but six years later prospectors had already located at least twenty-five deposits that exceeded 100,000 tons, and a few that were estimated to be a million tons. Lucky "uraniumaires" like Charlie Steen and Stella Dysart, whose colorful rags-to-riches stories fueled the "uranium fever," became heroes of the new era. Agricultural communities such as Moab, Utah, and Grand Junction, Colorado, were reincarnated as classic American mining towns, and older radium and vanadium mill communities, such as Uravan, Colorado, and Monticello, Utah, were reborn. Trading in low-cost uranium stocks gave Americans the chance to participate in the boom without suffering mining's hardships firsthand. Magazines offered "greenhorns" advice and outfitters sold them Geiger counters and army-surplus equipment. Hollywood made films and television episodes set in uranium country. Children practiced "duck and cover" atomic holocaust drills at school, but played with uranium-themed toys and games at home. The AEC's preference for an orderly and managed development of domestic uranium resources was swept aside by the frenzied dash for the riches made possible by the federal government's uranium program. In 1947, when the AEC began its procurement program, the domestic uranium industry was virtually non-existent—it produced only 67 tons of uranium oxide from two reconfigured vanadium mills. By the middle of the 1950s uranium was a $500 million industry.

Uranium excited the public's imagination and drove unprecedented mining activity across the American West, but the overriding objective of the AEC's procurement program was to produce concentrated uranium oxide, called yellowcake because of its canary color, which could be refined further to extract fissionable atoms for reactors and weapons. Yellowcake production required a massive expansion of milling capacity on the Colorado Plateau to process millions of tons of raw ore—by 1958, twenty-eight reduction mills were producing nearly 182,000 tons of uranium oxide. However, the construction costs for mills—up to $4 million each—added significantly to the costs of private uranium development. To stimulate mill construction, the AEC also subsidized the mills as another component of its uranium acquisition policy. A company interested in applying to the AEC for a mill license—often a company that already had significant projected ore production from its own mines—had to prove that it had the technical and financial capacity to operate the facility, and that there were sufficient ore supplies to support its construction. Once the company met these threshold conditions, the AEC contracted with it for the construction and operation of the mill. Since the AEC was the only customer for the mill's output, the agency determined the price for the yellowcake, taking into consideration variables including the cost to purchase ore, along with transportation, milling, and amortization costs. It also regulated nearly every aspect of mill operations. This public-private partnership meant that uranium milling, like uranium mining, was virtually guaranteed to be profitable.

The uranium boom, with all its passion, was also an incongruous blend of hope and fear. Uranium mining energized Americans' faith in their technical superiority, patriotic ideals, and national character, and it produced the raw material of atomic weapons that safeguarded that vision. The atom also threatened Armageddon. Yet Americans chose to focus more on uranium's utopian promise than its proven dangers, punctuated by smoldering Japanese cities and disfigured atomic survivors. The public uranium acquisition program, above all, reflected a deep-seated public trust in scientists and government technocrats to protect the nation from threats posed by the mysterious new energy. Too often, that faith was misplaced.[14]

ENVIRONMENTAL IMPACTS OF THE URANIUM BOOM
Mines

The "nuclear fuel cycle" has several stages: mining; milling ore to produce yellowcake; enrichment to separate and concentrate fissionable isotopes; fuel fabrication for reactors, weapons, or other purposes; and disposal. Uranium mining and milling, the "front end" of the fuel cycle, is a mechanical and chemical process similar to extracting and processing any other metallic ore—it is destructive and causes significant environmental changes. Domestic mining and milling for radium and vanadium during the opening decades of the twentieth century generated local pollution from the small mines and milling operations. Laboratories and factories that refined and sold radium products contaminated numerous communities throughout the nation. The rapid mid-century expansion of uranium mining and processing, however, amplified the industry's environmental impact. The boom eventually resulted in about 4,000 mines with documented production, though as many as 15,000 mine locations contained varying concentrations of uranium. At peak production, in 1961–1962, there were 925 active uranium mines.[15] The majority were located in Colorado, Utah, New Mexico, Arizona, and Wyoming, with three-quarters of them on federal or tribal lands. Most of them were underground mines, though a few were open-pit operations. Collectively, mines that had significant historical production generated about 3 billion metric tons of waste.[16]

As with any mining operation, overburden material and unmarketable ore formed massive tailings piles. Unreclaimed uranium and other hazardous materials commonly associated with uranium, such as arsenic, copper, phosphate, molybdenum, and vanadium, leached from tailings piles into nearby watercourses or blew off the piles as dust deposits that contaminated neighboring lands and communities. At the mills, processing chemicals leached from tailings and fouled rivers. Exhaust fans ventilated rock dust, radioactive gases, and reduction wastes from mines and mills into the atmosphere; tailings piles exhaled radon gas. Open-pit mine operations, which generate roughly forty-five times more mine waste than underground operations, destroyed vast areas at the mine itself and under the tailings piles. Boomtowns, with all their requirements for domestic urban infrastructure, blossomed in fragile desert ecosystems. And when demand for uranium waned and the mines and mills closed, tailings piles eroded, machinery and infrastructure decayed, and the once-thriving towns lost their vitality.[17]

Underground, miners faced additional environmental hazards. Most uranium mines were small operations located in remote areas, thus increasing the difficulties of risk management and safety oversight; about 60 percent of mines during the peak production years employed fewer than fifteen miners, and only about 13 percent employed more than fifty miners.[18] Risks typical of any mining operation were a constant part of their physically demanding and dangerous employment. Detonations filled the mines with dust and gas as miners blasted free the marketable ore. Mines were filled with machinery designed more for production than safety. Timbers and bolts stabilized walls and roofs, but collapses were a constant danger. Noise and engine exhaust underground added to the workday hardships. This hazardous work was done by thousands of white miners, many of them Mormon, and by a significant number of Native Americans, especially Navajos. In addition to these long-term regional residents, many miners and mill workers relocated to the Southwest seeking fresh opportunities. They were children of the Great Depression, and many were war veterans, all eager for well-paid physical labor that also served national interests. Uranium industry workers were usually conservative, nationalistic, vehemently anti-Communist, and often culturally and religiously insular, and they enthusiastically supported the nation's atomic defense policies. They were also happy with incomes that supported their families and communities in a historically poor region. Most also shared a faith that the architects of the nation's atomic policies were candid and accurate about the risks posed by the fast-growing uranium industry, and relied upon reassurances offered by federal atomic technocrats that their work was safe—or at least no more hazardous than any other mining.[19]

Uranium mining and processing, however, also generated hazards that were less obvious to uranium workers—radioactive contaminants. Drilling, blasting, crushing, and moving uranium ore released radioactive elements that became part of the atmosphere in the closed spaces of mines and mills. The cycle of radioactive uranium decay includes radon, an odorless and colorless gas. Radon atoms decay through a sequence of unstable elements, "radon daughters," with half-lives ranging from 160 microseconds to twenty-two years. With each radon decay step, these elements emit energized subatomic particles and ionizing radiation as the atoms shed their original unstable atomic mass of 222 to reach the stable mass of 206. These radon daughters, composed of various isotopes of polonium, lead, and bismuth, tended to cling to suspended nongaseous atmospheric particles such as cigarette smoke, dust, diesel exhaust, and water vapor, and also settled onto food and drinking water within the enclosed spaces of the mines and mills. Consequently, as workers breathed contaminated air, smoked cigarettes, or ate food, the radon daughters attached to their nose, throat, and lung tissues. When a step in the radon decay cycle occurred inside a miner's body, energized particles and ionizing radiation damaged nearby cells. The most common radon decay damage was to lung tissue, but another radon decay element, lead-210, with a half-life of twenty-two years, concentrated in human bones, where its slow decay damaged bone and marrow cells. Chronic cell damage from radioactive decay can eventually trigger malignancies, and uranium industry employees working in poorly ventilated mines and warehouses routinely received radiation doses that significantly increased their cancer risks.

Health physicists knew a great deal by the mid-twentieth century about the radiation risks in the uranium industry. Since the fifteenth century, European miners working in the Joachimsthal and Schneeberg mining districts, longtime sources for silver and pitchblende, had died from lung disease they called *Bergkrankheit,* or "mountain disease." In the late nineteenth century, doctors confirmed it was lung cancer, and eventually linked it to occupational radiation exposure. Health surveys demonstrated that those miners had about a twenty-year life expectancy after entering the mines, and as many as 75 percent died from lung cancer.[20] By the early twentieth century, European scientists had published extensively about the atmospheric hazards from radioactive materials in the confined spaces of mines. These studies were well known to American occupational hygiene and mine safety experts, and in the mid-1940s they had little doubt that radon significantly elevated lung cancer risk for miners, and that aggressive mechanical ventilation reduced the risk. European, Canadian, and African miners' mortality rates declined dramatically when their mines improved ventilation.[21]

In 1948, occupational safety and radiology experts within the AEC warned their colleagues about potentially high radiation levels in American uranium mines. They recommended that the agency include mine safety regulations in ore purchase contracts, especially mine ventilation standards to reduce radon levels, and heath monitoring of miners. Although the Atomic Energy Act empowered the AEC to oversee the uranium industry, in this instance AEC administrators interpreted the act narrowly to conclude that it did not authorize the agency to regulate mine safety, and consequently the commission did not implement comprehensive mine safety recommendations. In addition, despite decades of research and evidence linking radon exposure to lung disease, the agency also suggested that the causation mechanism for cancer in uranium miners was not definitive enough to justify interfering with private mining operations. Uranium mine safety, according to the AEC, was a matter for the states to address, many of which had developed a patchwork of industry regulations for conventional mines that were enforced sporadically by understaffed state mine inspectorates. States were also wholly unprepared to regulate radiological hazards in the uranium industry.[22]

The possibility of health problems among the thousands of uranium miners, coupled with the AEC's ambivalence about mine safety, spurred concerned state health officials and U.S. Public Health Service (USPHS) scientists to investigate radiation hazards in the uranium mines and mills without AEC assistance. They also planned research into mine ventilation techniques and worked to define safe radon exposure levels for the uranium industry. The results of such studies, they hoped, would provide the persuasive evidence needed to develop comprehensive safety recommendations for the uranium industry. The investigators, however, faced challenges. Unlike AEC officials, Public Health Service scientists had no legal authority to enter private mines without the owners' permission and, in order to gain admission to collect data, often had to agree not to inform miners of possible radon hazards. Medical consent forms for the health surveys could not mention health risks from mine radon. Because radon-induced lung cancer usually takes ten to fifteen years to develop, and the domestic uranium procurement program began in earnest only in the late 1940s, few miners would display symptoms until the late 1950s, further clouding the cancer causation issue. Miners, especially Navajos, were frequently transient laborers, worked in mines part-time, or simply disappeared for extended periods, making epidemiological research difficult. Nevertheless, despite mounting evidence of dangerously high radon levels in many mines, uranium companies maintained that their facilities were safe and resisted recommendations that they believed were too expensive and might interfere with production, and that the AEC did not require. Atmosphere safety guidelines proposed by health experts remained simply that—suggestions based on evidence—and they had no force unless a federal or state mine safety agency adopted them. Neither the uranium mining and milling companies nor the AEC enforced aggressive radon control protocols.[23]

Beginning in 1955, because of the AEC's ongoing reluctance to regulate radiation hazards in mines and mills, states developed their own safety standards based on the growing body of scientific evidence about the environmental risks in the industry. New Mexico began implementing radon safety standards in 1958, followed two years later by Colorado and Utah. These efforts had mixed success. The prevailing climate of national security secrecy and preemptive federal control of atomic energy made it difficult for state regulators to coordinate with the AEC to develop and enforce safety standards. State mine inspection agencies often lacked the finances to effectively monitor radon risks and were usually understaffed for the sort of comprehensive oversight program on the scale required to manage the burgeoning uranium industry. By 1960, when the USPHS offered its findings to the governors of uranium mining states about the radon risks in their uranium mines, evidence from epidemiological studies demonstrated that uranium miners had a 450 percent greater chance of contracting lung cancer than other miners and that the number of mines with unacceptably high radon levels was increasing. Finally, in 1967, after nearly seventeen years of buck-passing by private companies and federal and state agencies about uranium mine safety, the Department of Labor established the first federal standards for radon exposure in uranium mines. Rather than look to the Atomic Energy Act or its regulatory framework for that authority, the Labor Department relied upon the New Deal–era Walsh-Healy Act, which granted the federal government power to regulate safety and health conditions in businesses that had government contracts.[24]

Between 1950 and 1968, the entities that had the most control over the uranium industry—mining companies and the AEC—made little effort to minimize radon exposure among uranium workers or even to inform them about the possible radiation dangers they faced. As late as 1981, the American Mining Congress, the mining industry's official lobbying organization, still maintained that the necessity of ventilating uranium mines was not fully understood, despite nearly thirty years of American evidence and decades of European studies and practical experience linking radon exposure to fatal lung cancers. In some mines, the consequences of this safety failure were devastating. In the mines that had the highest radon concentrations, lung cancer mortality for miners was 50 percent by 1981. Another study completed in 1984 demonstrated a twelve-fold increase in mortality risk for uranium miners exposed to radon. The median age of Navajo uranium miners who died from lung cancer was nineteen years younger than the mortality age of nonminers in the community who died from lung cancers; among white miners, the radon-induced lung cancer mortality age was ten to fifteen years younger. By 1990, about 10 percent of uranium miners had died from lung cancer, a nearly 500 percent increase over lung cancer mortality among other miners.[25]

In 1979, uranium miners sued the United States for compensation in two class action lawsuits. Although the courts ultimately concluded that the doctrines of sovereign immunity and the discretionary functions of government agents under the Federal Tort Claims Act shielded the government from liability, the judges nevertheless found the miners' claims to be supported by evidence of decades of deliberate decision making that discounted their welfare.[26] They also agreed that the actions of the AEC and the USPHS reflected a national policy not to warn uranium miners of their employment risks. With the failure of adequate legal remedy for victims, their only recourse lay in the political process. Congress finally addressed the problem with the 1990 Radiation Exposure Compensation Act, which awarded up to $100,000 to uranium miners suffering from lung cancer or nonmalignant respiratory disease, provided they met threshold radiation exposure levels during their employment. While the legislation held out the promise of compensation, it also placed an extraordinary burden of proof on individual claimants. For many white miners, proving employment and radon exposure levels from boom-time employment decades earlier, in mines that kept poor or no records of radon levels, was challenging. For Navajo miners, many of whom never had occupational history records or legal proof of relationship status for next-of-kin claimants, the evidentiary hurdles were often insurmountable.[27]

The federal government, which legally dominated the uranium industry, failed to enforce mine safety regulations during the uranium boom years, or even adequately warn and educate miners about the radiation risk they faced, and as a direct result miners died. In the end, there was little evidence to support long-standing assumptions that mine safety protocols would compromise production, undermine national security, or impair the development of atomic power. It was, in short, a human disaster that could have been prevented. AEC policy makers, in particular, weighed miners' lives against the demand for domestic uranium and concluded that mine safety was less important than production. American uranium independence came at the cost of hundreds of deaths and was, as Judge William Copple wrote in his *Begay v. United States* opinion, "a tragedy of the atomic age."[28]

Mills

One of the earliest environmental problems from the front end of the nuclear fuel cycle that garnered widespread national attention was water pollution from uranium reduction mills. Mine wastewaters and tailings runoff contained metallic and radiological contaminants that found their way into surface waters and aquifers near the mines, but the chief source for water pollution from the uranium industry was milling. Mill tailings, around 99 percent of the original volume of ore delivered to the mill, contained nearly all their nonuranium radioactive materials, as well as chemicals used in the reduction process. Usually, tailings left the mills as semiliquid effluent called slime. In the early days of the uranium boom, mills occasionally discharged slime directly into rivers. More commonly, mills stored the viscous waste in containment ponds, where the liquids evaporated or soaked into the soil, leaving behind fine solid particulates. Uranium companies frequently relied upon the drying tailings to create their own dikes for the liquid effluent ponds, but with no structural support, containment dams occasionally failed, spilling chemical and radiological containments into local waterways.[29] Once dry, the sandy tailings were unstable and eroded easily from rain and wind. Spring runoff and changes in watercourses occasionally undercut the tailings that were located near rivers, causing tailings to collapse into the flowing water; a handful of mills even relied on seasonal runoff to help rid themselves of their tailings burden. Some mill tailings were located in remote areas away from densely populated communities, but others were located near urban centers such as Grand Junction and Durango, Colorado, and Salt Lake City, Utah, where a massive mill tailings pile could be seen from the steps of the state capitol.[30]

Beginning in the early 1950s, U.S. Public Health Service officials conducted tests of mill contamination of western rivers to determine the extent of water pollution from the uranium industry. Water samples taken below four mills located near rivers showed radium levels between 20 and 130 times greater than river samples taken above the mills. In-depth investigations of algae, fauna, and streambed materials downstream from eight uranium mills were even more alarming. Unlike the water samples, which were "snapshots" of moving water, the more comprehensive streambed analysis showed a cumulative radium contamination between 60 and 100 times greater than normal background levels, as well as the presence of chemicals used in the milling process.[31] One USPHS bioassay of the Animus River below the mill at Durango, Colorado, showed that the chemical and radiological contamination was so extensive that the lead investigator characterized it as a "biological desert" for fifty miles downstream from the mill—where nearly 30,000 people relied upon the Animas for domestic and agricultural water.[32]

Yet, despite evidence pointing to chemical and radiological pollution levels in surface waters, experts remained uncertain as to what levels of radiological contamination posed a health threat. Biologists with the National Committee for Radiation Protection, for example, considered the pollution a possible long-term threat and urged a cautious approach to determining safe exposure levels. The AEC, on the other hand, maintained that its power to control uranium ore entering reduction mills did not require regulation of "unimportant" radioactive materials, such as radium, that could not be used for nuclear fuel. The nation's atomic agency instead deferred oversight responsibility for mill tailings to other federal and state agencies. As was the case with mines, however, state and federal public health agencies were unprepared to manage water contamination generated by the uranium industry.[33]

In 1954, after a revision of the Atomic Energy Act designed to promote civilian atomic power, and in response to mounting pressure from state public health and environmental protection agencies and the overwhelming success of the AEC's uranium acquisition program that eased concerns that regulation might impede uranium production, the commission began to address the issues of mill water pollution. It was a slow conversion. Federal and state public health experts wanted the agency to exercise its considerable power to curtail mill water pollution. But the AEC's competing mandates to both promote and regulate the uranium industry led it to sympathize more with the industry than with cautious experts who urged greater agency control over polluters. Finally, in 1959, after considerable state and national political criticism and public outrage over the agency's halfhearted regulatory actions, the AEC ordered millers to assure that concentrations of radioactive materials in wastewaters be brought within "permissible limits." In 1960, the agency also reversed its long-standing position that it did not have jurisdiction over radium, the most potentially hazardous radionuclide in the tailings. It now concluded that it could enforce radiological pollution standards at operating mills that it licensed, because tailings were an integral part of mill operations and were consequently subject to mill licensing requirements. Over the next few years, operating mills undertook abatement programs that significantly lowered radiation contamination of local watersheds.[34]

Although the AEC eventually undertook measures to require its mill licensees to reduce pollution from operating mills, tailings from closed uranium mills continued to contaminate watercourses. Once again, jurisdictional disputes between federal and state agencies amplified the environmental impacts of uranium tailings. The AEC maintained that once a uranium mill closed and the agency no longer issued an operating license for the facility, it had no responsibility or authority to manage mill tailings. In 1965, the agency also reasserted its long-standing claim that uranium tailings contained so little uranium after ore processing that they did not constitute an unreasonable environmental risk and, consequently, tailings did not require agency oversight.[35] State health and environmental officials recognized the irony: the AEC had already conceded that tailings from operating mills were a sufficient environmental threat to force millers to manage their tailings to reduce pollution, but the agency now claimed that those same tailings at closed facilities were not sufficiently hazardous for the agency to regulate them directly. Instead, the agency favored control of the abandoned tailings by some other, preferably state, entity. Federal authorities conceded that while tailings were sometimes a public nuisance because of blowing dust, they were not a greater environmental threat than effluent from other types of ore processing and could be handled and regulated in a similar fashion. As one AEC official put it, managing the tailings at abandoned mills was "a matter of good housekeeping prudence [. . .] but aesthetics [were] not within [AEC] jurisdiction."[36]

In contrast to the atomic agency's short-term assessment of radiation hazards from abandoned tailings, federal and state public health authorities remained concerned about potential harms from long-term exposure to tailings and the likelihood that radiological contaminants would eventually erode from the piles and pose a threat to downstream water users.[37] The AEC's reluctance to address tailings threats also ignited a national controversy among a public increasingly sensitized to such dangers by the environmental movement of the 1960s. Reporter Terry Drinkwater captured the mood when he reported on the *CBS Evening News with Walter Cronkite*: "If nothing is done to cover up or remove these tailings, and if government health officials are right, then another generation may well look at these radioactive man made mountains as monuments to the carelessness of this generation, man's carelessness in the early years of the nuclear age." Senator Edmund S. Muskie convened a Senate subcommittee hearing on the matter of abandoned tailings in 1966, and testimony revealed that only one tailings pile generated by thirty-four mills in the western states had been adequately stabilized to limit water and wind erosion.[38]

By late 1966, public criticism, pressure from western lawmakers, and a growing body of evidence about tailings risks prompted uranium states' environmental and health agencies to move ahead with their own uranium tailings management policies. Regulations promulgated by Colorado, for example, required millers to submit plans to state regulators for stabilization of their tailings at their closed mills, including controlling tailings erosion, limiting public access to the piles, preventing tailings runoff from flowing into local waterways, and generally maintaining the piles to limit radiological contamination of local environments. Other western states quickly followed suit. It was, in one sense, a victory for the AEC, which had successfully shifted the regulatory burden for tailings at closed facilities onto states. Yet the new state regulatory schemes validated the states' environmentally cautious approach to long-term radiation contamination. The new assertive state position about tailings also reflected a growing public skepticism about the AEC's commitment to radiation safety within the uranium industry, skepticism that influenced subsequent policy debates about management of the nation's atomic trash.

The new regulatory structure and the resolution of water pollution concerns originating from the historically careless management of uranium tailings reduced environmental hazards. AEC licensing requirements for active mills, coupled with state regulations for tailings at inactive mill sites, forced the uranium industry to stabilize or relocate tailings piles away from nearby watercourses. State-mandated monitoring programs kept careful watch on water contamination. Radioactive wastes that had so concerned environmental and health safety experts seemed, finally, to have been contained and the tailings crisis resolved.

But water pollution was only one of the environmental threats posed by the uranium mills. Crushing machines filled the air with pulverized rock dust and radioactive materials; yellowcake containing uranium and other radioactive trace elements and chemical contaminants dusted mills' packaging rooms. Mills often vented their buildings to the outdoors, elevating the concentrations of chemical and radioactive contaminants in the mills' immediate vicinity. Radon gas exhaled from decaying radium in tailings escaped into the atmosphere, and the dry tailings had for years blown from the piles and coated nearby communities with a fine dust that environmental and health experts feared might result in harmful radiation exposure. Although decades of evidence had established that radon in enclosed spaces, such as poorly ventilated mines or mills, elevated the risk of cancer and death, few in the AEC or industry were concerned about atmospheric radon emissions. The agency assumed that radon would quickly dilute in the atmosphere and therefore posed little atmospheric radiation risk beyond the immediate vicinity of the tailings. Consequently, the agency declined to take aggressive action to compel the uranium industry to reduce atmospheric radon and chemical pollution, and the uranium industry was reluctant to voluntarily undertake what might prove to be costly mitigation measures for a problem that may not exist.

In the communities surrounding the mines and mills, residents were not convinced by reassurances that radiation hazards beyond the tailings piles were minimal. Public and industrial health experts, as usual, urged a cautious approach that included extensive air monitoring near the mills to determine atmospheric radon levels. By 1969, such research confirmed initial assumptions that the radon from tailings piles posed a minimal atmospheric threat to local communities. The studies also revealed that no significant public radiation exposure had resulted from the radon exhaled from tailings piles: the worst fears about atmospheric radon pollution, while grounded in legitimate concern about chronic radon exposure in closed spaces, were unwarranted when radon was diluted in the atmosphere. Nevertheless, the atomic agency's earlier experience with public criticism and state pressure to manage tailings prompted it to use its licensing power to encourage millers to keep the tailings dust from blowing off-site, not because it was a health hazard but because the agency considered limiting dust to be good public relations.[39]

By the mid-1960s, ongoing water and atmospheric monitoring demonstrated that the uranium industry had significantly reduced its chemical and radiological pollution. The AEC's Peter Morris confidently testified before the Senate in 1966, "We find it difficult to conceive of any mechanism whereby the radioactive material which is now so widely disbursed could become so concentrated as to exceed current applicable standards for protection."[40] Less than a year later, however, state and federal health officials discovered that just such a mechanism was ongoing and had existed for over a decade.

As tailings accumulated at mills, new state and federal regulations made the uranium industry even more conscious of the need to manage its bulky wastes. One solution was to encourage alternative uses for tailings, especially for construction. Tailings were an ideal building material because milling pulverized ore into a uniform, sandy consistency in preparation for chemical processing; and there were tens of thousands of tons if it available to meet the construction demands of growing uranium boomtowns. Millers were happy to give their tailings away, and local contractors used the material as backfill against building foundations and for streets, sidewalks, and railroad beds. The fact that most tailings were slightly acidic also made them ideal soil additives for gardeners and farmers. The Federal Water Pollution Control Agency even encouraged construction with tailings, believing that disbursing tailings underground would isolate the waste from the atmosphere, prevent them from eroding and contaminating waterways, and generally reduce environmental risks.[41] Thus repurposed, tailings became a serious source of radiological contamination in several uranium communities.[42]

The experience of Grand Junction, Colorado, exemplifies the health risks from the indiscriminant use of tailings. Beginning in 1956, the

Climax Uranium Company permitted builders to take its tailings for free. Over the following decade, as much as 300,000 tons of tailings went into highway and road construction, sidewalks, sewer projects, and residential and commercial structures. In January 1966, investigators from the Colorado Department of Health and the U.S. Public Health Service stumbled upon the then-routine practice of releasing tailings into the Grand Junction community. They also immediately recognized the possibility of health threats from elevated radon and gamma radiation levels in contaminated structures. Their fears were confirmed when preliminary monitoring over the following months established that radon levels were significantly higher than background in their small sample of polluted structures. Citing that research, in the summer of 1966, Colorado ordered an immediate halt to the distribution of tailings for construction.[43]

Further state testing through 1967 confirmed that tailings were a widespread problem in Grand Junction. Colorado public health officials sought technical and financial assistance from the federal government to conduct a comprehensive inventory of contaminated structures. The AEC rejected the request, asserting that the radon from construction tailings posed no greater threat than naturally occurring radon and therefore was not a significant health risk. The agency agreed that it had the authority to oversee tailings at the mills, because it determined that tailings were an integral part of milling operations subject to the agency's licensing power. But once tailings were removed from the mill and transferred to third parties, such as builders and home owners, the agency asserted, the tailings were no longer within its jurisdiction. Tailings contamination beyond the mills, the AEC maintained, was a state responsibility. Once again, the irony was not lost on Colorado health officials and politicians. The allegation that radon inside buildings reflected background levels relied on scant data about interior radon levels—data that could best have been generated by federal authorities who resisted thorough investigation of the issue. Moreover, it seemed illogical that tailings in the possession of mills and on mill property were hazardous enough to justify forcing the uranium industry to prevent tailings escaping from the mills' properties into the environment, but once the tailings were deliberately distributed off-site throughout the community, they were not sufficiently dangerous to warrant oversight by the nation's leading atomic experts. Disappointed but undeterred, Colorado public health officials proceeded as best they could on their own with their survey of the contaminated city. The ongoing state investigation revealed that tailings contamination was far more wide-spread than originally estimated and that radon levels in some buildings approached those of poorly ventilated mines. By 1971, 2,870 structures in Grand Junction were identified as contaminated, and other mill towns and cities discovered that they, too, had polluted buildings.[44]

While governments disagreed about responsibility for remedial action, few public health officials doubted the potential hazards of radon in closed living spaces. It was also quickly becoming apparent that decontamination would be expensive. Property owners demanded financial assistance because most believed that the atomic agency's programs had caused the pollution problem and that it was the federal government's responsibility to solve it. The growing scope of the indoor tailings pollution also attracted nationwide public attention as residents of the affected communities, their political leaders, and much of the national public questioned the AEC's position that the federal government bore no responsibility for the problem. Popular media often sensationalized the pollution, but for years Grand Junction remained emblematic of the risks that ordinary people, not only uranium industry workers, faced as a consequence of the nation's drive for uranium independence. By the early 1970s, the issue of indoor tailings pollution was rapidly eroding public faith in the AEC to such a degree that even its own employees began to question openly the agency's position. Critics argued that it was unfair that residents should shoulder a disproportionate share of the harms associated with the nation's drive for uranium independence. The states' struggles to understand the nature and extent of the pollution, and the AEC's reluctance to cooperate with state managers, reinforced the federal agency's image as out of step with the environmental risks of the uranium industry.[45]

In February 1972, Colorado politicians introduced identical bills in the U.S. House and Senate to support decontamination of Grand Junction. The following summer, the final version of the law allocated $5 million for the program, with the federal government shouldering 75 percent of the costs, and the states expected to cover the remainder. Perhaps more important, the states and the AEC defined their separate governmental responsibilities: the AEC would provide scientific and technological expertise, while the states coordinated the on-the-ground work and interaction with claimants. The new law soundly rejected the traditional AEC position that it had no responsibility for the contamination, and, instead, the federal government assumed the primary financial and technical—and, many believed, moral—obligations for the pollution that the community demanded.[46]

AEC staff concluded that with the new legislation and the Grand Junction cleanup under way, public hostility toward the agency would ease. But the agency underestimated the depth of public suspicion it faced, especially among western lawmakers and the western public. Now that Colorado politicians had succeeded in securing funding for Grand Junction, other states, most forcefully Utah, pressed their own demands in Washington to cope with tailings pollution in their communities. In November 1977, western politicians introduced four House bills intended to force the federal government to solve the problems of tailings once for all: tailings were to be managed with the goal of long-term environmental safety, and the federal government would pay for the majority of cleanup costs. Collectively, these bills, combined with the earlier Grand Junction remedial action legislation, were a forceful reproach to the historically sweeping authority of the federal government in matters of atomic energy. The proposals, hearings, and ongoing research also revealed that the AEC's preferred approach to tailings pollution management—state action—was untenable in practice. Remedial action for the growing number of contaminated properties, as well as for the tailings piles themselves, was far beyond the states' expertise and financial capability. As of 1978, twenty-five states had assumed some responsibility for tailings oversight, but standards varied from state to state and were sometimes less stringent than what public health experts considered safe. It was clear that the scope of the tailings contamination and the technical expertise that remedial action demanded were so great that it could best be managed by comprehensive national legislation. During the summer of 1978, tailings remedial legislation supporters labored to reconcile the separate bills into a single proposal, and in the hectic closing days of the Ninety-Fifth Congress, Congress passed the Uranium Mill Tailings Radiation Control Act.[47]

Supporters of UMTRCA routinely characterized the law as a matter of equity—uranium production benefited the entire nation, but the harms fell on local residents. The act directed the Department of Energy to undertake remedial action at twenty-four inactive uranium tailings sites and nearly 5,200 vicinity properties, and encouraged federal, state, tribal, and industry cooperation. It also mandated interagency cooperation within the federal government. The federal government agreed to pay 90 percent of the remedial costs, with the states paying the balance. No provision was made for tailings at active mills, reflecting congressional intent that the uranium industry should shoulder remedial costs whenever possible. The law's drafters also took pains to prevent UMTRCA from becoming a model for future environmental remedial action, stressing the unique historical and environmental circumstances of the tailings contamination. In addition, the law mandated that remedial action achieve a permanent solution to the tailings problem that would require no, or at least minimal, ongoing maintenance. For active mills, it directed the Nuclear Regulatory Commission to develop tailings disposal standards that would be enforced through the mill-licensing process. It was a challenging mandate: roughly 39 million cubic yards of tailings and abandoned and contaminated buildings and equipment covered some 3,900 acres, and thirteen sites required removal of tailings to off-site internment locations. All plans also had to comply with the National Environmental Protection Act.

The law's supporters had little to fear that UMTRCA would be a bellwether for future federal remedial pollution action—the law was hardly a model of environmental legislation. Passed in the closing days of the congressional session, it was hastily drafted, had internal inconsistencies that made the law difficult to implement, and involved a vast number of governmental agencies that generated bureaucratic congestion and delay. Imposing costly new environmental standards on operating facilities without killing the already-declining uranium industry proved nearly impossible. Historically, states had been the most responsive to environmental concerns, but the law's dominance and the need to develop comprehensive remediation policies virtually eliminated state autonomy in managing uranium hazards. Conversely, reduced state participation weakened the states' traditional enforcement mechanisms at a time when stakeholders had little confidence that the federal agencies would aggressively develop and enforce remedial standards and programs.

Yet, despite its shortcomings, UMTRCA was a milestone in the history of uranium mining in the United States. It was foremost a tangible improvement in the way the nation managed its uranium trash. It was also a political rebuke of the decades-long tradition of bureaucratic delay and denial and marked the first time that the federal government acknowledged its obligation to take the lead in resolving an environmental pollution problem that it had been instrumental in creating. It settled the long-standing intergovernmental legal disputes that had hampered efforts to study and manage uranium industry contamination. Perhaps above all, the law helped to end a governmental culture that placed domestic uranium production ahead of environmental and health concerns. Since the end of World War II, the government had campaigned to convince Americans of the benefits and safety of the nation's atomic pursuits. UMTRCA was a moment when growing public skepticism about atomic goals was validated—when the nation finally admitted that those goals were achieved with baleful environmental and health consequences, and that those costs must be borne by the nation.[48]

CONCLUSION

The human and environmental damage caused by the uranium industry in the mid-twentieth century was rooted not in a failure of science and technology, but in the shortcomings of atomic technocrats and policy makers, most notably within the AEC, in adequately addressing the risks of an industry that they both nurtured and regulated.

The challenges of managing a hazardous industry considered essential to national security and future prosperity, whether it was producing radium for medicine or uranium for weapons and reactors, began with scientists' and policy makers' conceptualization of the resource. Once uranium was defined as necessary for the public good, policy development and implementation placed a premium on its acquisition at the expense of other imperatives. The unprecedented control the federal government exercised over the postwar uranium industry, combined with deferential oversight from Congress and the executive, resulted in a policy climate that discounted both short- and long-term environmental consequences of uranium production. Federal atomic bureaucrats, who had considerable scientific expertise and political sophistication, were motivated by national security and an atomic-powered utopian idyll and were consequently reluctant to acknowledge the environmental costs associated with achieving their goals. Secrecy prevailed within both the military and the civilian atomic bureaucracy, and when environmental concerns conflicted with national security and dreams of national prosperity, disputes were usually resolved in favor of aggressive uranium acquisition regardless of its environmental or health consequences.

Fearing that a labor backlash might imperil vital uranium production if the true extent of the risks were acknowledged, administrative watchdogs downplayed or dismissed environmental concerns raised by health and safety experts, both within and outside the atomic bureaucracy and industry, and remained hostile to criticism that they believed challenged the nation's atomic priorities. Atomic technocrats' actions, and inactions, intended to promote the national interest weighed uranium acquisition against the well-being of laborers, their communities, and regional ecosystems and found resource acquisition to be paramount. The Atomic Energy Commission was also shielded from tort liability for actions it undertook, or failures to regulate hazards, that resulted in civilian injuries or deaths.[49] Consequently, risk analysis of the nation's uranium program that appeared to challenge basic assumptions was discounted by those in the best positions to implement remedial action. Critics were marginalized and hazardous conditions were a hallmark of the uranium industry from the 1940s through the 1970s. Policy makers' zeal to promote atomic energy exposed thousands of uranium industry workers, countless boomtown residents, and the regional environments to conditions that mining engineers, atomic scientists, occupational health and safety experts, the military, and AEC officials themselves knew, or had reason to know, elevated the risk of harm to human and environmental health.

Moreover, while such activities may have been legal under the terms of legislation crafted to incentivize uranium production, they raised serious moral questions. At a historical moment when few outside the atomic community understood the technology and its risks, deception and prevarication in the name of national security that threatened the environment and human life eventually eroded public confidence in the integrity of atomic policy makers. Above all, atomic technocrats were convinced that they knew better than anyone the risks associated with atomic technology, and they convinced the nation that they had fully anticipated and resolved radiation-related environmental dangers. On their optimistic balance sheet, their vision of an atomic future simply outweighed calls for a comprehensive recognition of risk and environmental stewardship. Over the sixty-year history of uranium mining in the West, that approach left a legacy of toxic mine scars, tainted water, remediation costs, dead miners, and legal obligations to former industry workers. Eventually, the nation's uranium program, born in Cold War atomic attitudes that promoted acquisition at the expense of environmental quality, was confronted by a new and growing public appreciation that environmental issues were also issues of equity and social justice, and that the pursuit of national goals had to take into consideration the consequences of those actions for ecological and human health.

NOTES

1. Silvestre Vélez de Escalante, *The Domínguez-Escalante Journal: Their Expedition through Colorado, Utah, Arizona, and New Mexico in 1776* (d. 1792), trans. Fray Angelico Chavez, ed. Ted J. Warner (Salt Lake City:

University of Utah Press, 1995); Herbert E. Bolton, *Pageant in the Wilderness: The Story of the Escalante Expedition to the Interior Basin, 1776* (Salt Lake City: Utah State Historical Society, 1951, 1972). Despite historical and translation errors, *Pageant* remains the most comprehensive account of the Domínguez-Escalante expedition.

2. An example of this historical narrative of uranium mining is presented at the New Mexico Mining Museum, Grants, New Mexico. See, generally, Terre Ryan, *This Ecstatic Nation: The American Landscape and the Aesthetics of Patriotism* (Amherst: University of Massachusetts Press, 2011).

3. Earle R. Caley, "The Earliest Known Use of a Material Containing Uranium," *Isis* 38:3–4 (February 48): 190–93; Herman Fleck, "A Series of Treatises on the Rare Metals," *Proceedings of the Colorado Scientific Society*, vol. 11 (Denver: Colorado Scientific Society, 1916); L.O. Howard, "Development of Our Radium Bearing Ores," *Salt Lake Mining Review* 15 (February 28, 1914); Herman Fleck and William G. Haldane, "A Study of the Uranium and Vanadium Belts of Southern Colorado," *Report of the Colorado State Bureau of Mines for the Years 1905–6* (Denver: Colorado State Bureau of Mines, 1907); Thomas M. McKee, "Early Discovery of Uranium Ore in Colorado," *Colorado Magazine* 32 (July 1955); Larry L. Meyer, "The Time of Great Fever: U-Boom on the Colorado Plateau," *American Heritage* 32 (June–July 1981); United States Vanadium Company, *Mesa Miracle in Colorado, Utah, New Mexico, Arizona* (New York: United States Vanadium Company, Union Carbide and Carbon Corporation, 1952); H.C. Hodge, J.N. Stannard, and J.B. Hursh, eds., *Uranium, Plutonium, and Transplutonic Elements: Handbook of Experimental Toxicology*, vol. 36 (Berlin: Springer-Verlag, 1973), 5–12. Eugène-Melchior Péligot produced the first sample of metallic uranium in 1841. Joseph J. Katz and Eugene Rabinowitch, *The Chemistry of Uranium (Part 1): The Element, Its Binary and Related Compounds* (New York: McGraw-Hill, 1951), 122.

4. Donna Strahan, "Uranium in Glass, Glazes, and Enamels: History, Identification, and Handling," *Studies in Conservation* 46:3 (2001): 181–95; Dan Kline and Ward Lloyd, eds., *The History of Glass* (London: Orbis, 1984), 174–75; Richard B. Moore and Karl L. Kithil, "A Preliminary Report on Uranium, Radium, and Vanadium," *United States Bureau of Mines Bulletin, No. 70* (Washington, DC: U.S. Government Printing Office, 1913), 58; Richard B. Moore, "Uranium and Vanadium," in *The Mineral Industry: Its Statistics, Technology, and Trade during 1920*, G.A. Rouch (New York: McGraw-Hill, 1921), 708; McKee, "Early Discovery of Uranium Ore in Colorado," 192; Francis L. Pittman, "The Direct Production of Uranium Steel" (MS thesis, Colorado School of Mines, 1914), 22; Carrington H. Bolton, "Index to the Literature of Uranium, 1789–1885," *Annual Report of the Board of Regents of the Smithsonian Institution, Part I* (Washington DC: U.S. Government Printing Office, 1886), 922; *Salt Lake Mining Review* (February 28, 1914), 14; J. Baxeres de Alzugaray, "Manufacture and Metallurgy of Ferro-vanadium, *Mining World* (June 24, 1905), 659.

5. United States Geological Survey, *Mineral Resources of the United States* (Washington, DC, 1901–11); Don Sorensen, "Wonder Mineral: Utah's Uranium," *Utah Historical Quarterly* 31 (Summer 1963); Moore and Kithil, "A

Preliminary Report on Uranium, Radium, and Vanadium"; Fleck and Haldane, "Study of Uranium and Vanadium Belts of Southern Colorado."

6. "The Biological Effects of Radium," *Science* 33 (June 30, 1911): 1001–5; Carroll Chase, "American Literature on Radium Therapy Prior to 1906," *American Journal of Roentgenology and Radium Therapy* 8 (1921): 766–67. See, generally, Rober Abbe, "Subtle Power of Radium," *Transactions of the American Surgical Association* 22 (1904); Louis Wickham and Paul Degrais, "Radium: Its Uses in Cancer and Other Diseases," *Contemporary* 8 (August 1910): 174–88; Emile F. Krapf, "Recent Investigations on the Use of Radium for Malignant Diseases," *Radium* 1 (May 1913): 10–13.

7. Kathleen Bruyn, *Uranium Country* (Boulder: University of Colorado Press, 1955), 42–43, 57.

8. Fleck, "A Series of Treatises on the Rare Metals," 174–75; U.S. Congress, House, Committee on Mines and Mining, *Radium Hearings on H.J. Res. 185 and 186*, 63d Cong., 2d sess., January 19–28, 1914, 166; Charles L. Parsons, "Our Radium Resources," *Science* 38 (October 31, 1931), 617.

9. Bruyn, *Uranium Country*, 41–59; Waren Bleeker, "Private or Governmental Radium Production," *Engineering Magazine* 49 (April 1915): 102–3; Bleeker, "The Production of Radium in Colorado," *Science* 42 (August 6, 1915): 184–85; Bleeker "On Extraction of Radium by the U.S. Bureau of Mines," *Journal of Industrial and Engineering Chemistry* 8 (May 1916): 469–73; Charles H. Viol, "The Production of Radium," *Science* 43 (June 2, 1916): 778–79; Viol, "Radium Production," *Science* 49 (March 7, 1919): 227–28; Viol, "Radium Production," *Science* 49 (June 13, 1919): 564–66; Herbert T. Wade, "Extracting Radium from American Ores," *Scientific American* 114 (February 19, 1916), 194–95.

10. John S. MacArthur, "The Radium Industry and Reconstruction," *Engineering and Mining Journal* 107 (April 5, 1919): 605–6; A. T. Parsons, "Radium, with Special Reference to Luminous Paint," *Journal of the Oil and Colour Chemists' Association* 12 (January 1929): 3; James E. Lounsbury, "Famous Pittsburgh Industries: The Standard Chemical Company of Pittsburgh, Pa.," *Crucible* 22 (June 1938), 134; U.S. Bureau of Mines, *Annual Report to the Secretary of the Interior for Fiscal Year 1918* (Washington DC: U.S. Government Printing Office, 1919), 78; Charles H. Viol and Glenn D. Krammer, "The Application of Radium in Warfare," *Transactions of the American Electrochemical Society* 32 (1918): 381–90.

11. Richard B. Moore, "Radium," in Rouch, *The Mineral Industry*, 615–19; Frank L. Hess, "Radium, Uranium and Vanadium," in Rouch, *The Mineral Industry*, 601; Viol, "Radium Production," (March 7, 1919); MacArthur, "The Radium Industry and Reconstruction"; H.E. Bishop, "The Present Situation in the Radium Industry," *Science* 57 (March 23, 1923): 341–45; Bishop, "Radium Ore in Africa," *Literary Digest* 76 (January 13, 1923), 23; Camille Matignon, "The Manufacture of Radium," *Annual Report of the Board of Regents of the Smithsonian Institution, 1925* (Washington, DC: U.S. Government Printing Office, 1926), 233.

12. Jonathan E. Helmreich, *Gathering Rare Ores: The Diplomacy of Uranium Acquisition, 1943–1954* (Princeton: Princeton University Press, 1986).

13. James R. Newman, "The Atomic Energy Industry: An Experiment in Hybridization," *Yale Law Journal* 60 (December 1951); Harold P. Green and Alan Rosenthal, *Government of the Atom* (New York: Atherton Press, 1963); Frank G. Dawson, *Nuclear Power: Development and Management of a Technology* (Seattle: University of Washington Press, 1976); Mark Hertsgaard, *Nuclear, Inc.: The Men and Money behind the Nuclear Power Industry* (New York: Pantheon Books, 1983); Richard S. Lewis, *The Nuclear Power Rebellion* (New York: Viking Press, 1972); Philip Mullenbach, *Civilian Nuclear Power* (New York: Twentieth Century Fund, 1963); Morgan Thomas, *Atomic Energy and Congress* (Ann Arbor: University of Michigan Press, 1956); James M. Jasper, "Nuclear Policy as Projection: How Policy Choices Can Create Their Own Justification," in *Governing the Atom: The Politics of Risk,* ed. John Byrne and Steven M. Hoffman (New Brunswick, NJ: Transaction, 1996); Gerald H. Clarfield and William M. Wiecek, *Nuclear America: Military and Civilian Nuclear Power in the United States, 1940–1980* (New York: Harper and Row, 1984); Richard O. Niehoff, "Organization and Administration of the United States Atomic Energy Commission," *Public Administration Review* 8 (May 1948); United States Congress, Joint Committee on Atomic Energy, *Atomic Power and Private Enterprise: Hearings before the Joint Committee on Atomic Energy,* 81st Cong., 1st sess., 1949; United States Congress, Joint Committee on Atomic Energy, *Atomic Power and Private Enterprise* (Joint Committee Print, 1952); United States Congress, Joint Committee on Atomic Energy, *Hearings Before the Joint Committee on Atomic Energy on Atomic Power Development and Private Enterprise,* 83rd Cong., 1st sess., 1953.

14. Meyer, "The Time of Great Fever"; George Dannenbaum, *Boom to Bust: Remembrances of the Grants, New Mexico, Uranium Boom* (Albuquerque: Creative Designs, 1994); "Uranium Grows Up—Big Business Now," *U.S. News and World Report* (April 6, 1956); "History's Greatest Metal Hunt," *Life* 38 (May 23, 1955); "Colorado Plateau Uranium Population Doubles in 2 Years," *Denver Post* (February 6, 1955); Michael A. Amundson, *Yellowcake Towns: Uranium Mining in the American West* (Boulder: University Press of Colorado, 2002); Stephen I. Schwartz, *Atomic Audit: The Costs and Consequences of U.S. Nuclear Weapons since 1940* (Washington, DC: Brookings Institution Press, 1998); Arthur R. Gomez, *Quest for the Golden Circle: The Four Corners and the Metropolitan West, 1945–1970* (Albuquerque: University of New Mexico Press, 1994); Raymond W. Taylor and Samuel W. Taylor, *Uranium Fever, or No Talk under $1 Million* (New York: Macmillan, 1970); Perrin Stryker, "The Great Uranium Rush," *Fortune* (August, 1954); Raye C. Ringholz, *Uranium Frenzy: Boom and Bust on the Colorado Plateau* (Albuquerque: University of New Mexico Press, 1991); Maxine Newell, *Charlie Steen's Vi Vida* (Moab, UT: Moab's Printing Place, 1992); Jesse C. Johnson, "The Romance of Uranium Mining," *Science Digest* 40 (September 1956); Burt Meyers, "Uranium Jackpot," *Engineering and Mining Journal* 154 (September 1953); Elizabeth Pope, "The Richest Town in the U.S.A.," *McCall's* (December 1956).

15. Howard Ball, *Cancer Factories: America's Tragic Quest for Uranium Self-Sufficiency* (Westport, CT: Greenwood Press, 1993), 37.

16. Environmental Protection Agency, Office of Radiation and Indoor Air, Radiation Protection Division (6608J), *Uranium Location Database Compilation,* EPA 402-R-05–009 (August 2006).

17. See, generally, Homer Aschmann, "The Natural History of a Mine," *Economic Geography* 46:2 (April 1970): 172–89.

18. Ball, *Cancer Factories,* 37.

19. Amundson, *Yellowcake Towns;* Leonard J. Arrington, *Great Basin Kingdom: An Economic History of the Latter Day Saints, 1830–1900* (Champaign-Urbana: University of Illinois Press, 2004); James H. McClintock, *Mormon Settlement in Arizona* (Whitefish, MT: Kesslinger, 2010); William S. Abruzzi, *Dam That River* (Lanham, MD: University Press of America, 1993); Richard H. Jackson, ed., *The Mormon Role in the Settlement of the West* (Provo, UT: Brigham Young University Press, 1980); William S. Abruzzi, "Ecology, Resource Redistribution, and Mormon Settlement in Northeastern Arizona," *American Anthropologist,* New Series, 91:3 (September 1989): 642–55; D.W. Meinig, "The Mormon Culture Region: Strategies and Patterns in the Geography of the American West, 1847–1964," *Annals of the Association of American Geographers* 55:2 (June 1965): 191–220; Milton R. Hunter, "The Mormons and the Colorado River," *American Historical Review* 44:3 (April 1939): 549–55; Ruth M. Underhill, *The Navajos* (Norman: University of Oklahoma Press, 1983); Clyde Kluckhohn and Dorothea Leighton, *The Navajo* (Cambridge, MA: Harvard University Press, 1992); Laura Gilpin, *The Enduring Navajo* (Austin: University of Texas Press, 1987); John U. Terrell, *The Navajos: The Past and Present of a Great People* (New York: HarperCollins Perennial Library, 2000).

20. Robert N. Proctor, *Cancer Wars: How Politics Shapes What We Know and What We Don't Know About Cancer* (New York: Basic Books, 1995), 186.

21. Ball, *Cancer Factories,* 36. See, for example, Sigismund Peller, "Lung Cancer among Mine Workers in Joachimsthal," *Human Biology* 11 (1939): 130–43; William C. Hueper, *Occupational and Environmental Cancer of the Respiratory System* (Springfield, IL: Charles C. Thomas, 1942), 435–56; William C. Hueper, *Occupational Tumors and Allied Diseases* (Springfield, IL: Charles C. Thomas, 1942); Egon Lorenz, "Radioactivity and Lung Cancer: A Critical Review of Lung Cancer in the Miners of Schneeberg and Joachimsthal," *Journal of the National Cancer Institute* 5 (August 1944): 1–15; E. Cook, "Ionizing Radiation," in *Environment,* ed. W.W. Murdock (Sunderland, MA: Sinauer Associates, 1975), 304; J. Newell Stannard, *Radioactivity and Health: A History* (Oak Ridge, TN: Office of Scientific and Technical Information, 1988), 131–32; *Radiation Exposure of Uranium Miners, Part One: Hearings before the Subcommittee on Research, Development, and Radiation of the Joint Committee on Atomic Energy,* 90th Cong., 1st Sess. (1967); Duncan Holaday, Chief, Occupational Health Field Station, Public Health Service, "Employee Radiation Hazards and Workmen's Compensation," Joint Committee on Atomic Energy, 86th Cong., 1st Sess. (1959), 190.

22. Merril Eisenbud, *An Environmental Odyssey* (Seattle: University of Washington Press, 1990); *Health Impact of Low-Level Radiation: Joint Hearing before the Subcommittee on Health and Scientific Research of the Senate Committee on Labor and Human Resources and the Senate Committee on the Judiciary,* 96th Cong., 1st Sess. (1979), 40–41; Ball, *Cancer Factories,* 47.

23. *Sylvia Barnson v. United States*, 630 F. Supp. 418 (1985), 420–21; Proctor, *Cancer Wars*, 44; Wilhelm C. Hueper, "Adventures of a Physician in Occupational Cancer: A Medical Cassandra's Tale" (1976), unpublished autobiography, Hueper Papers, Series I, National Library of Medicine, National Institutes of Health, Bethesda, MD; *John H. Begay v. United States*, 591 F. Supp. 991 (D. AZ, 1984).

24. Duncan Holaday, *Federal Radiation Council, Preliminary Staff Report, No. 8: Radiation Exposure of Miners, Part One* (Washington, DC: Joint Committee on Atomic Energy, 1967), 89.

25. Ball, *Cancer Factories*, 52–53.

26. *Begay v. United States*; *Sylvia Barnson et. al v. United States*, 531 F. Supp. 614 (1982); *Barnson v. United States* (1985). *Begay* was the Navajo miners' litigation against the federal government and mining companies addressing conditions in uranium mines; the *Barnson* cases dealt with non-Indian miners.

27. Pub. L. 101–426, 4 STAT. 920; Public Law 101–426, 101st Congress, October 15, 1990. The act created three payment categories: $50,000 to Nevada Test Site "downwinders"; $75,000 for workers involved with atmospheric nuclear weapons testing; and $100,000 for uranium miners, millers, and ore transporters. As of the spring of 2013, 27,246 claims had been approved, and 10,348 denied, for total award payments of $1.8 billion. U.S. Department of Justice, Civil Division, "Radiation Exposure Compensation System Claims to Date: Summary of Claims Received by 5/28/2013; All Claims," http://www.justice.gov/civil/omp/omi/Tre_SysClaimsToDateSum.pdf.

28. *Begay v. United States*, 1013; Ball, *Cancer Factories*, 65–94.

29. The most serious containment dam failure occurred at the Kerr-McGee uranium mill at Shiprock, New Mexico, August 22–23, 1960, which released as much as 780,000 gallons of radioactive and toxic organic compounds into the San Juan River. The spill boosted radiation levels to about twenty times what the USPHS deemed permissible for drinking water. Further investigations of the riverbed revealed that the high radiation levels were not attributable to the spill alone, but reflected long-term seepage from the tailings impoundment ponds. At the time, thousands of people relied on the San Juan River for agricultural and domestic water. Other dam failures occurred at the Union Carbide mill at Green River, Utah, August 19, 1959; another Union Carbide mill at Maybell, Colorado, December 6, 1961; the Mines Development mill at Edgemont, South Dakota, June 11, 1962; the Climax mill at Grand Junction, Colorado, July 2, 1967; the Petrotomics mill, Shirley Basin, Wyoming, February 16, 1971; the Western Nuclear mill, Jeffrey City, Wyoming, March 23, 1971; another Kerr-McGee mill at Churchrock, New Mexico, April 1, 1976; the Homestake mill at Milan, New Mexico, February 1, 1977; the United Nuclear mill at Churchrock, New Mexico, July 16, 1979. U.S Department of Health, Education, and Welfare, Public Health Service (Region VIII, Denver, Colorado), *Shiprock, New Mexico, Uranium Mill Accident of August 22, 1960*, Colorado River Basin Water Quality Control Project (January 1963); U.S. Nuclear Regulatory Commission, *Regulatory Guide 3.111.1, Rev. 1: Operation Inspection and Surveillance of Embankment Retention Systems for Uranium Mill Tailings* (October 1980).

30. Robert C. Merritt, *The Extractive Metallurgy of Uranium* (Golden: Colorado School of Mines Research Institute, 1971); Katz and Rabinowitch, *The Chemistry of Uranium (Part 1)*, 111–32; Thomas C. Hollocher and James J. MacKenzie, "Radiation Hazards Associated with Uranium Mill Operations," in Union of Concerned Scientists, *The Nuclear Fuel Cycle* (Cambridge, MA: MIT Press, 1975); U.S. Congress, Subcommittee on Air and Water Pollution of the Senate Committee on Public Works, *Radioactive Water Pollution of the Colorado River Basin*, 89th Cong., 2nd Sess., May 6, 1966.

31. "Transcript of Conference on Interstate Pollution of the Animas River, Colorado–New Mexico," Santa Fe, New Mexico, April 29, 1958, New Mexico Environment Department; R.F. Poston, "Uranium Milling Waste Studies— Colorado and Utah" (unpublished memorandum), Western Gulf and Colorado Basin Office, U.S. Public Health Service (USPHS) (March 19, 1951); E.C. Tsivoglou, A.F. Bartsch, D.L. Rushing, and D.A. Holiday, *Report of Survey of Contamination of Surface Waters by Uranium Recovery Plants* (Cincinnati, OH: USPHS, Robert A. Taft Sanitary Engineering Center, 1956).

32. Colorado Department of Public Health, *Uranium Wastes and Colorado's Environment*, 2nd ed. (Denver: Colorado Department of Public Health, 1971); USPHS Technical Report W62–17 (Cincinnati, OH: Robert A. Taft Sanitary Engineering Center, 1962). Aleck Alexander to Division of Sanitary Engineering Services et al., memorandum, February 20, 1958, Box 4, Accession No. 90–62A-672, U.S. Public Health Service; and Murry Stein to James Harlan, June 22, 1959, Box 1, Accession No. 90–62A-121, U.S. Public Health Service, National Archives and Record Administration, Washington, D.C. U.S. Congress, Subcommittee on Air and Water Pollution of the Senate Committee on Public Works, *Radioactive Water Pollution of the Colorado River Basin*; Hollocher and MacKenzie, "Radiation Hazards Associated with Uranium Mill Operations."

33. Joseph F. Hennessey to Glenn Seaborg et al., memorandum, June 9, 1966, Box 184, U.S. Atomic Energy Commission, Glenn Seaborg Collection, U.S. Department of Energy Archives, Germantown, MD.

34. Hollocher and MacKenzie, "Radiation Hazards Associated with Uranium Mill Operations"; H. Peter Metzger, *The Atomic Establishment* (New York: Simon and Schuster, 1972); Tsivoglou et al., *Report of Survey of Contamination of Surface Waters by Uranium Recovery Plants*; Colorado Department of Public Health, *Uranium Wastes and Colorado's Environment*; "Transcript of Conference on Interstate Pollution of the Animas River, Colorado"; W.B. Harris et al., *Environmental Hazards Associated with the Milling of Uranium Ore, HASL-40* (New York: U.S. Atomic Energy Commission, Health and Safety Laboratory, Operations Office, June 1958).

35. U.S. Congress, Subcommittee on Air and Water Pollution of the Senate Committee on Public Works, *Radioactive Water Pollution of the Colorado River Basin*.

36. Ibid.; Colorado Department of Public Health, "Report on Control of Uranium Mill Tailings," Occupational and Radiological Health Section, October 28, 1966, Box 9937, Colorado State Archives, Denver; R.E. Hollingsworth

and Harold L. Price to Glenn Seaborg, memorandum, December 2, 1965 (AEC-R 18/32), Accession No. 9210120297, Nuclear Regulatory Commission Papers, National Archives and Records Administration, Washington, DC.

37. U.S. Department of Health, Education, and Welfare, *Disposition and Control of Uranium Mill Tailings Piles in the Colorado River Basin* (Denver: Federal Water Pollution Control Administration, Region VII, March 1966).

38. See, for example, "West Slope Studying Pollution by Uranium," *(Denver) Rocky Mountain News,* December 14, 1965; "Uranium Mystery on the Colorado Basin," *New Republic* 154 (March 5, 1966): 9; U.S. Congress, Subcommittee on Air and Water Pollution of the Senate Committee on Public Works, *Radioactive Water Pollution of the Colorado River Basin.*

39. Betty L. Perkins, *An Overview of the New Mexico Uranium Industry* (Santa Fe: New Mexico Energy and Minerals Department, 1979); Ken Silver, "The Yellowed Archives of Yellowcake," *Public Health Reports* 111 (March–April 1996); Richard Waxweiler et al., "Mortality Patterns among a Retrospective Cohort of Uranium Mill Workers," *Proceedings of the Sixteenth Midyear Topical Meeting of the Health Physics Society* (Albuquerque, 1983); Ringholz, *Uranium Frenzy*; U.S. Public Health Service, *Evaluation of Radon 222 near Uranium Tailings Piles,* DER 69-1 (Rockville, MD: U.S. Dept. of Health, Education, and Welfare, 1969).

40. U.S. Congress, Subcommittee on Air and Water Pollution of the Senate Committee on Public Works, *Radioactive Water Pollution in the Colorado River Basin,* 20.

41. U.S. Department of Health, Education, and Welfare, *Disposition and Control of Uranium Mill Tailings Piles in the Colorado River Basin,* 8.

42. Congress, House, Subcommittee on Energy and the Environment of the Committee on Interior and Insular Affairs, *Uranium Mill Tailings Control: H.R. 13382, H.R. 12938, H.R. 12535 and H.R. 13049,* 95th Cong., 2nd sess., June 26–27, July 10, 17, 1978; "Landscaping (Industrial Strength)," *Nuclear Energy* (2nd Quarter, 1993), 9–11; J.M. Costello et al., "A Review of the Environmental Impact of Mining and Milling of Radioactive Ores, Upgrading Processes, and Fabrication of Nuclear Fuels," in *Nuclear Energy and the Environment,* ed. Essam E. El-Hinnawi (Oxford: Pergamon Press, 1980); "Management of Inactive Uranium Mill Tailings," *Journal of Environmental Engineering* 112 (June 1986); D. Lush et al., "An Assessment of the Long Term Interaction of Uranium Tailings with the Natural Environment," *Proceedings of the Seminar on Management, Stabilization, and Environmental Impact of Uranium Mill Tailings* (Albuquerque: OCED Nuclear Energy Agency, July 1978).

43. U.S. Congress, Joint Committee on Atomic Energy, Subcommittee on Raw Materials, *Use of Uranium Mill Tailings for Construction Purposes,* 92nd Cong., 1st sess., October 28–29, 1971; Colorado Department of Public Health, *Uranium Wastes and Colorado's Environment*; Colorado Department of Public Health, "Report on Control of Uranium Mill Tailings"; Robert N. Snelling and Robert D. Seik, "Evaluation of Radon Film Badge" (Southwestern Radiological Health Laboratories, USPHS, and Colorado Department of Public Health, April 1968); Metzger, *The Atomic Establishment,* 171; Frank E. McGinley to

Elton A. Youngberg, memorandum, November 16, 1971, Accession No. 9210120418, U.S. Nuclear Regulatory Commission, Public Documents Reading Room, Washington, DC.

44. *Tailings for Construction Hearing*; Colorado Department of Public Health, *Uranium Wastes and Colorado's Environment*; U.S. Public Health Service, Department of Health Education and Welfare, *Evaluation of Radon 222 near Uranium Tailings Piles*; H. Peter Metzger, "Dear Sir: Your House Is Built on Radioactive Uranium Waste," *New York Times Magazine*, October 31, 1971; Stephen H. Greenleigh to Legal Files, memorandum, December 19, 1970, Nuclear Regulatory Commission, Accession No. 9210120373, Public Documents Reading Room, Washington, DC. Other communities with significant tailings contamination include Durango, Colorado; Salt Lake City, Utah; Tuba City, New Mexico; and Canonsburg, Pennsylvania (the site of a radium-era uranium reduction mill that later also processed ore for the AEC). Ellen Wilson, "Some Like It Hot," *Environmental Action* 17 (November–December 1985); Ralph Haurwitz, "Families Cry for Radiation Park Action," *Pittsburgh Press*, October 26, 1980.

45. *(Denver) Rocky Mountain News* (September 1, 1966); *Washington Daily News* (September 1, 1966); J. Samuel Walker, *Containing the Atom: Nuclear Regulation in a Changing Environment* (Berkeley: University of California Press, 1992), 262; Robert Saile, "Radon Gas Found in Junction Homes," *Denver Post* (December 19, 1969); *(Grand Junction) Daily Sentinel* (January 13, March 25, 26, 29, 1970); Roger Rapoport, "The Trouble with 90.5 Million Tons of Radioactive Tailings," *Los Angeles Times West Magazine* (April 12, 1970); N. Wood, "America's Most Radioactive City," *McCalls* 97 (September 1970); Tom Rees, "Committee Recommends Uranium Tailings Action," *(Denver) Rocky Mountain News* (September 22, 1971); "Radon? Sure. So What Else Is New? Ask the Folks in Grand Junction," *Nuclear Industry* 18 (October–November 1971).

46. P.L. 92–314, Title II, 86 Stat. 222 (June 16, 1972); U.S Comptroller General, *Report to Congress: Controlling the Radiation Hazard from Uranium Mill Tailings* (Washington, DC: General Accounting Office, May 1975); *Denver Post* (September 6, 1971); Renee Baruch and Madonna Ghandi, "Radioactive Waste: A Failure in Governmental Regulation," *Albany Law Review* 37 (1972): 97–134.

47. U.S. Congress, Joint Committee on Atomic Energy, Subcommittee on Raw Materials, *S. 2566 and H.R. 11378: Uranium Mill Tailings in the State of Utah*, 93rd Cong., 2nd Sess., March 12, 1974; U.S. Congress, House, Committee on Interior and Insular Affairs, Subcommittee on Energy and the Environment, *Uranium Mill Tailings Control: H.R. 13382, H.R. 12938, H.R. 12535, H.R. 13049*; Congressional Quarterly, *Weekly Report* 36 (August 19, 1978); P.L. No. 95–604 (November 8, 1978), 92 Stat. 3021 et. seq., 42.

48. Jay B. Bell and Richard E. Turley, *The Need for Remedial Action and Federal Participation in the Case of the Abandoned Vitro Uranium Mill Tailings Located in Salt Lake County, Utah* (Salt Lake City: Office of the State Science Advisor, March 7, 1974), Utah State Records, Salt Lake City; U.S. Congress, JCAE, Subcommittee on Legislation, ERDA, *Authorizing Legislation, Fiscal*

Year 1976, 94th Cong., 1st Sess., February 18 and 27, 1975; U.S. Congress, JCAE, Subcommittee on Raw Materials, *S. 2566 and H.R. 11378*; U.S. Congress, House, Committee on Interior and Insular Affairs, Subcommittee on Energy and the Environment, *Uranium Mill Tailings Control: H.R. 13382, H.R. 12938, H.R 13049*; U.S. Department of Energy, *Inactive Uranium Mill Tailing Remedial Action Program: Salt Lake City (Vitro) Site Offsite Decontamination Program Survey; Compilation of Candidate Properties for Remedial Action* (Albuquerque: UMTRCA Project Office, November 18, 1983); H. Peter Metzger, "AEC vs. The Public: The Case of the Uranium Tailings," *Science News* 106 (July 13, 1974); "EPA Finds 'Intolerable' Radioactivity in Drinking Water near Uranium Mines," *Environmental Reporter* 6 (August 22, 1974); "Report Says Radon Exposure Major Hazard in Living near Uranium Tailings Deposits," *Environmental Reporter* 7 (May 7, 1976); "NRC Radon Impact Estimates 'Grossly' in Error, Says Oak Ridge Official," *Environmental Reporter* 8 (November 2, 1977); Congressional Quarterly, *Weekly Reports* 36 (August 19, 1978); Chris Shuey, "Bringing Tailings under Control," *Workbook* 10 (1985); Elizabeth V. Scott, "Unfinished Business: The Regulation of Uranium Mining and Milling," *University of Richmond Law Review* 615 (1984); John D. Collins, "Uranium Mine and Mill Tailings Reclamation in Wyoming: Ten Years after the Industry Collapsed," *Land and Water Law Review* 26 (1991); Mary Boaz, "Retroactive Liability for Clean-Up of Hazardous Waste in *Atlas v. United States*: The Nuclear Industry's Failed Attempt to Make the Government Pay," *Journal of Mineral Law and Policy* 6 (1990–91).

49. See Federal Tort Claims Act of 1946. The "discretionary function" exception, one of twelve such exceptions to the act that limits the ability of private citizens to seek redress in federal courts for government action, invalidates "any claim based upon any act or omission of any employee of the Government, exercising due care, in the execution of a statute or regulation, whether or not such statute or regulation be valid, or based on the exercise or performance or the failure to exercise or perform a discretionary function or duty on the part of a federal agency or any employee of the government, whether or not the discretion involved be abused."

CHAPTER 9

A Comparative Case Study of Uranium Mine and Mill Tailings Regulation in Canada and the United States

ROBYNNE MELLOR

The U.S. regulatory system in the atomic age, according to prevailing scholarly opinion, was a complete failure in the uranium mines. Unlike other countries, the United States let companies pollute land, water, and air indiscriminately in a drive to achieve Cold War goals. Moreover, there is often the implication—or the blunt statement—that the U.S. government, in an almost conspiratorial manner, sacrificed the lives of miners.[1]

In Canada, the situation was quite different. Far fewer miners contracted serious cases of lung cancer, and there was little or no widespread and vocal civil backlash against the mining companies or the government. From this set of circumstances, one might suppose that the better health and safety record in Canada was due to better regulatory oversight. This supposition, however, is wrong. In fact, Canada had an arguably worse regulatory record than the United States.

When addressing the history of regulation of mines and mining, environmental justice is a major issue, both in U.S. historiography and in the little scholarship about uranium mining in Canada. Scholars of U.S. history of uranium mining focus specifically on the injustice done to Native Americans in the American Southwest who worked on and lived near the uranium mines. Thus the argument that the U.S. government was exceptional in its regulatory neglect has an additional layer of meaning: discrimination against Native Americans.[2] This chapter does not deny the injustice that working-class miners and Native Americans—especially the Navajo Nation—endured, but will argue rather that it was unexceptional. I seek to show that the environmental justice issues related to uranium mining were widespread and not contained within U.S. borders.

This chapter uses two case studies—one in Elliot Lake, Ontario, and the other in Grants, New Mexico—to compare the regulatory oversight of uranium mining and milling in Canada and the United States. I argue that the Canadian regulatory system was not better than its counterpart in the United States, and in some respects was worse.

This argument, however, presents a paradox: Canada had a regulatory system just as bad—if not worse—than that of the United States, but the United States had a greater public backlash against, and more health problems associated with, its uranium-mining program. Why? At the end of this essay, I present a few hypotheses that might resolve the paradox.

THE BEGINNING OF THE ATOMIC AGE
AND INITIAL PROCUREMENT

Following World War II, a new conflict emerged between the United States and the Soviet Union in which uranium played an important role. Their Cold War competition prompted the two countries to stockpile ever-growing numbers of nuclear weapons. Thus, uranium ore became essential to U.S. national security strategy from the late 1940s. The weapons race also caused profound feelings of insecurity among Canadian policy makers. Nuclear weapons posed a threat not just to the United States, Canadian policy makers concluded, but to North America as a whole.[3]

In 1946, both the American and Canadian governments created similar federal entities to dictate nuclear policy in the postwar world. The Atomic Energy Act created the U.S. Atomic Energy Commission (AEC; see chapter 8, this volume), and in Canada, the Atomic Energy Control Act established the Atomic Energy Control Board (AECB). Both acts addressed uranium mining in a general way, with only broad statements as to how the governments would regulate raw materials.[4] The generality with which the documents treated the uranium mines led to similar regulatory problems in Canada and the United States.

Due to Cold War pressures, a single concern dominated the newly created AEC in its early years: procurement. The uranium industry grew exuberantly in the 1950s. (For more detail on procurement programs within the United States, see chapter 8, this volume.) But between 1955 and 1958, the U.S. situation changed dramatically and unexpectedly; uranium ore poured into U.S. government stockpiles, with the most significant sources located in the Grants region of New Mexico. Similarly, the Elliot Lake region of Ontario began to expand uranium production in the same period, becoming Canada's most important producing region. The common features of the uranium procurement programs would impact people and landscapes in Ontario and New Mexico for decades to come, but for some reason, it would impact them differently. Below I show that difference was not due to better Canadian regulation.

ELLIOT LAKE, ONTARIO

Elliot Lake is a small town some twenty miles north of the northern end of Lake Huron. It lies on the Canadian Shield's thin soils, amid rolling hills and a multitude of lakes. Before 1955, the land was sparsely populated because it was not conducive to agriculture. It remained part of the traditional fishing and trapping grounds of the Anishinaabe, or Serpent River First Nation, population.

Most of the uranium mines in the Elliot Lake mining region are essentially located beneath Quirke Lake, the region's largest lake. The uranium ore is uniform and continuous. Although the quality of uranium in the region is not exceptionally rich (usually around 0.20% uranium oxide) its formation in long veins makes it amenable to mass mechanized removal.[5] Most mines in the postwar period operated mills to refine the ore through acid treatment. At the height of operation in the late 1950s, there were ten mills, with approximately each mine running its own mill. A secondary industry, a sulphuric acid plant, was established to support acid leaching in the mills.[6] Mines, mills, and acid plants all emptied their waste into the Serpent River Watershed, the complex waterway system in which the Elliot Lake mining region is located.

By 1959, Elliot Lake mines produced 74 percent of Canada's total milled uranium oxide, or yellowcake—about 11,000 tons. To produce 11,000 tons of yellowcake, the companies working in the Elliot Lake region also produced millions of tons of waste. In 1959, the *daily* milling capacity for the region was 34,000 tons.[7] The 99.8 percent of the ore not converted to yellowcake was left to seep into the many lakes and streams surrounding the mines, raising the water's acidity as well as releasing harmful radioactivity.

Although it flowed silently into local water supplies in enormous amounts, radioactive waste was not a concern for those who lived in

Elliot Lake, nor did it alarm the Canadian government. A boom mentality seized people at both the local and federal levels. Pamphlets and articles from the late 1950s promoted the brand-new mining town as a uranium utopia in harmony with nature and peopled with "Atom Age pioneers." The rush for uranium was comparable to the Klondike gold rush, but at the same time thoroughly modern and unlike anything that came before, according to its proponents.[8] Foreign Minister Lester Pearson compared the importance of uranium miners to Canada's Cold War pilots in Europe, stating they both had "a direct and vital relationship to security against aggression." Elliot Lake, he claimed, was responsible for the safety of the free world.[9]

Aware of the importance of Elliot Lake's uranium industry to global security, Canadian authorities wanted nothing to slow it. The Serpent River First Nation had no say in the development of the land, and those among them who worked on the region's development and in the mines faced daily discrimination.[10] Pearson noted that an "old Indian" lamented: "'When the white man came here first, he killed our animals and sold the furs. Then he cut down our trees and sent them off as lumber. And now—never satisfied—he is taking away our rocks.'" And what was Pearson's reply to this painful recounting of First Nations history? "He [the white man] is certainly doing his best."[11] The atomic age pioneers began to take not only the rocks but also the lakes, streams, and fields, effectively rendering much of it unusable for hunting and fishing for at least the next few decades, if not millennia.

The industry's impacts on the landscape were far from the minds of officials, however. Uranium euphoria resounded in the reports of the Canadian Atomic Energy Control Board during the 1950s. Records from 1954 to 1958 reveal unchecked optimism about the increased production in the mines. In 1958, the AECB reported that Canada had exported ore worth $131 million, or about one-third of the AECB's budget for an entire year in 1957.[12] None of the AECB budget went to regulating mines, and there was no Canadian counterpart to the U.S. Public Health Service studies (about which, see below) conducted in the United States during the 1950s. American officials, even if they were not doing much, were doing more than Canadians.

In the early days at Elliot Lake, according to mining engineer R. W. Thompkins, production overrode all other concerns, including safe working environments for the miners. From the 1950s through the 1970s, in fact, the health and safety conditions within the mines were not the responsibility of the AECB, but rather of contractors, such as Thompkins, whom individual companies, such as Denison Mines in Elliot Lake, hired. The federal government expected the Ontario Department of Mines, which had no experience with radiation, to oversee health and safety concerns and keep companies and engineers accountable.[13]

But lack of regulatory oversight was not the only obstacle preventing companies from operating safe mines. In his memoir, Thompkins notes differences of opinion between physicists and engineers about how best to control what he calls the little-understood problem of radiation. Even though engineers and scientists were aware that radiation could be harmful, the issue of how to regulate and protect against it was still complex. Engineers understood mining, and scientists understood the physics of radiation, but it was difficult to connect the two issues. Application of safety standards was also problematic. Although the International Radiation Protection Committee had established a safe level of radiation for uranium miners in the postwar period, enforcing it was a much more difficult task. Thompkins reports that measurement of radiation levels in the mines was not an effective way of monitoring or preventing exposure, as it did not address the source of radon gas. He concludes that "although staffs did their best to correct bad situations, it can be called, in retrospect, only a fumbling best effort."[14]

Thompkins maintains that despite the lack of help from the government, the companies provided as much funding as needed and that all uranium mines in Elliot Lake during the period "opened with more than double the volumes of air circulating than other mines in Ontario of a comparable size or tonnage."[15] But good ventilation monitoring in one mine did not necessarily indicate proper ventilation in another. If a certain company had a stubborn manager or an incompetent engineer, a system could potentially be installed that increased, rather than decreased, radiation risks within the mine. According to Thompkins, that was the case with the Stanrock mine, where the extant system of radiation gravely compromised the health of the miners.[16] Good ventilation had nothing to do with superior government regulation in Canada. Companies and engineers, if they could and if they wished, provided it.

If no one fully understood issues concerning the underground mining environment, there was even less understanding about the effects of tailings on the surrounding landscape. Correspondence from the Pronto mine in Elliot Lake shows that prior to 1954, companies could use any land they owned as they saw fit without any government oversight. In addition, the companies and officials viewed the uranium mining waste as no different from tailings at other, nonuranium mines. The mine operators considered the acidity of the tailings to be within acceptable levels, and seemed to give no thought to the unique problems of radioactive contamination.[17] In 1954, the province of Ontario did create a pollution authority to monitor dumping of contaminants into lakes and streams, but the only concerns it raised for Pronto officials were pH and iron levels in the water.[18] Monitoring the watersheds with an eye to radiation—though still without governmental guidelines—began in 1958. The reason for this beginning is not entirely clear, but it is certain that it did not come from federal prompting.[19]

At this time, officials had little idea of how the process of uranium mining affected the environment. The waste produced in the mining and milling of uranium in Elliot Lake was similar to the situation in the United States (see chapter 8, this volume). It had three pathways by which it affected the surrounding environment: solid waste rock and tailings piles, liquid effluents, and gaseous emissions. Although the contamination pathways are similar to those in other forms of mining, the presence of radiation is a unique and long-lasting problem. Radon daughters are the biggest airborne problem associated with uranium extraction, and they can be transferred into the surrounding environment from all of the stages of processing: extraction, milling, and tailings storage. Liquid effluents in Elliot Lake contained progeny radionuclides as well as sulphuric acid and sulphates from the leaching process. Radionuclides have the potential to move up through a food chain, potentially accumulating in harmful amounts in larger organisms.[20] During the 1950s, officials and companies at Elliot Lake did not study or attempt to ameliorate any of these problems.

Just as Canada began to settle into its lucrative role as one of the world's largest uranium producers, the U.S. buying situation shifted. In 1959, Washington pulled out of its contracts due to a glut of ore. The 1959–60 AECB report noted that the United States would not buy any uranium after 1966.[21] The Canadian uranium industry was losing its biggest customer.

The uranium bust hit the Elliot Lake region hard. Deliveries of yellowcake to the United States dropped from 15,909 tons in 1959 to 5,630 tons in 1964.[22] By 1964, only four mines remained open in the Elliot Lake region.

Although waste production also decreased, the mines still produced a significant amount of tailings. Even in the slow year of 1964, the mines in Elliot Lake milled 12,000 tons per day.[23] The Industrial Hygiene Branch of the Ontario Department of Health estimated that the accumulated amount of waste tonnage from uranium mining activity in Elliot Lake from 1957, the year mining and milling operations began, to 1964 exceeded 50 million tons.[24]

A year later, in 1965, as the danger of radionuclides became better known, the Ontario Committee of Deputy Ministers—not the AECB—recommended that the pollution of the uranium mining and milling industry in Canada be investigated in detail. Two government organizations responded to this task: the Ontario Water Resources Commission (OWRC) and the Radiation Protection Service, a part of the Ontario Department of Health. Between May 1966 and June 1969, these groups investigated how the mining process affected the bodies of water surrounding the mines.[25] This new interest in radioactive pollution was likely related to increased international discussion in 1959 and 1960 about safe levels of radiation exposure.[26] During this period, the Federal Radiation Council in the United States, the AECB in Canada, and the International Commission on Radiological Protection produced occupational safety standards. Neither Canada nor the United States applied these standards to uranium mines, but they do indicate increased scientific interest and understanding of radionuclides.

When Ontario government officials tested water from the Elliot Lake region, they received a rude surprise. Measurements revealed that some of the water that people used for drinking, washing, and cooking in the region came from bodies with higher radiation measurements than were internationally acceptable. Officials discovered radium-226 levels five or six times the contemporary (unofficial) safe drinking level of 6 picocuries per liter.[27] Officials had admitted earlier that they did not know what was safe versus unsafe in terms of radioactive content. Nonetheless, the OWRC finally released water safety standards relating to radioactivity in 1965. It determined that drinking water is safe only when it has fewer than 10 picocuries of radium-226 per liter, and water is safe to swim in only with under 40 picocuries per liter. The organization was confident that swimming in water with forty times its natural radium-226 content did not present "any hazard to health."[28]

The OWRC's environmental studies in Elliot Lake from the late 1960s to the early 1970s present a bleak picture. By this time, many companies had abandoned their deposits, mines, and wastes. The mines lucky enough to continue operation continued to discharge tailings into the surrounding lakes. One mine, Stanrock, caused the "impairment" of three lakes and was partially responsible for issues with four others, including Quirke Lake.[29] Provincial government studies in the early

1970s revealed radionuclides in many contaminated lakes between 50 and 200 times the normal background levels. The presence of sulphates and ammonia from the uranium reduction process also caused acidity problems. The environment the mines produced caused fish populations to plummet.[30]

In addition to normal waste disposal, dam failures plagued Elliot Lake. In the spring of 1964, two environmental disasters hit Stanrock. In April, a dam collapsed, releasing 9,000 tons of tailings into the water system. In June, the decant structure—a sort of drain meant to remove cleansed water from the rest of the tailings pond—failed, and another 250 tons of tailings spewed into the Serpent River Watershed.[31] These disasters hinted at the problem of poor construction that characterized most dams at the Stanrock mine. A provincial government report found that all dams at Elliot Lake were either leaking or deteriorating.[32]

So what did the OWRC do in response to these alarming results? It concluded it could not undo the past, and so officials decided to minimize future damage as much as possible.[33] The watershed, it seemed, was beyond redemption. The OWRC placed the blame on Stanrock. The AECB—the country's central atomic regulatory body—remained aloof from the issue.[34] There is no evidence the federal government took notice.

Records from the 1960s indicate that official communication of water- and airborne risk to workers, families, and communities living in Elliot Lake and the Serpent River Watershed was nearly non-existent. One OWRC employee stated that the only people concerned with pollution were old, drunk, or delusional.[35]

The 1970s brought profound change. The 1973 and 1979 oil crises prompted energy concerns that proved a boon to nuclear power interests as well as uranium mining. Construction, mining, and drilling resumed, and in one year alone, Denison Mines' profits increased by 56 percent.[36] The press predicted a second life for uranium mining in Elliot Lake.[37] At the same time, two momentous events prompted the federal and provincial governments to introduce relevant environmental and occupational safety legislation.

The first event was the birth of the modern environmental movement. Nuclear, water, and air pollution issues dominated the Canadian environmentalist counterculture, including Canada's ecological crusaders known as Greenpeace.[38] But Canada's environmental movement was not as successful as its U.S. equivalent for several reasons. First, it focused on international as opposed to national problems.[39] Second, the Department of the Environment, a federal body established in the

1970s, left most of the issues of regulation to the provinces. The budgets passed in Parliament did not provide it with sufficient resources, and it experienced problems maintaining leadership. These issues, combined with the pro-business, and pro-mining, tendency of the Canadian federal government, meant that Ottawa did not introduce effective environmental legislation for the mines.[40] Thus, because of the weakness of environmental legislation in Canada, the start of the environmental movement was of little consequence for Elliot Lake in the 1970s.

The second important event was miner protests. There is no evidence that company or government officials informed workers of the potential dangers to their health in the 1950s or 1960s. Although union leaders were aware of the dangers as early as the 1950s, there is no clear indication that they conveyed this information to the workers until the 1970s.[41] The United Steelworkers, which represented the miners, was aware of health risks related to mining as early as the 1950s, but it did not call for a health survey of all 1,500 miners employed in Elliot Lake until 1973. This study showed a higher incidence of silicosis among these miners than in control populations.[42] This was the first study of Canadian uranium workers' health—more than twenty years after the first American study of uranium workers' health. In 1974, Canadian miners held an unofficial strike to protest the lack of health and safety guidelines, provoking a government response in the form of the Royal Commission on the Health and Safety of Workers in Mines, often referred to as the Ham Commission. Though the commission covered all areas of health and safety in all types of mining, it addressed many issues related to the uranium mines.[43]

Again, government intervention prompted little immediate change. Even after the Ham Commission was formed, health and safety regulation remained haphazard. The AECB still left the regulation of occupational health and safety to the Ontario Department of Mines and to the mining companies. Because of the lack of government oversight, mine safety continued to depend on mine managers and engineers. Thompkins, the mining engineer, argues in his memoirs that because of his work at Denison there was a never a "radiation crisis."[44] Despite such efforts, by one account more than one thousand people died from "industrial diseases directly linked to uranium mining."[45] Unfortunately, what these diseases were is unclear, as is their link to uranium mining specifically. This distinction is important, as the government still was not able to control levels of silica dust during the same period, and silicosis was widespread.[46] Strangely, there is no evidence of public backlash against government treatment of Canadian uranium miners and no widespread reports of miners dying from lung cancer.

The Department of Mines released the first government health and safety guidelines in 1967, but radiation levels were not consistently brought below guidelines until 1976.[47] The AECB did not involve itself in radiation control in the mines until 1978, over a decade after the American AEC released its first radiation guidelines for miners. It was in 1978 that the federal regulatory body finally considered uranium miners as federal radiation workers, and thus covered under its health and safety guidelines.[48] The regulatory record—or lack thereof—of the AECB proves that it was not superior regulatory oversight that reduced health and safety risks or addressed environmental issues in the Canadian uranium mines.

GRANTS MINERAL BELT, NEW MEXICO

Just as the Elliot Lake region in Ontario was crucial to the Canadian ore procurement program, so was the Grants Mineral Belt in northwestern New Mexico vital to the U.S. demand for uranium. It produced almost half of the total U.S. uranium in 1975. Between the 1950s and 1970s, mining techniques included underground and open-pit mining, and the mills used either acid-leach or alkaline-leach methods.[49] By the mid-1970s, there were three active and three inactive mills in the region.[50]

In stark contrast to Elliot Lake, there is little surface water in the Grants Mineral Belt. Because of the lack of surface water, groundwater is especially important. The climate in Grants is semiarid to arid and, in springtime, prone to strong winds, which is an important factor when considering the spread of dry uranium tailings.[51] Before uranium mining, the main industries in Grants were farming and ranching, and Navajos largely constituted the population. They used this land for grazing their livestock, and used the valleys for growing vegetables with the help of irrigation.[52]

But in the 1950s, uranium prospectors flocked to New Mexico. Mining and milling quickly dominated the economy of the Grants region, and many secondary industries developed to support the mining industry, including the production of chemicals, construction, and housing. Mining and related activity, however, did not wipe out stock raising and agriculture. Rather, the two land uses coexisted in the Grants Mineral Belt.[53]

The mining and milling system in the Grants region differed from that in the Elliot Lake region. In Elliot Lake, as noted above, the same company often ran both the uranium mine and mill, and almost every mine had its own mill. In the Grants region, the system was more complex. Deposits in the U.S. Southwest were both small and large. And because not all deposits were deep underground, some uranium was cheaper to extract than at Elliot Lake. Independent miners could lease land and equipment from larger companies to extract uranium from smaller deposits or to mine small deposits themselves. These less expensive mining operations were mixed with deeper, more expensive deposits that large mining companies operated. Thus, while many companies operated both mines and mills, there were also independent producers of uranium feeding their ore to the mills.[54]

Between 1953 and 1970, the mills at Grants processed 40,689,945 tons of ore and produced about 80,300 tons of yellowcake.[55] In 1959, the average amount of ore milled per day would have been around 9,000 tons, which was over 60 percent of the entire amount that the AEC purchased that year, but less than one-third of the amount that Elliot Lake produced.[56] The amount of waste in this process was around 99.8 percent.

Although information on the monitoring and control programs of the mills in the Grants region is scarce, pollution studies from better-documented regions in the United States show that there was no concern about, or testing for, radioactive particles during the 1950s. And testing remained haphazard through the 1960s.[57]

Mine safety was usually a state issue, but the states in which uranium mining occurred did not produce any regulatory guidelines during the 1950s. Moreover, the AEC did not believe it had the legal authority to regulate the mines, and so it removed itself from the issue, just as the AECB did in Canada. Although troubling evidence emerged about miner death rates, little action to improve mine safety resulted.[58]

There were a few notable exceptions to the lack of accountability in the United States. Certain governmental bodies conducted important epidemiological work during this period. The state of Colorado and the U.S. Public Health Service (USPHS) began collaborative epidemiological studies of uranium miners and conducted air sampling in uranium mines in 1950. One engineer with the USPHS, Duncan Holaday, called for better control of radon within the mines.[59] Still, federal bodies—including the AEC, the Department of Labor, and the Bureau of Mines—insisted that the health of the miners was the bureaucratic responsibility of the states, and the state governments continued to do nothing.[60]

The regulatory situation in U.S. mines was much more complex than in Canada due to the aforementioned mix of large and small mines. In Grants and other southwestern mining regions, there were many small "doghole" or "gopher hole" mines alongside the larger mines. These small operations did not require much money or time to make a profit, so they often opened and closed quickly without proper regulation. They had little to no ventilation, and miners worked closely with the radioactive ore, blasting, sorting, and carrying it to the mills.[61] Thompkins, whom Kerr-McGee contracted to work on its ventilation in Grants, describes these small mines as very unsafe.[62]

Larger mines were not always better, and Thompkins notes the limitations of the measurement techniques used. Levels of radiation could vary greatly from one part of a mine to another because of the circulating air.[63] As in Canada, companies did not fully understand radiation's behavior or its movements, nor did the institutions tasked with regulating it.

In the early 1960s, radiation joined the everyday concerns of the American public. Fallout controversies related to testing—the evacuation of the Marshallese and the radiation exposure of those aboard the *Lucky Dragon*—caused greater awareness.[64] It was public attention to the fallout controversy, and not the plight of the miners, that prompted U.S. president Dwight Eisenhower to look seriously into occupational safety hazards. In 1959, in response to this controversy, Eisenhower created the Federal Radiation Council (FRC). Comprising the AEC chairman and several cabinet secretaries, the FRC did not begin to examine safety within the mines until 1961.[65] In that year, the council listed radiation regulation for miners as one of its future goals,[66] but nuclear fallout issues continued to supersede problems within mines.[67] Nevertheless, the U.S. government recognized the issue of mining health and safety in a way that had no Canadian equivalent, and acknowledged the problem much earlier than its northern neighbor.

Just as environmental and health regulation began, the above-noted sudden glut of ore caused the AEC to revise its earlier buying program. In fact, it was the Ambrosia Lake mine in the Grants region that caused, and was subsequently deeply affected by, this change. The name Ambrosia Lake is misleading, for the mine is not a lake at all, but rather a massive deposit of uranium located deep underground with a grade of about 0.3 percent uranium oxide. Whereas lakes and streams surrounded the mines at Elliot Lake, the land surrounding the Ambrosia Lake deposit was arid, with the main water sources coming from underground.

As the AEC's director of raw materials announced in 1958, Ambrosia Lake was the reason for "the very rapid increase in our [U.S.] overall reserves."[68] Immediately following a full survey of the deposit, the AEC announced the reduction in its domestic buying program. During this period, the mills had to revise their buying contracts, causing half of the mills to shut down in the 1960s. This shutdown left regulators with two separate problems: the control of active mill tailings and the control of abandoned mill tailings.

But into the 1960s, the AEC claimed there was nothing to regulate with regard to abandoned tailings piles. During the slowdown of the buying program, the commission faced mounting piles of tailings. Approximately 60 million tons of solid tailings across the country by 1965 faced impending abandonment. AEC officials claimed, however, that they had no jurisdiction over tailings piles because the amount of radioactive material contained therein did not fall under the commission's regulatory jurisdiction. Moreover, the AEC concluded that the tailings posed no public health risk.[69]

The year 1967 proved a watershed for health and safety regulation in U.S. mines, partly due to public attention to the problem. That year, the *Washington Post* exposed the unsafe working conditions within uranium mines and the effect they had on the miners.[70] In April, the AEC was brought before the Joint Committee on Atomic Energy to hold hearings about the risks associated with uranium mining. During the hearings the FRC finally submitted regulation guidelines for work within the mines, which the president then approved.[71] These regulations were inexcusably late—twenty years late, in fact. But they were still eleven years earlier than those put in place in Elliot Lake.

In the early 1970s, the need for nuclear power and the growing environmental movement changed the mill waste situation in Grants. By 1967 in Grants alone, there were 41,334,000 tons of active and inactive tailings that were almost entirely unregulated. This would soon change with the emergence of the then-popular environmental movement, which was much more nationally significant in the United States. In 1970, 20 million people gathered to celebrate the first Earth Day. It was the beginning of what one scholar deemed the Green Decade, during which the government enacted new environmental legislation and created new environmental organizations.[72]

One such organization was the New Mexico Environmental Improvement Agency (NMEIA). The NMEIA realized that there was a paucity of information about the effects of the uranium mining industry on surface water and groundwater and approached the recently created U.S. Environmental Protection Agency (EPA) to study the region's water resources.[73] New Mexico's plea for help prompted the EPA to conduct a monitoring program in Grants for six months in 1975. The EPA produced three reports on this program which revealed that the tailings regulatory system in New Mexico was at best disorganized and ineffective, and at worst nonexistent.[74]

In terms of regulation, the EPA used several benchmarks to assess the quality and potential danger of the water. In New Mexico, the NMEIA set a basic water quality standard at 30 picocuries per liter of radium-226. With regard to potable water, the regulatory standard under the 1974 Safe Drinking Water Act was 3 picocuries per liter, the same as at Elliot Lake at the time.[75]

As they tested different water sources around the mines, the EPA scientists found gross violations of the drinking water guidelines. At one of the mills in the area, Kerr-McGee, water that workers used for washing and drinking contained fifty times more picocuries per liter than the federal government officially deemed potable. Near the Churchrock mine, EPA testing of the workplace water found it to contain over four times the picocuries per liter that was then deemed safe. The EPA stumbled upon yet another horrifying water situation around the Churchrock mine. A water source that contained almost 40 picocuries per liter of radium-226, thirteen times the safety threshold, supplied water for a mobile home where the wife and three young children of one miner lived.[76]

The results of the EPA's groundwater-monitoring project were slightly more encouraging. EPA officials concluded that groundwater pollution was not a pervasive problem in the study area. Though more encouraging than surface water measurements, the lack of egregious radiation violation was not due to the foresight of the mining and milling companies, but rather the absorbent quality of the soil between the tailings and the groundwater.[77]

With regard to self-monitoring, the records of the mining companies operating in the region were abysmal. Often, neither groundwater- nor surface water–monitoring systems existed at all, as at the Jackpile and Paguate open-pit mines. These two mines produced tailings with a radium-226 content of nearly 190 picocuries per liter. Examining those mining company monitoring systems that did exist, the EPA concluded that they did not collect samples from proper locations, or did so too rarely to be effective. In addition, the mining companies were not using any of the latest geophysical or geohydrologic techniques.

The EPA did not place the blame for this deficiency solely on the companies. Its reports also concluded that Atomic Energy Commission reporting requirements and inspections were "incomplete and disorganized." The EPA blamed this failure on the commission.[78] Essentially, the mining and milling monitoring systems were ineffective, and the companies were not held accountable to any state or federal body for their shoddy monitoring activity. This self-regulation was certainly inadequate, but not unique to the United States.

At this time, the mid-1970s, U.S. federal and state agencies conducted studies of abandoned tailings in order to assess them for future cleanup projects. At Ambrosia Lake, officials reported that the only barrier between the tailings and living organisms were barbed-wire fences and a few signs warning of radiation. Cattle still grazed nearby. Population in the area was sparse, with only ten residences and one school in the vicinity. Officials judged that the stability of the piles without human intervention was quite good. The piles formed an outer crust that prevented wind erosion, and water erosion was not a large problem. They also noted that the shale beneath the pile was impermeable, and so tailings would not affect the groundwater. Stabilization and fencing were considered adequate until more effective measures could be taken.[79]

In November 1978, the U.S. Congress passed the Uranium Mill Tailings Radiation Control Act, but legislation took a long time to travel from Washington, D.C., to Grants, New Mexico. During the period between its enactment and implementation, disaster struck. Equivalents of the shoddy dams observed in Elliot Lake existed in Grants. On 16 July 1979, a dam near Churchrock burst, spilling 93 million gallons of liquid tailings and 1,100 tons of solid tailings into the Rio Puerco via a small streambed. The tailings continued downstream for one hundred miles, through Navajo grazing land, and into Arizona.[80]

Although this accident garnered much less attention than Three Mile Island—the power plant in which a partial meltdown had occurred three months prior—it was indicative of the terrible record of regulation within the mining and milling industry. There is likely another reason that this incident is often overlooked: it primarily affected Navajo land.

CONCLUSION

I argue in this essay that the regulatory oversight in Elliot Lake was at best equal to the oversight in Grants, and often worse. If more stringent regulation in Elliot Lake does not explain why it was safer and radiation caused less public backlash than at Grants, then what does explain the difference? There are three hypotheses that might, with further research, answer this question.

First is the human hypothesis. It holds that the difference in safety within a given mine depended on different individuals working in different mines. Thompkins, in his memoir, notes two examples of this difference. One is that an oil man who had an insufficient knowledge of how to work a mine ran Kerr-McGee, and the other is the Stanrock manager, who sabotaged the work of mining engineers because he did not understand it. The safety of a mine depended on who ran it. The drawbacks of this hypothesis are that it is based on a memoir of a mining engineer, who obviously has his own biases about who was best suited to the job. To test this hypothesis would require careful assembly of data on mine safety records and mine management personnel at many mines.

The remaining hypotheses fall more in line with Timothy LeCain's materialist argument that matter itself has historical agency and power. He argues that matter works in cooperation with humans, and is not simply acted upon (see chapter 6, this volume). I call the second hypothesis the ore hypothesis. It holds that the difference in the formation of the uranium deposits in Elliot Lake versus Grants led to different environmental effects. In Grants, ancient bodies of water deposited scattered amounts of low-grade uranium ore across the landscape. It was widely dispersed in different formations across the Four Corners region in many different deposits, small and large. The location and size of the deposits meant that small groups of miners could extract uranium using setups such as the aforementioned doghole mines alongside larger mining companies. Ventilation was often inadequate in these smaller mines as they were temporary and had less capital investment. Due to slack government regulation, these smaller mines were rarely tested for safe radiation levels, and they could not afford protection. Because of the transient and short-term nature of these mining jobs, the same workers who were exposed to high levels of radiation would later work in the large company mines.[81] In Elliot Lake, in contrast, glaciers shaped and reshaped the region over 500 million years, leaving deposits of minerals in their wake, including veins of uranium. These veins were concentrated in specific areas and were located far enough underground that they required significant capital investment to mine. Therefore, in Elliot Lake there were only large mining companies, which were more likely to have more sophisticated ventilation. The ore hypothesis holds that it was the way the ore was deposited that led to different systems of extraction with differing levels of safety. Thus the impact of weak regulation—common to both Grants and Elliot Lake—led to different outcomes based on the character of the ore itself.

The third and final hypothesis is the mechanization hypothesis, and it is also related to the way the ore was originally deposited. In Elliot Lake, it was easier for companies to mechanize the extraction process because of the uranium's position in veins.[82] In the smaller mines in the American Southwest, according to one source, the way the ore was deposited—thinly and irregularly—meant that it was often sorted by hand.[83] Again, this was especially true in smaller mines.

A limitation of the latter two hypotheses, however, is that large companies also ran most of the uranium mining in Grants. The smaller mines were more common in Utah, Arizona, and Colorado, where deposits were located closer to the surface. But miners certainly traveled among the mines in the Four Corners region, and there were still independent, small-scale producers in Grants, while Elliot Lake had none.

Testing the ore hypothesis and the mechanization hypothesis would require collecting data on safety and radiation exposure in specific mines in Grants. Did the smaller, independent mines genuinely have a worse record, as some veterans of the industry recall? This research would not be easy. The small, independent mines never underwent regular testing, and so their records are sparser than those of the larger mines. Furthermore, testing in the mines in the 1950s, 1960s, and 1970s was not always an accurate measure of the radon levels within them, as the level varied greatly from one area of a mine to another and from minute to minute.

Any of these three hypotheses, or all three together, might explain why the Grants radiation safety record was poorer than that of Elliot Lake. But superior Canadian regulation of uranium mines does not explain it.

Arguing that the lack of regulatory oversight in Elliot Lake was on the same level as the situation in Grants is not to make an excuse for Washington, D.C., or to make Ottawa the new cautionary tale of Cold War zeal. Instead of weakening the environmental justice argument connected to U.S. uranium mining, I hope this chapter strengthens it, by showing how uranium ore procurement affected working-class and indigenous populations outside the United States. The repercussions were international. Moreover, I hope it brings the research closer to understanding how the individuals involved in uranium mining—and how the rock itself—contributed to lasting effects of Cold War uranium procurement. I hope it helps ask the right questions about the exploitation of people who exploited the ore.

NOTES

1. For examples of this argument, see Doug Brugge, Timothy Benally, and Esther Yazzie-Lewis, eds., *The Navajo People and Uranium Mining* (Albuquerque: University of New Mexico Press, 2006); Peter H. Eichstaedt, *If You Poison Us: Uranium and Native Americans* (Santa Fe: Red Crane Books, 1994); Barbara Rose Johnston, ed., *Half-Lives and Half-Truths: Confronting the Radioactive Legacies of the Cold War* (Santa Fe: School for Advanced Research, 2007); Judy Pasternak, *Yellow Dirt: An American Story of Poisoned Land and a People Betrayed* (New York: Free Press, 2010); Raye C. Ringholz, *Uranium Frenzy: Saga of the Nuclear West* (Logan: Utah State University Press, 2002); and Stewart L. Udall, *The Myths of August: A Personal Exploration of Our Tragic Cold War Affair with the Atom* (New York: Pantheon Books, 1994).

2. All books listed in note 1 address the political and moral aspects of uranium mining. See also Advisory Committee on Human Radiation Experiments, *Final Report* (1995), http://www.hss.doe.gov/healthsafety/ohre/roadmap/achre/chap12_2.html (accessed May 5, 2015). For an example of this argument in Canadian historiography, see Laurel Sefton MacDowell, "The Elliot Lake Uranium Miners' Battle to Gain Occupational Health and Safety Improvement, 1950–1980," *Labour/Le Travail* 69 (Spring 2012): 91–118; and Lorraine Rekmans, Keith Lewis, and Anabel Dwyer, eds., *This Is My Homeland: Stories of Nuclear Industries by People of the Serpent River First Nation and the North Shore of Lake Huron* (Cutler, Ontario: Anishinabe Printing, 2003).

3. Reginald Whitaker and Gary Marcuse, *Cold War Canada: The Making of a National Insecurity State, 1945–1957* (Toronto: University of Toronto Press, 1994), 138–57.

4. U.S. Senate, 79th Congress, *Public Law 585: Atomic Energy Act of 1946* (Washington, DC: U.S. Atomic Energy Commission, 1965); J. Samuel Walker, *Containing the Atom: Nuclear Regulation in a Changing Environment, 1963–1971* (Berkeley: University of California Press, 1992), 258; *Atomic Energy Control Act, 1946*, 10, Geo. VI, Ch. 37, August 31, 1946; Gordon H.E. Sims, *A History of the Atomic Energy Control Board* (Hull, Quebec: Canadian Government Publishing Centre, 1981), 78–79.

5. Ira U. Cobleigh, "Vision at Blind River," *Commercial and Financial Chronicle* (October 7, 1954), Franc R. Joubin Fonds—Elliot Lake Uranium, Vol. 41, File 3: Algoma District, Blind River Boom, Articles, Library and Archives Canada, Ottawa, Ontario (hereafter cited as LAC).

6. Rekmans, Lewis, and Dwyer, *This Is My Homeland*, xv.

7. J.W. Griffith, *The Uranium Industry: Its History, Technology, and Prospects* (Ottawa: Queen's Printer and Controller of Stationary, 1967), 95.

8. "A City Is Born . . . Elliot Lake," Research and Development Studies, Mining—Elliot Lake Area, Vol. I, Part I, Archives of Ontario, Toronto, Ontario

(hereafter cited as AO); "What It's Like to Live in Ontario's Atomic City," *(Toronto) Star Weekly* (February 11, 1956); "Rush to New Uranium Camp like Klondike on Wheels," *Toronto Daily Star* (April 3, 1957); "Elliot Lake—Pioneer Boom Town of the 20th Century," *Canadian Homes and Gardens* (March 1958), Franc R. Joubin Fonds—Elliot Lake Uranium, Vol. 46, File 18: Elliot Lake Development, Clippings, 1955–1959, LAC; "Town in a Billion" *Weekend Magazine* 6:36 (1956), Vol. 44, File 8: Algom Uranium Mines Ltd., Press Clippings, Articles, LAC.

9. Lester Pearson "Canada's Uranium Miners Are Vital as Jet Pilots in Free World's Defence," *Toronto Daily Star* (October 5, 1957), Vol. 43, File 6: Uranium, Power Prospects, Articles, LAC.

10. For a detailed account of this racism, see Jean-Marie Barsalou, *Lure of a Dream: Remembrance of a Life's Experience* (Sudbury, Ontario: privately printed, 1994): 55–59.

11. Pearson, "Canada's Uranium Miners Are Vital as Jet Pilots in Free World's Defence."

12. Yearly reports of the Canadian Atomic Energy Control Board (hereafter cited as AECB), 1954–1958, Canadian Nuclear Safety Commission, http://nuclearsafety.gc.ca/eng/resources/canadas-nuclear-history/index.cfm (accessed September 20, 2015).

13. R.W. Thompkins, *The Richest Canadian: The Life and Career of a Canadian Mining Engineer* (Toronto: Whistler House, 1994), 130–39.

14. Ibid.; Brian Walker, "Government Regulation of Health Hazards in the Ontario Uranium Mining Industry, 1955–1976," in *At the End of the Shift: Mines and Single-Industry Towns in Northern Ontario*, ed. Matt Bray and Ashley Thomson (Toronto: Dundurn Press, 1992), 130–31.

15. Thompkins, *The Richest Canadian*, 131.

16. Ibid., 137.

17. Paul E. Young to L.F. Labow, memorandum, Re: Tailings Disposal Area, May 21, 1954, and R.P. Ehrlich to F.R. Joubin, Re: Pronto Tailings, May 17, 1954, Franc R. Joubin Fonds—Elliot Lake Uranium, Vol. 45, File 15: Pronto Uranium Mines Ltd. Mill Tailings Pollution, Correspondence, Memos, LAC.

18. R.P. Ehrlich to R.M. Ennis, interoffice memo, July 20, 1954, Franc R. Joubin Fonds—Elliot Lake Uranium, Vol. 45, File 15: Pronto Uranium Mines Ltd. Mill Tailings Pollution, Correspondence, Memos, LAC.

19. One Health Department employee attributed the advent of watershed monitoring to "curiosity [more] than anything else." L.B. Lepperd, "MPC (Maximum Permissible Concentration) Objectives for Drinking Water Contaminated by Certain Uranium-Thorium Daughter Mixtures," reproduced in Ontario Water Resources Commission (hereafter cited as OWRC), "Industrial Wastes Survey of Stanrock Mines Limited, Elliot Lake Ontario," 1969, AO, 94–95.

20. Earle A. Ripley, Robert E. Redmann, and Adèle A. Crowder, *Environmental Effects of Mining* (Delray Beach, FL: St. Lucie Press, 1996), 206–21.

21. Yearly reports of the AECB, 1959–1960.

22. Yearly reports of the AECB, 1959–1965.

23. Griffith, *The Uranium Industry*, 95.

24. L. B. Lepperd, "MPC (Maximum Permissible Concentration) Objectives for Drinking Water Contaminated by Certain Uranium-Thorium Daughter Mixtures," 94–95.

25. OWRC, "Water Pollution from the Uranium Mining Industry in the Elliot Lake and Bancroft Areas, Volume 1—Summary," October 1971, AO, 1.

26. Ibid., 2; Lepperd, "MPC (Maximum Permissible Concentration) Objectives for Drinking Water," 94–95.

27. OWRC, "Industrial Wastes Survey of Stanrock Mines Limited, Elliot Lake Ontario," 1969, AO, 94–105.

28. "Report of Radiological Water Pollution in the Elliot Lake and Bancroft Areas—1965," reproduced in part in OWRC, "Industrial Wastes Survey of Stanrock Mines Limited, Elliot Lake Ontario," 1969, AO, 104–5.

29. OWRC, "Industrial Wastes Survey of Stanrock Mines Limited, Elliot Lake Ontario," 1–2.

30. Ripley, Redmann, and Crowder, *Environmental Effects of Mining*, 213–15; OWRC, "Water Pollution from the Uranium Mining Industry in the Elliot Lake and Bancroft Areas, Vol. 1—Summary," 3–11.

31. OWRC, "Industrial Wastes Survey of Stanrock Mines Limited, Elliot Lake Ontario," 37–39.

32. Ibid., 70–74.

33. Ibid., 120–21.

34. Ibid., 125–26.

35. Albert A. Visentin to S. E. Salbach, July 29, 1966, Box B281661, Research and Development Studies, Mining—Uranium, Elliot Lake Field Reports, Vol. 1, AO.

36. Elliot Lake Secondary School Students of English, *Jewel in the Wilderness: A History of Elliot Lake* ([Elliot Lake]: self-published, n.d.), 108–9.

37. "Elliot Lake—The Town That Refused to Die" *The Standard* (October 16, 1975); "Elliot Lake—The Ghost Town Is Making a Comeback," *Toronto Star* (August 20, 1976); "The Uranium Capital That's Well on Its Way to a Second Prosperity," *(Toronto) Globe and Mail* (March 24, 1977), Franc R. Joubin Fonds—Elliot Lake Uranium, Vol. 47, File 1: Elliot Lake Development, Clippings, 1959–1983, LAC.

38. Benjamin Kline, *First along the River: A Brief History of the U.S. Environmental Movement* (Lanham, MD: Rowman and Littlefield, 2011), 87–92; Laurel Sefton MacDowell, *An Environmental History of Canada* (Vancouver: University of British Columbia Press, 2012), 243–46. For more on Greenpeace, see Rex Weyler, *Greenpeace: How a Group of Journalists, Ecologists, and Visionaries Changed the World* (Emmaus, PA: Rodale, 2004), and Frank Zelko, *Make It a Green Peace! The Rise of Countercultural Environmentalism* (Oxford: Oxford University Press, 2013).

39. This argument is especially true regarding Greenpeace. Weyler, in *Greenpeace*, mentions Canadian uranium only a few times, and usually only in regard to nuclear proliferation concerns. Only once, in passing, does he mention the plight of miners (p. 563). Zelko, *Make It a Green Peace!* does not mention the issue at all.

40. Macdowell, *An Environmental History of Canada*, 253.

41. Walker, "Government Regulation of Health Hazards in the Ontario Uranium Mining Industry," 131; Henri Groulx, "'Nobody ever told us that uranium was hazardous to your health,'" in Rekmans, Lewis, and Dwyer, *This Is My Homeland*, 85.

42. Silicosis is an occupational lung disease caused by silica dust. Walker, "Government Regulation of Health Hazards in the Ontario Uranium Mining Industry," 132.

43. Ibid.; MacDowell, "The Elliot Lake Uranium Miners' Battle," 91–92.

44. Thompkins, *The Richest Canadian*, 145, 155.

45. MacDowell, "The Elliot Lake Uranium Miners' Battle," 116.

46. Walker, "Government Regulation of Health Hazards in the Ontario Uranium Mining Industry," 131.

47. Ibid., 130–31.

48. Ibid., 133.

49. Both alkaline and acid leaching work in similar ways, except different leaching agents are used. The most common leaching agent is sulfuric acid. Leaching is used to refine uranium to yellowcake.

50. R. F. Kaufman, G. G. Eadie, and C. R. Russell, "Effects of Uranium Mining and Milling on Ground Water in the Grants Mineral Belt, New Mexico," *Ground Water* 14:5 (1976): 296–97.

51. R. F. Kaufmann, G. G. Eadie, and C. R. Russell, *Summary of Ground-Water Quality Impacts of Uranium Mining and Milling in the Grants Mineral Belt, New Mexico*, Technical Note ORP/LV-75-4 (Las Vegas, NM: U.S. Environmental Protection Agency, Office of Radiation Programs, August 1975), 13–21; U.S. Environmental Protection Agency, *Impacts of Uranium Mining and Milling on Surface and Potable Waters in the Grants Mineral Belt, New Mexico* (Dallas, TX: National Enforcement Investigations Center and USEPA Region VI September 1975), 1–3, 11–14.

52. Kaufmann, Eadie, and Russell, *Summary of Ground-Water Quality Impacts of Uranium Mining and Milling in the Grants Mineral Belt, New Mexico*, 13–21.

53. Ibid.

54. For a detailed description of working as an independent miner, see oral history interview of Hanson L. Bayles conducted in 1970 by Dorothy E. Erick, OH 329, Uranium History Project, Center for Oral and Public History, California State University, Fullerton (hereafter cited as COPH-CSUF).

55. Kaufmann, Eadie, and Russell, *Summary of Ground-Water Quality Impacts of Uranium Mining and Milling in the Grants Mineral Belt, New Mexico*, 13–21.

56. The amount of ore produced is an approximation based on average production figures from the mills during their entire periods of operation. Mill capacities and the numbers of mills open at a given time changed over the years. I have done my best to be as accurate as possible in this approximation based on which mills were open and what their individual operating capacities were in 1959. Numbers are based on the following sources: File: "Summary History

of Domestic Uranium Procurement Under U.S. Atomic Energy Commission Contracts Final Report," prepared by Holger Abrethson Jr. and Frank E. McGinley, Grand Junction, CO, October 1982, Box 2, Collection: Project Histories, Office of Administration Histories, Related Documents, Publications and Appraisal Files, 1943–1996, Record Group (RG) 434, National Archives and Records Administration, Rocky Mountain Region, Denver (hereafter cited as NARA-RMR). AEC purchase numbers are based on Nielsen B. O'Rear, "The Domestic Uranium Program, 1946–1966" (September 16, 1966), File: History of Raw Materials Program (3 of 3), Box 2, Collection: Project Histories, Office Administration Histories, Related Documents, Publications and Appraisal Files, 1943–1996, RG 434, NARA-RMR.

57. Multiple examples of the absence of testing for radon may be found in the following files: Legal 11 Litigation, Galigher Company Folder #1, through 3/25/55, Folder 1 of 1; Legal 11 Litigation, Galigher Company, Folder #2, through 10/27/55; and 11 Litigation, Galigher Company, Folder #3, through 9/10/56, Folder 1 of 1, all in Box 8, Collection: Litigation Case Files, National Lead Co., 1956–1982, Correspondence, 1982–1985, RG 434; NARA-RMR.

58. Walker, *Containing the Atom*, 233–37.

59. Advisory Committee on Human Radiation Experiments, "Chapter 12: Observational Data Gathering—The Uranium Miners," in *Final Report* (1995); Trial—Testimony of Duncan Holaday, in Joint Committee on Atomic Energy, *Radiation Exposure of Uranium Miners: Part 1*, 90th Cong., 1st Sess., May 9, 10, 23, June 6, 7, 8, 9, July 26, 27, and August 8, 10, 1967, p. 601, archived at https://collections.stanford.edu/atomicenergy/bin/page?forward=home (accessed September 20, 2015).

60. Eichstaedt, *If You Poison Us*, 74–75.

61. Bayles oral history interview, COPH-CSUF.

62. Thompkins, *The Richest Canadian*, 163.

63. Ibid.

64. J. Samuel Walker, *Permissible Dose: A History of Radiation Protection in the Twentieth Century* (Berkeley: University of California Press, 2000), 18–19.

65. Walker, *Containing the Atom*, 238–39.

66. Federal Radiation Council—Minutes and Records of Action, Meeting of May 17, 1961, Papers of the Federal Radiation Council, RG 412 (Records of the EPA), National Archives and Records Administration, College Park, Maryland (hereafter cited as NARA-CP).

67. Minutes and Records of Action Meetings, 1961–1968, Papers of the Federal Radiation Council, RG 412, NARA-CP.

68. Joint Committee on Atomic Energy, *Problems of the Uranium Mining and Milling Industry*, 85th Cong., 2nd Sess., February 19, 24, 25, 1958, p. 6, archived at https://collections.stanford.edu/atomicenergy/bin/page?forward=home (accessed September 20, 2015).

69. AEC, "Ultimate Disposition of Uranium Mill Tailings (Discussion Paper)," November 30, 1965, archived at https://collections.stanford.edu/atomicenergy/bin/page?forward=home (accessed September 20, 2015).

70. George T. Mazuzan and J. Samuel Walker, *Controlling the Atom: The Beginnings of Nuclear Regulation, 1946–1962* (Berkeley: University of California Press, 1984), 241; J. V. Reistrup, "Hidden Casualties of Atom Age Emerge," *Washington Post* (March 9, 1967).

71. Joint Committee on Atomic Energy, *Radiation Exposure of Uranium Miners: Summary Analysis of Hearings*, 90th Cong., 1st Sess., May 9, 10, 23, June 6, 7, 8, 9, July 26, 27, and August 8 and 10, 1967, 5, archived at https://collections.stanford.edu/atomicenergy/bin/page?forward=home (accessed September 20, 2015).

72. Kline, *First along the River*, 95–111. For a more detailed discussion of Earth Day and its connection to the rise of the environmental movement, see Adam Rome, *The Genius of Earth Day: How a 1970s Teach-In Unexpectedly Made the First Green Generation* (New York: Hill and Wang, 2013).

73. Prior to NMEIA's work with the EPA, New Mexico public health officials conducted a few studies, none of which were published, as well as some studies of surface water associated with the Colorado River Basin. These studies had more information about groundwater than surface water, and they did not lead to any significant legislation concerning the tailings. USEPA, *Impacts of Uranium Mining and Milling on Surface and Potable Waters in the Grants Mineral Belt, New Mexico,* 3–4; Kaufman, Eadie, and Russell, "Effects of Uranium Mining and Milling on Ground Water in the Grants Mineral Belt, New Mexico," 296.

74. U.S. Environmental Protection Agency, *Impact of Uranium Mining and Milling on Water Quality in the Grants Mineral Belt, New Mexico*, EPA 906/9-75-002 (Dallas, TX: USEPA Region VI, September 1975), 1.

75. USEPA, *Impacts of Uranium Mining and Milling on Surface and Potable Waters in the Grants Mineral Belt, New Mexico,* 16–18.

76. Ibid., 37–39.

77. The conclusions about lack of groundwater contamination indicated another issue to the EPA scientists: if the soil absorbed or adsorbed the radionuclides, the soil, instead of the water, was contaminated. Because the study was about groundwater, however, soil contamination was only a hypothesis. Moreover, although the mining companies were not polluting the deep aquifers, the EPA scientists feared that the companies were dewatering the aquifers at a rapid rate, contaminating the water, and then discharging it as highly contaminated surface water. Kaufmann, Eadie, and Russell, *Summary of Ground-Water Quality Impacts of Uranium Mining and Milling in the Grants Mineral Belt, New Mexico,* 17–21.

78. Ibid., 22–25; USEPA, *Impact of Uranium Mining and Milling on Water Quality in the Grants Mineral Belt, New Mexico,* 8–9.

79. "Summary Report—Phase 1 Study of Inactive Uranium Mill Sites and Tailings Piles" (October 1974), File: 10/74 Report of Inactive Uranium Mill Sites, Folder 1 of 1; "Report on Conditions of Uranium Millsite and Tailings at Ambrosia Lake, New Mexico," File: Report on Conditions of Radioactive Sands at Lowman, Folder 1 of 2, both in Box 12, Collection: AEC Airborne Anomaly Location Files, 1953–1971, Site Information Packets, 1950–1980, RG 434; NARA-RMR.

80. Oversight Hearing before the Subcommittee on Energy and the Environment of the Committee on Interior and Insular Affairs, *Mill Tailings Dam Break at Church Rock, New Mexico*, 96th Cong., 1st Sess., October 22, 1979.

81. Eichstaedt, *If You Poison Us*, 44–45; Thompkins, *The Richest Canadian*, 163.

82. Cobleigh, "Vision at Blind River."

83. Eichstaedt, *If You Poison Us*, 43.

CHAPTER 10

The Giant Mine's Long Shadow

Arsenic Pollution and Native People in
Yellowknife, Northwest Territories

JOHN SANDLOS AND ARN KEELING

In a report from 2002, Canadian government economists Warwick Bullen and Malcolm Robb reviewed the production history of three major gold mines near the small northern Canadian city of Yellowknife: Con (1938–2003), Discovery (1950–68), and Giant (1948–2004). These mines, the reader learns, contributed an estimated 13.5 million ounces of gold and employment for Yellowknife citizens over sixty-five years of operation. Of these, Giant mine was the largest producer in the Northwest Territories, generating just over 7 million ounces of gold and just over $3.3 billion (adjusted to 2013 values) over the mine's life. Despite the fact that all the mines have now closed, Bullen and Robb suggest that Yellowknife's gold mines represent an example of sustainable development, because private wages flowed through the economy and because the generation of public wealth with mining taxes and royalties built more enduring infrastructure such as roads, schools, and hospitals.[1]

What this accounting failed to acknowledge, however, was the massive environmental and economic legacy of arsenic contamination at Giant mine (and to a lesser extent Con), an omission that neglects the fact that these gold mines caused low-level and sometimes acute poisoning of Yellowknife and nearby Aboriginal communities for more than a half century. Beginning in the late 1940s, the highly toxic compound arsenic trioxide traced a pathway outward from the mines via the tailings ponds and roaster stacks at Con mine and Giant mine. Falling from the air as a light dust or leaking as slurry from tailings ponds, the arsenic dispersed through the region's rivers, streams, lakes, air currents, and surface soils, eventually working its way into the bodies of fish, terrestrial wildlife, plants, and humans. Although the arsenic represented a danger to the entire Yellowknife population, the toxic fallout from the mine represented a particularly dire threat to the Yellowknives Dene First Nation, located adjacent to Yellowknife, because the poison was deposited on local berries and vegetables and also in snow used as a water supply in winter. By 1951, several cases of acute arsenic exposure prompted a public health campaign over the following twenty-five years to reduce arsenic levels in the local environment and within the bodies of individuals to so-called safe threshold level values (TLVs).

If these pollution control efforts reduced acute water and air pollution problems in the short term, the techno-fix approach that was adopted produced additional problems in the long term through the disposal of arsenic trioxide dust. In 1951, Giant mine began to pump arsenic dust collected in a Cottrell electrostatic precipitator (ESP) into mine chambers, placing 237,000 tons of the material underground by 2004. This created the potential for massive toxic seepage as water tables began to rise and permafrost failed to reestablish itself (as mine officials had originally hoped). Currently the federal government's Indigenous and Northern Affairs Canada (INAC) is undertaking a remediation program involving surface restoration and the freezing and stabilization of underground arsenic. For the Yellowknives Dene, the remediation program has raised profound issues not only about the perpetual care issues associated with the so-called frozen block method of stabilization, but also about the historical environmental injustices associated with gold mining in Yellowknife.[2] These legacies of arsenic pollution at Giant mine run counter to the triumphal tone of local Yellowknife histories and commemorative activities equating gold mining with the advance of modern civilization in northern Canada.[3]

In a broader context, the history of arsenic loading at Yellowknife resonates with environmental histories of industrial mining and pollution at other sites in North America. The long struggle for adequate pollution control at Giant mine reflects the weak regulatory regimes surrounding mining pollution generally in Canada, the United States, and Mexico before the 1970s. Fearful of jeopardizing jobs and investment, state, federal, and territorial governments tended to accommodate industrial interests and addressed pollution problems only when forced to do so by the courts, cross-border disputes, or catastrophic pollution episodes.[4] For its part, the industry responded to environmental regulation with political and legal resistance and, when pressed to deal with pollution, turned to "technological fixes" that often deferred or displaced contamination rather than preventing it. Timothy LeCain's work on the limits of environmental technologies such as the Cottrell ESP (which may remove toxic particulate matter such as arsenic from the air, but then produces new problems associated with the disposal of the collected material) finds ample reinforcement in the chambers under Giant mine containing thousands of tons of arsenic dust.[5] The critique of the techno-fix approach highlights both the politics of regulation surrounding industrial pollution and the persistent materiality of contaminants generated by industrial mining and dispersed into local human and animal bodies and environments.[6]

Another key element in this story, common to North American mining controversies, was the contested knowledge surrounding environmental pollution and human exposures. The works of several historians, principally Linda Nash and Nancy Langston, but also many of the contributors to forums on the historical dimensions of industrial toxins in the journals *Environmental History* and *Osiris*, have pointed to the historically constructed nature of scientific knowledge within the fields of toxicology and public health.[7] In particular, they have argued that the public health responses to crises analogous to the arsenic problem in Yellowknife have been bound up with the idea that science can determine safe TLVs for toxic loading in individuals and safe levels of toxins that can be consumed in food and water. These determinations ignore the complex interactions of bodies with the full panoply of chemicals in ecosystems and fail to account for the effects of chronic low-dose exposure to toxins over long periods of time.[8] More recently, historians Brett Walker and Joy Parr have expanded this discussion of toxic histories beyond the application of science to the issue, instead emphasizing the importance of sensory experiences such as pain or, in Parr's case, shifting soundscapes and smells that accompany human encounters with industrial development.[9]

All of these themes are clearly echoed in local perceptions of disease and sickness within bodies and local landscapes in Yellowknife. Nonetheless, the issue of arsenic pollution in Yellowknife finds its most relevant frame outside the field of environmental history, within the political ecology and anthropology literature tracing the environmental injustices and colonial dimensions of mining on indigenous lands throughout the globe.[10] Although much of this literature focuses on the developing world, broadly similar processes of "industrial colonization" are evident in the recent history of mining-induced landscape and economic change in the Canadian north.[11] Indeed, the story of gold mining at Yellowknife is important precisely because it illustrates how industrial polluters have served as agents of colonial dislocation and dispossession in hinterland environments, appropriating local land and water as an industrial pollution sink and transforming the material subsistence base of Native communities into a threatening and hazardous landscape.

MINING GOLD AND POISON

In the 1930s, the Yellowknife Bay area became the epicenter of the second gold rush to hit Canada's territorial North, after the Klondike rush of the 1890s. The conservative mindset of Depression-era capitalists had created a high demand for gold, one of the only secure investments—and stable mining sectors—during this period of global economic upheaval. The Yellowknife region in the Northwest Territories (see fig. 10.1) was even more remote and difficult to access than the Klondike, however, with the former's gold deposits inconveniently encased in hard rock rather than fluvial deposits. Thus, from the very beginning large companies dominated gold production rather than hordes of individual miners. The result was industrial-scale mines that, due to the expense and difficulty of transport, contained processing facilities on site that completely transformed solid rock into finished gold bricks.

In 1935, Canadian government geologists discovered gold on the east side of Yellowknife Bay, and the Canadian mining giant Consolidated Mining and Smelting (CM&S, later known as Cominco) staked a large series of claims in 1936 near the shores of Great Slave Lake and quickly opened the Con mine (with the adjacent Rycon property forming part of the same operation) in September 1938. A new company, Negus Mines Ltd., opened a mine of the same name one year later adjacent to Con. In 1935 two prospectors with Burwash Yellowknife Mines staked the first twenty-five Giant claims. After a lengthy period of exploration and development, the new company, Giant Yellowknife Gold Mines, Ltd. (GYGML), began full production of gold at the head of Yellowknife Bay in May 1948.[12]

Arsenic pollution did not present a problem during the earliest years of gold production at Yellowknife. At Con and Negus, the first exploited ore bodies contained no arsenic, and the companies used the common process of dissolving gold in cyanide and separating the solution from the surrounding waste rock. In 1940, however, miners at Con discovered

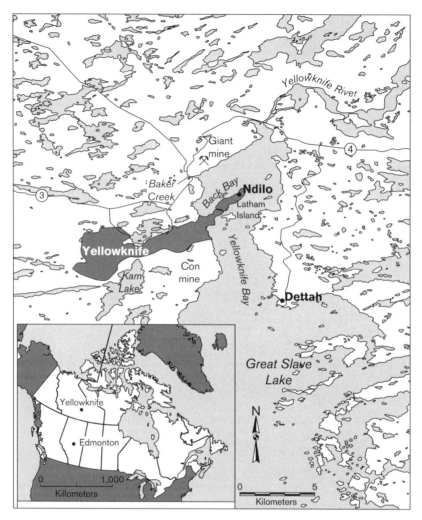

FIGURE 10.1. The mines, settlements, and water bodies around Yellowknife, Northwest Territories. Map by Charlie Conway.

gold within arsenopyrite formations, a type of ore in which sulfides prevent the cyanide from accessing the gold and leaching it from surrounding material. To burn off sulfur, such ores require roasting at extremely high temperatures (usually above 260 degrees Celsius) prior to cyanidation. The roasting process also produces the highly toxic compound arsenic trioxide in the form of a fine white dust that condenses as it cools.[13]

FIGURE 10.2. Aerial view of the Giant mine headframe and buildings in Yellowknife, Northwest Territories, 1955. Courtesy of Busse/NWT Archives/N-1979–052–1947.

By the late 1940s the city of Yellowknife and its environs were sitting on a ticking arsenic pollution time bomb. The roasting facility at Con mine, shut down in 1942 due to wartime restrictions after only six months of operation, was reopened in 1948. One year later, GYGML opened a roasting facility at the new mine and began to send arsenic dust into the air around Yellowknife. Although the mining companies did not formally monitor emissions from the roaster stacks until 1954, a government report from the 1960s estimated total arsenic dust output at Con and Giant at 22,000 pounds per day between 1949 and 1953. While Con Mine had a higher mill capacity than Giant at this time, arsenic emissions at Giant accounted for the lion's share of this total (approximately 16,500 pounds per day) because 100 percent of its gold deposits were contained in arsenopyrite ore, while only 20 percent of the ores at Con required processing in the roaster. Regardless, in their rush to process gold as quickly as possible, both gold mines sent arsenic dust up their roasting stacks without any pollution control equipment, Con from 1948 to 1949 and Giant from 1949 to 1951.[14]

Arsenic was hardly a new poison in the late 1940s. Although much of its popular notoriety comes from its history as a murder weapon, it also caused sickness and death within many nineteenth-century homes in Europe and North America due to its use in the pigmentation of wall-paper. By the early twentieth century, arsenic pollution was a focal point for disputes over the impact of smelter smoke from large-scale facilities such as the Washoe smelter in southwestern Montana. More recently, wells contaminated by arsenic from surrounding rock in South and Southeast Asia have provoked broad-scale public health crises and campaigns for safe drinking water.[15]

The health impacts of arsenic exposure depend in part on the dose ingested. We know that arsenic trioxide kills quickly with a single dose of 70–180 milligrams. Short-term exposure to levels just below the lethal threshold may cause abdominal pain, muscle pain, vomiting, diarrhea, skin rashes, burning or tingling feelings in the extremities (paresthesias), and thickening of the skin in hands and feet (keratosis). Lower-dose exposure over longer periods of time (over ten years) may also cause black spots to appear on the skin (hyperpigmentation). Long-term exposure to arsenic may also produce nausea and diarrhea and disrupt normal heart rhythm and blood vessel circulation, as well as brain and kidney function. It may also produce potentially fatal blad-der, liver, skin, and lung cancers. The human body has some defenses against arsenic: it is absorbed only through ingestion of contaminated water, food, or dust and cannot be absorbed through the skin while swimming or being exposed to dust. Human bodies can also jettison the material over time through urine and sweat, and it does not readily bioaccumulate, limiting risks from eating fish and terrestrial mammals exposed to arsenic. Although the United States and Canada have desig-nated a concentration of 0.01 parts per million (PPM) as a safe level of arsenic in drinking water, recent studies of the long-term exposure risks suggest that a value of zero is the only safe threshold level.[16]

In Yellowknife, the 11 tons of untreated arsenic dust emitted from the Giant and Con roasting stacks each day in 1949 spread widely with even the lightest breezes. When it settled in water, the arsenic was read-ily soluble in its compound form of arsenic trioxide, posing potential health risks to all organisms that lived in or drank from a contaminated water source. It also settled on soil and snow, concentrating particularly well on the latter during the long winter months because there was no rainfall to wash it away. In spring, meltwater suddenly mobilized months of accumulated arsenic, contaminating local waterways with dangerous levels of the toxin. Poor tailings management at Giant and Con compounded water pollution problems, as archival documents suggest several spills of arsenic-laden slurry and waste rock into Baker

Creek, Great Slave Lake, and other, smaller water bodies in the 1950s and 1960s. Arsenic trioxide dust also inevitably settled on local food sources, including berries, vegetables, and the leafy browse and grasses favored by wildlife.[17]

The impacts of such intensive arsenic loading in the Yellowknife area became apparent almost immediately in the late 1940s. In 1949 two workers at the Akaitcho mine exploration site just north of Giant mine received treatment in hospital after using melted snow as drinking water.[18] The most acute human health issues occurred at the Yellowknives Dene settlement on Latham Island (now called Ndilo). This community is located across a narrow channel to the northwest of Yellowknife's Old Town and, when winds blew out of the north, directly downwind from the Giant mine roaster stack. While not officially a Native reserve, the congregation of Dene on Latham Island reflected discrimination by local non-Native settlers who objected to Native people moving into town.[19] Another settlement, located on the east side of Yellowknife Bay at Dettah, predated the town of Yellowknife as a seasonal camping site, but became one of the two main Native settlements when the Yellowknives Dene were pressured to settle permanently after 1959.[20] These communities remained unserviced long after sewerage and water supplies were provided to Yellowknife proper in the 1940s (at a new town site, located farther from the mine), and residents relied on polluted snow and lake water rather than municipal supplies.

The location of these underserviced Dene communities proved dangerous, even deadly, in the face of widespread arsenic pollution. In April 1951, arsenic deposition from the gold mines killed a two-year-old Dene boy on Latham Island. A coroner's inquest ruled that the boy had died from "acute gastroenteritis caused by arsenical poisoning administered by unknown means."[21] Subsequent reports clarified the precise cause of death. Superintendent of Indian Affairs I. F. Kirkby reported that the boy had died from contaminated drinking water.[22] The minutes of a meeting of government officials held in June 1951 to assess the arsenic situation suggest that inspectors of the Department of Resources and Development had noticed large concentrations of arsenic in snow in the Yellownife area, "particularly at the northern end of Latham Island." The inspectors contacted Dr. O. L. Stanton, the Yellowknife medical health officer, who placed advertisements in the local paper (*News of the North*) and signs around the area warning people to be cautious with their use of water during spring runoff. The Yellowknife Indian agent had previously warned the local chief of the arsenic danger, but government officials at the June 1951 meeting suggested that "in spite of these precautions certain Indians living on the north end of Latham Island used the water in the vicinity, with the result that a number of them had to be given hospital treatment and one died."[23]

This was a fairly transparent attempt on the part of government officials to deflect blame from the mining company and local health officials. It is not clear how the Yellowknives Dene were supposed to respond to the posted warnings and determine which sources of water were dangerous, especially since arsenic is tasteless and odorless. Barriers of literacy and language (at least some Yellowknives elders did not speak English fluently) also likely reduced the effectiveness of the local advertising campaign, as did its limited scope. Through all of 1951, for example, only five small advertisements appeared in the back pages of the *News of the North*. In addition, the release date of April 6 for the first advertisement likely came far too late for a community that had undoubtedly melted snow as a water source all through the long Yellowknife winter. Remarkably, the local press published no news stories on arsenic poisoning or the death of the young Dene boy in 1951, failing to inform the public of the severe health crisis in its community.[24] Giant Yellowknife Gold Mines, Ltd. did not completely absolve itself of responsibility for this crisis, however. In August, the mine compensated the family of the boy who had been poisoned with $750 for the loss of their son.[25]

Existing records do not make clear how many other Yellowknives Dene were sickened by arsenic after April 1951. Nonetheless, a November 1953 memo describes one additional possible case associated with an "indigent Indian" named Henry Lafferty. In 1977 a Yellowknife doctor, recalled treating a "middle aged Indian" in winter 1957–58 for several arsenic-related anemia and skin conditions.[26]

The severity of arsenic pollution around Yellowknife can also be traced through local ecological changes and impacts on domesticated animals. A Yellowknives Dene community history refers to elders' stories about sled dogs, cattle, and chickens dying in large numbers during Giant mine's early years. In retrospect the elders believe the animals died from drinking water contaminated with arsenic.[27] Oral histories that focus on Yellowknife's gold-mining glory days contain very little discussion of arsenic, but some testimony supports the Yellowknives' observations of animal mortality and suggests at least some formal medical study of the issue stemming from a legal dispute. Laurie Cinnamon recalled that her father's horse team died in spring 1950 because it "got arsenic poisoning from drinking the spring run-off water lying about in puddles."[28] Similarly, Barbara Bromley remembers that fresh milk deliveries came to an end in 1951 because "they [the cows] died and I think he had an investigation and stated that they were poisoned by arsenic and that Cominco was to blame. I don't remember the outcome, although I believe there was a court case."[29] Helen Kilkenny, a farmhand at the Bevan family farm from 1947 to 1951, similarly describes the "hard luck" that plagued the farm a result of arsenic: "We watered the cows from Kam Lake about 500 yards from the barn. In winter we would cut a hole in the ice for the cows to drink. In the summer the cows would feed along the road and in the grassy places in the rocks. But after four years the cows got arsenic poisoning and they all died."[30] These incidents raise questions about the effectiveness of warning signs and advertisements. Government officials clearly identified Kam Lake as an arsenic hot spot at a June 1951 meeting, a situation that arose because CM&S officials had admitted to previously dumping arsenic-laden slurry from the company's stack scrubbers directly into Pud Lake, which drained directly into Kam. The fact that a local farmer deliberately cut holes in the ice to water his cattle in a water source that industry and government officials knew was contaminated further illustrates the lax pollution control and public information regime regarding the dangers of arsenic in 1950s Yellowknife.[31]

How did federal government and mining company officials respond to the arsenic crisis in Yellowknife? At the previously mentioned meeting in June 1951, A. K. Muir and A. C. Callow, the general manager and secretary treasurer, respectively, of Giant Yellowknife Mines, Ltd., met with a dozen government officials in Ottawa. The discussion focused on a three-pronged approach to mitigating arsenic impacts: continuation of the public information campaign warning about the dangers of consuming local water, fruit, and vegetables; a stepped-up program of testing and monitoring of public water and food sources as well as taking urine samples from humans; and most important, the installation of pollution control technology on the Con and Giant stacks. On the last-named issue, federal officials heavily criticized the impinger (or scrubber) method of pollution control CM&S had developed, noting that because the technology was based on cleaning the arsenic gas with water spray, the resulting contaminated slurry might leak from containment ponds and pollute local water sources, including Great Slave Lake. Government officials, along with Chief Medical Officer Stanton, were unanimous in preferring Giant's proposed approach—the installation of a Cottrell electrostatic precipitator with the storage of dry arsenic underground—to Con's wet method.[32] Remarkably, however, Muir reported that, due to shipping delays, his company had not yet installed the precipitator at Giant mine. At the meeting and in subsequent correspondence, government officials expressed only mild concern, and the Cottrell ESP was not operational until October 29, 1951.[33] Giant mine was thus permitted to operate at full tilt for more than six months with no pollution control equipment on a roaster stack that had poisoned a small boy to death and sickened an unknown number of his fellow community members.

Over the next three decades, the efforts of the federal government and the company to contain arsenic pollution remained intermittent at best. Officials reacted erratically to the issue, acting only with the emergence of public controversy and internal scares about spikes in contamination levels, rather than consistently applying program principles. Perhaps the most damning evidence of a lax regulatory regime is the fact that Giant mine's Cottrell ESP was not a particularly effective form of pollution control on its own. Tests from 1954 to 1958 revealed a relatively large percentage reduction in total Yellowknife arsenic emissions from an estimated 22,000 to 7,250 pounds per day due to the installation of a second Cottrell ESP. In absolute terms, however, a large amount of toxic material was still being loaded into the local environment through much of the 1950s, the majority of which originated with Giant mine. The installation of a baghouse (a secondary treatment method that captured arsenic dust in a large filter) at Giant between 1957 and 1959 resulted in a much more dramatic absolute reduction in total arsenic emissions to 695 pounds per day.[34]

Through the 1950s and 1960s, the federal government continued to monitor the impact of arsenic on water and food supplies in the region. Dr. Kingsley Kay, chief of the federal government's Industrial Health Laboratory, led a survey team to Yellowknife to conduct water and vegetable testing in the fall of 1951, while archived results suggest ongoing monitoring to 1960.[35] At the initial June 1951 meeting called in response to the Yellowknife arsenic crisis, federal officials adopted the Canadian standard of 0.05 PPM as the concentration of arsenic considered a safe threshold level for drinking water (this is five times the current threshold level in Canada). If higher concentrations were detected, warning signs would be posted around the contaminated body of water.[36] On a broad scale, the adoption of the 0.05 PPM threshold clearly indicates that federal officials were steeped in the "dose makes the poison" thinking that prevailed at the time, failing to account for or consider possible harms that might accrue from continued exposure to arsenic over long periods of time.[37]

In any case, the federal government failed to enforce even this standard. Data from the late 1950s reveal consistent spikes in arsenic above the 0.05 threshold for drinking water in Yellowknife and at the mines.[38] In other words, despite the strong rhetoric surrounding pollution control and public health after the crisis of 1951, roaster emissions and tailings spills from the mines into Back Bay resulted in the contamination of the Yellowknife tap water supply with arsenic concentrations above an already–dangerously high threshold level. One retrospective report suggested that the Yellowknife water supply contained arsenic levels above the acceptable limit of 0.05 PPM approximately 15 percent of the time between 1951 and 1960.[39] Similarly, reports from the 1950s and 1960s on arsenic contamination of vegetables and grass samples revealed staggeringly high levels of arsenic contamination, with mean values ranging from 18 PPM to 2,228 PPM over eight years. Federal officials acknowledged that such contamination levels were many orders of magnitude greater than the U.S. Public Health Service–recommended value of 1 PPM.[40] Clearly the installation of the Cottrell ESP at Giant mine failed to prevent the ongoing arsenic contamination of food and water in the Yellowknife area through the 1950s, with tests merely confirming that status quo mining operations continued to poison local residents.

If federal and Yellowknife public health authorities seem to have adopted a relatively passive approach to the arsenic issue through the late 1950s and early 1960s, public health officials nevertheless did begin to focus again on the issue in 1965. One official from the Indian and Northern Health Services Division declared on December 10: "I have recently discovered that the problem of arsenic pollution at Yellowknife is far from solved," based on tests showing 40–50 PPM of arsenic on lettuce and cabbage leaves in the area and on the fact that roaster stacks were still pumping out 300 to 400 pounds of the material every day.[41] Public health officials from across several divisions of the Department of National Health and Welfare called for a study of the issue.[42] Dr. G. C. Butler, the department's regional director, alerted Ottawa to the fact that local doctors were reporting high rates of anemia, an indicator for low-level arsenic poisoning, among female patients who had moved to Yellowknife more than four months previously.[43] Arsenic contamination of the local water supply in Back Bay continued to be a problem through the 1960s, one that public officials tried to solve in 1969 by moving the pipeline intake further upstream from pollution sources to the mouth of the Yellowknife River.[44] This solution failed to help Native residents of Latham Island and Dettah, who were not connected to the new system. Many continued to collect ice and water from the bay because they could not afford the cost of water trucked to their community. In 1973 Latham Island resident Michel Sikyea wrote to Jean Chretien, minister of Indian affairs and northern development, to protest the fact that the city of Yellowknife had "been after us to pay" for the past few years and had threatened to cut deliveries of trucked water. Sikyea wrote,

> Why should we pay others who poisoned our water? Most of the people in the valley don't get enough welfare to have food; we can't pay $5.00 a month for water too. None of us have the money, and even if we did, we should not be forced to pay for all this trouble. Starting tomorrow people will begin having to drink the water from Yellowknife Bay and soon our people will be sick and maybe some will die again.[45]

The federal government's response to these ongoing concerns about the danger of arsenic was tentative at best, reflecting a tendency to control and limit the flow of public health information. Between 1966 and 1969, Dr. A. J. de Villiers from the Biomedical Unit of the Department of National Health and Welfare led a comprehensive study, but information on results was not forthcoming until 1971, even for other federal officials who made repeated requests for further information.[46] In 1967 a health engineer of the Northern Health Service raised the possibility that current arsenic levels in Yellowknife's water could be carcinogenic, but the director general of medical services suggested dismissively that the claim was based on a single article from the United States, and wrote that "we are unwilling to assume, without some further evidence, that 0.1 ppm of arsenic in the water would be as toxic in a cold climate as in a warm one since in the former the water consumption per capita would be far less."[47] Regional Director Butler replied derisively that there were extensive references pointing to the possible carcinogenic properties of arsenic and that northerners did tend to drink lots of water in the form of tea, coffee, and alcohol with a water mix.[48]

Three years later, Dr. Butler had managed to pry enough information from his department's arsenic survey to determine that rates of contamination on vegetables had declined, but were still in most cases 0.05 PPM to 3 PPM above the allowable limit. Butler threatened to commission an independent investigation of the arsenic situation from a university or a provincial government (one that might actually be completed), and warned he would issue a public statement informing Yellowknife residents not to eat locally grown vegetables, a move "likely to make headlines."[49] The commissioner of the Northwest Territories requested that Dr. Butler contact the mines to see what preventative action could be taken before making a public statement and "allowing this problem to be blown out of proportion."[50] Two cabinet ministers—Jean Chrétien, minister of Indian affairs and northern development (and a future prime minister), and John Munro, minister of health and welfare—agreed to order Butler to remain silent on the issue. Munro suggested that the large reduction in arsenic levels on vegetables "indicate[s] that the problem is under reasonable and practical control," even if some samples showed contamination rates four times the federal government's own safety standards.[51]

Such attempts at secrecy came back to haunt the federal government in the 1970s, prompting the third wave of public concern about the arsenic issue. One measure of this concern is the large number of angry requests the federal government received for more information on the 1966 survey from the city of Yellowknife, the Indian Brotherhood of the Northwest Territories (a Native advocacy group), and the Canadian Broadcasting Corporation. In 1975 CBC radio's *As it Happens* brought the issue to national attention by suggesting that the results of the de Villiers report had been suppressed, particularly the sections pointing to high rates of lung cancer in Yellowknife due to long-term arsenic exposure and the fact that some Yellowknife residents still used the water from Back Bay.[52] In response to the CBC story, Health Minister Marc Lalonde issued a statement in 1975 claiming that the de Villiers report contained no data on links between arsenic and cancer rates in Yellowknife, but promising to conduct a study on arsenic rates in Yellowknife residents as a precaution.[53] The federal government proceeded with these public health studies in 1975, testing arsenic rates in human hair and urine samples and finding elevated levels only in mill workers at Giant mine, results that public health officials interpreted as a minor localized workplace matter rather than a widespread health issue.[54]

The media sensation associated with charges of a cover-up reflected, in part, the massive shift in Canadian attitudes toward pollution, northern development, and Native rights in the 1970s. As in the United States, the 1970s ushered in an era of environmentalism in Canada, with pioneering groups like the University of Toronto's Pollution Probe bringing issues, such as phosphate loading in the Great Lakes and urban air pollution, to public prominence. Federal and provincial governments created new environment departments and issued regulations aimed at curbing pollution, including the federal Arctic Waters Pollution Prevention Act (1970).[55] Between 1974 and 1977, Justice Thomas Berger's high-profile inquiry into the proposed Mackenzie Valley Pipeline frequently brought northern Native criticisms of development and environmental destruction to the front pages of Canadian newspapers. During this period, the arsenic issue became another important focus for heightened Aboriginal activism and the increasing concerns of southern Canadians over the environmental impacts of northern development. Almost unique to the case of arsenic at Yellowknife is the fact that environmentalism and Native activism briefly coalesced with the labor and occupational health movements to challenge the federal government's declarations that pollution levels remained safe in the local area.[56]

Indeed, Native and labor groups remained so concerned about the safety of arsenic emissions that they joined forces to produce their own health and environmental research on the issue. In 1975 the National Indian Brotherhood (NIB) conducted a small hair-sampling study of eighteen Native people in the Yellowknife area and arranged for the samples to be analyzed in a laboratory at the University of Toronto's Institute for Environmental Studies. Although high arsenic levels were found in samples from children, the government refused direct requests from the NIB to conduct testing on this seemingly at-risk group. In response, the NIB joined forces in 1977 with the United Steelworkers union and researchers at the University of Toronto to release a comprehensive hair study of local Native children and Giant mill workers. The results showed arsenic rates greater than 10 PPM in 30 percent of the study group and greater than 5 PPM in 50 percent of the samples tested. Robert Jervis, a professor at the Institute for Environmental Studies, suggested that results above 5 PPM were extremely rare in Canada and that none of the samples collected from a control group study in Whitehorse showed arsenic levels above this level. In Jervis's analysis, the NIB and United Steelworkers hair samples "clearly demonstrate a very high degree of exposure to arsenic for Indian children living at Yellowknife."[57]

Health Minister Marc Lalonde faced intense media scrutiny almost immediately after the NIB and United Steelworkers went public with their findings. Not only did his department's previous research efforts come under fire, but accusations of a cover-up persisted because the department failed to release a major environmental study of arsenic in the Yellowknife area completed in 1975, and because a confidential memo was leaked suggesting that an impending recommendation for dramatic reductions in arsenic levels in air should be kept from the public "as it may cause undue concern."[58] In response to the mounting criticism, Lalonde ordered a new independent study be conducted by the nonprofit Canadian Public Health Association (CPHA).[59] Much of the CPHA's work focused on urine samples from workers and local Native people. After extensive testing, the CPHA concluded that the impacts were largely confined to the workplace; arsenic levels in the general population remained below threshold safety levels. The CPHA's final report recommended ongoing monitoring of arsenic levels, careful washing of vegetables and berries, and the trucking of water to Ndilo and Dettah in winter, with warnings to locals not to use snow as a source of drinking water as studies still indicated high levels of concentration in this source.[60]

If the CPHA's report repeated the federal government's earlier claims that arsenic did not constitute a public health crisis, its release did not blunt criticism and concern. Dr. Hector Blejer, an occupational health expert from the United States who had been appointed as advisor to the task force at the request of the United Steelworkers and the NIB, suggested that the task force focused too narrowly on the threat from short-term arsenic poisoning while ignoring the increasingly well-established lung and skin cancer threat from long-term chronic exposure.[61] The final task force report ignored these criticisms, but the NIB and steelworkers union mentioned them liberally in public comments. The two groups questioned the idea that arsenic levels in Yellowknife were safe, because, as Dr. Blejer had suggested, safe levels simply do not exist for substances that cause cancer.[62]

The arsenic issue at Yellowknife did ultimately fade from public prominence in the 1980s. Official voices had declared that arsenic levels at Yellowknife were safe, while improvements to the arsenic collection technology for water and air emissions produced further dramatic reductions in pollution. At Giant mine, stack emissions fell from 850 pounds per day in 1973 to 29 pounds per day by 1979.[63] Although the substance may be retained in soils for long periods of time, there was a marked decline in arsenic on surface environments. One study indicated an 80 percent drop in arsenic trioxide in snow core samples from 1976 to 1986.[64] Combined with improvements to tailings storage and treatment (including the construction of an effluent treatment plant at Giant in 1981),[65] these results suggest that after nearly three decades, the federal government finally managed to mitigate the problem of acute arsenic pollution problems in Yellowknife.

Concerns about arsenic resurfaced in the 1990s, however, when Yellowknife environmental activists Chris O'Brien and Kevin O'Reilly requested an investigation of the environmental and health impacts of arsenic and sulfur dioxide from Giant mine under the NWT Environmental Bill of Rights, focusing in particular on the fact that regulators were permitting a known carcinogen to be emitted in close proximity to an urban area. Subsequent investigation concluded that emissions were within safe limits, but the two activists continued to raise concerns about the issue in local media. In the end, the question of arsenic releases into the Yellowknife environment was not solved for good until the Giant mine closed in 2004.[66]

THE AFTERLIFE OF ARSENIC

The controversy over arsenic disposal at Giant mine continues unabated, as the abandoned mine presents a massive contemporary environmental liability. In the early 1950s, Giant general manager A.K. Muir and several government officials were confident that arsenic trioxide dust deposited underground would be contained as permafrost became reestablished in the mine, though Muir suggested that cold air might have to be pumped in to counteract the heat rising from deeper tunnels.[67] Even as mine operations were winding down in the 1990s, it became apparent that permafrost had not reestablished itself and the water table was rising dangerously toward the chambers storing the arsenic dust. Government memoranda obtained by the National Indian Brotherhood in the 1970s revealed mounting concern about the potential mobilization of this highly toxic stored arsenic in groundwater.[68] In 2002 the federal government's commissioner of the environment and sustainable development highlighted Giant mine as one of four abandoned mines in northern Canada where massive public investment was required to clean up severe toxic legacies.[69] Three years later the federal and territorial governments signed an agreement to develop a remediation plan for the site.[70] By 2006 the Department of Indian and Northern Affairs Canada developed a plan involving the use of thermosyphon technology (passive heat exchange using natural convection) to freeze the arsenic underground permanently. After freezing, the site will require ongoing monitoring and maintenance, with water pumped from the mine indefinitely. Such a perpetual-care scenario prompted the city of Yellowknife to request an

FIGURE 10.3. Thermosyphons at Giant mine remediation test plot, Yellowknife, Northwest Territories, May 2011. Photograph by John Sandlos.

environmental assessment (EA) through the Mackenzie Valley Environmental Impact Review Board.[71] Although the EA resulted in new commitments from the federal government to fund research toward a permanent solution to the underground arsenic problem within a century, local concerns remain over the long-term risk of maintaining 237,000 tons of toxic material should there be no lasting resolution of the issue.[72]

The environmental assessment has also provided a forum for the Yellowknives Dene to highlight their historical memory of cultural and environmental loss associated with gold mining in the area. Primary sources tracking Native reactions to the introduction of mining at Yellowknife in the 1930s and subsequent issues with arsenic are difficult to find as the archival record belongs exclusively to the voices of government and mine company officials. Nonetheless, recent forums such as the environmental assessment and a public workshop on perpetual care, and older sources such as a Yellowknives Dene Community History and a set of hearing transcripts from 1995 on revisions to the Canadian Environmental Protection Act (CEPA), offer a clearer window into the historical impacts of Giant mine on the Yellowknives Dene.

Many elders clearly see the mines as the central agent of colonialism in the Yellowknife region, a progenitor of social, economic, and ecological changes that dramatically altered the Yellowknives' way of life based on hunting and trapping. At the 1995 CEPA hearings, elder Michel Paper described that way of life as it was before the mine and suggested how little contact the Yellowknives had with non-Native southerners:

> At that time, trapping was the way the people survived. Caribou was another source [. . .] by which the people survived. All the fish that were available were known by the people. The people lived a very healthy life by hunting for wildlife. All year long we would follow the caribou, and at that time we did not have to pay for wood. We did not pay for the food we gathered. We travelled by dog team only. When the firewood ran low in the camping area, we would move on to another place where there was plenty of wood. We did not pay for the firewood. That was the way our people lived in the past.
>
> In 1934–35 we heard news that the white people had arrived in the Yellowknife area. It was at Burwash Point. We travelled at night by dog team back to Yellowknife and we could see Burwash Point lit up from a distance. We heard that the white people had arrived, and we were afraid of them so we travelled back around the way of Dettah. At that time, the white people were also afraid of us.[73]

A community history prepared by the Yellowknives Elders Advisory Council in 1997 describes the impact of these new arrivals on patterns of subsistence in the local area:

> Explosions of dynamite by prospectors, air traffic, the development of a town and mines, the building of commercial fish plants, a prison, and roads, and the use of the land and waters for recreation. These developments contributed to the gradual withdrawal of moose and other animals, and to caribou changing their migration route through the area. In spring, Weledeh Yellowknives Dene used to wait for caribou returning north where the Prince of Wales Northern Heritage Centre now sits on Frame Lake. Although now it is rare to see moose near the Weledeh, these animals used to be common and could be relied on by Weledeh Yellowknives for food and clothing.[74]

The Yellowknives were not passive in the face of such dramatic changes. Paper and elder Isadore Tsetta suggested that they and twelve to fifteen other Yellowknives Dene found work at the mines (often hauling lumber), but such adaptations to the new mining economy did not erase from the elders' memories the mines' deleterious impacts.

Indeed, memories of the arsenic crisis of the early 1950s are widely recalled in the communities of Dettah and Ndilo and form the core narrative of the Yellowknives' encounter with the gold mines. The number of the dead and dates often vary according to the speaker (and may indicate that more fatalities occurred beyond that of the young boy described in the archival record). Regardless, Yellowknives elders and community leaders continually point to the tragic death and sickness of the 1950s as the most profound injustice associated with the gold mines. At the 1995 hearings, then-chief Fred Sangris told the legislators:

> The first case of death within our community came in 1959, when three children in the same family died in the Yellowknife Bay area. At that time, there was no adequate water delivery from either the government or DIAND [Department of Indian Affairs and Northern Development]. You had the responsibility to look after first nations, because there was that fiduciary obligation to do so. The family that lost the three children were compensated $1,000. That's all they were given. They were told, "Here, take the money, and forget about everything." Eventually, this person and his wife got into drinking because they couldn't deal with that. A lot of the first nations in this area did the same thing. One person mentioned here that they were powerless; yes, it's true.[75]

The community history produced by the Yellowknives Elders Advisory Council tells the poisoning story in this way:

> The people were never warned about the impacts and risks of living near mines. In late December of 1949, a massive emission from the Giant mine dispersed huge amounts of arsenic into the air, settling into the ice and snow. Melting snow in the spring of the following two years was so toxic that notices were printed in Yellowknife newspapers warning people not to drink or use the meltwater. Few Weledeh Yellowknives Dene could read the notices. Anyone who washed their hair with arsenic-laden meltwater in the next two springs went bald. [. . .] But the greatest tragedy occurred in spring 1951: four children in family camps in Ndilo died. The mine owners gave their parents some money, as if it could compensate for the loss. Women stopped picking medicine plants and berries, which used to grow thickly in the area of Giant mine. The people moved away, avoiding the mine area for some years, although it had once been so important to them.[76]

Statements such as these suggest that the Yellowknives Dene experienced a toxicity-induced alienation from the land that had once sustained them. Anthropologist Stuart Kirsch has described this process as the cultural loss associated with the weakening of ties to local landscapes and ecologies that have been made dangerous through industrial development or military activity.[77] The memories of many Yellowknives elders focused on the broad-scale impacts of toxic loading on humans, animals, water, and local food sources and the broken relationships among these overlapping ecologies. In an interview transcribed as part of a submission to the Giant Mine Remediation EA, elder Joseph Charlo described the impact of arsenic pollution:

> Ever since it started, I have never heard one good thing about mining: it destroys the land. We survive by the animals: all our ancestors lived by the animals on the land, and the animals were healthy. If we don't take care of the animals, if the mining starts up and the animals get contaminated, the people will also. They [i.e., the mining companies] should be careful as to how they work with the Dene and how they should work to protect the environment. And my wife, she remembers when she used to go berry picking in the Giant Mine area; she used to go there with her grandmother. Right now, you can't put anything in your mouth from that area: everything is contaminated. It's as if they've killed everything around here. We need to make a statement that we don't want to destroy anything on this land.[78]

Rachel Ann Crapeau further explained the sickness that spread from the land and water into the bodies of her people:

> Before the Yellowknives Dene understood what arsenic was, they were aware of changes that made them wary of the water, fish, berries and plants near the mine sites. When land users took their sled dogs through the tailings ponds that crossed their traditional trails, the dogs would lose the fur on their paws within a day or two. The Elders can recall people falling off their sled into the tailings ponds, which stayed open year-round, and becoming ill, losing their hair soon after. After many of their sled dogs died without obvious cause, dog owners stopped feeding them fish from Weledeh. People, too, started dying from cancer at a rate previously unknown to Yellowknives Dene.[79]

Isadore Tsetta suggested that that even in 1995 his community remained wary of the dangers associated with subsistence activity on the land, a loss of an economic base for which, he argued, the Yellow-knives should be compensated:

> We do not know what to do with the contaminated water now. We cannot use the water now. After the land is spoiled, plants cannot grow in the contaminated soils. That is the situation with us now. If justice was done [. . .] we should be compensated somehow for the contaminated water. Giant Mines and Con Mines have ruined the water and we cannot use it any more. We were here first, before the white people arrived and the mining started. We all know how the land was.[80]

At a public workshop on perpetual care held in 2011, Michel Paper suggested much the same idea when he said very simply, "People love the land but mining has changed the land and made it dangerous."[81] Through sixty years of gold mining in Yellowknife, the area's Native inhabitants have experienced not only pain associated with sickness and death but also fear associated with the hazardous nature of the land.

CONCLUSION: THE SHADOW OF THE GOLD MINES

Standing on the eastern shore of Yellowknife Bay in the community of Dettah, the huge Robertson shaft headframe at the Con mine once towered above all other city buildings located almost directly across the bay. When people from Dettah traveled around Back Bay to Yellowknife, the Giant mine's old roaster stack and headframes loomed over the narrow highway. Although these landmarks have recently been demolished as part of site reclamation plans, for decades the residents of what became Dettah and Ndilo (across the bay) lived almost literally within the shadows of Yellowknife's gold mines. These shadows had a long reach, causing one of the worst cases of industrial poisoning in the history of northern Canada. Although the Canadian government did mobilize public expertise and action in the face of the crisis, the campaign's narrow focus on data-driven threshold values for drinking water and air quality failed to account for the full range of ecological impacts that the production of a toxic landscape had on the Yellowknives Dene. The public health campaign was also limited by the fact that, for company and public officials, continued operation of the mines always remained sacrosanct regardless of the potentially dangerous material being loaded into the air and water bodies of the Yellowknife region. Never once did the federal government attempt to establish strict regulations for the amount of arsenic coming out of the gold mines' smelter stacks or contained in tailings effluent, so long as the local environment was able to assimilate and dilute the pollution to levels reasonably close to prevailing safe standards. Technological fixes such as the Cottrell ESP, the baghouse, and the Con impinger likely prevented further deaths in the short term, but government and company officials failed to consider mounting evidence suggesting the medical risks of low-dose arsenic exposure over long periods for mine workers and Yellowknife residents. Although no historical epidemiological study has traced these impacts in Yellowknife, certainly archival and oral records describing high rates of exposure among mill workers and First Nations people, and evidence of low dose impacts from other jurisdictions, all suggest that arsenic levels were not as safe as government officials claimed.[82]

The actual spread of arsenic also furthered the colonial advance of an industrial economy in the Yellowknife region. As a by-product of gold mining, arsenic emissions instigated material environmental changes and toxic exposures that reinforced the inequities associated with colonial resettlement of Dene territory.[83] The suddenness and severity of these ecological changes suggest that one need not rely on naïve notions of pristine nature or a static view of culture to acknowledge that arsenic poisoning precipitated a major disruption in the ties that bound the Yellowknives to their land. If the land and water became less toxic as public health initiatives reduced the amounts of arsenic in the local environment, it was as difficult for the Yellowknives to regain trust in a poisoned landscape as it was to reestablish fully the subsistence land base that had been invariably changed through related developments such as urban growth in Yellowknife and the expansion of surface impacts (including tailings ponds, infrastructure, and four open pits at Giant) associated with the mines.[84] The public health campaign also failed to find a long-term solution to the problem of arsenic disposal at Giant mine; the prospect of another mass poisoning from the many tons of arsenic trioxide stored in the mine continues to propagate the idea among the Yellowknives that the land is both sick and dangerous. For the Yellowknives, gold mining, environmental disaster, and cultural loss are all synonymous with one another, and are part of a history of environmental injustice and dispossession that continues to shape contemporary responses to mining in the Northwest Territories.

NOTES

This research was supported by Social Sciences and Humanities Research Council of Canada research grant number 866–2008–0016.

1. Warwick Bullen and Malcolm Robb, "Social-economic Impacts of Gold Mining in the Yellowknife Mining District," http://www.miningnorth.com/_rsc /site-content/library/Socio-Economic%20Impacts%20of%20Gold%20 Mining%20in%20Yellowknife%202002.pdf (accessed January 10, 2017). By comparison, Bullen and Robb suggest that Con Mine produced 5.5 million ounces of gold, or just over $2.5 billion in revenue. Though smaller than the Timmins, Ontario, gold-mining complex, where Canada's biggest producer, the Hollinger mine, yielded close to 20 million ounces of gold, the Yellowknife gold-mining complex remains historically significant because it provided the economic basis for the development of Yellowknife, the major modern settlement in the Northwest Territories.

2. For an overview of the remediation project, including links to a historical overview of the arsenic issue, see "Giant Mine Remediation Project," Aborigi-

nal Affairs and Northern Development Canada (AANDC), http://www.aadnc-aandc.gc.ca/eng/1100100027364 (accessed March 12, 2012).

3. See Susan Jackson, ed., *Yellowknife, NWT: An Illustrated History* (Yellowknife: Nor'West, 1990); *Yellowknife Tales: Sixty Years of Stories from Yellowknife* (Yellowknife: Outcrop, 2003).

4. In the United States and Mexico, mine smelter emissions largely remained unregulated until the U.S. Clean Air Act of 1967 (and amendments to the act in 1970 and 1977), the creation of the U.S. Environmental Protection Agency in 1970, and the U.S.-Mexico La Paz agreement of 1983. For an overview, see John D. Wirth, *Smelter Smoke in North America: The Politics of Transborder Pollution* (Lawrence: University Press of Kansas, 2000). For further examples illustrating the unregulated nature of air and water pollution caused by mining in North America, see Robert V. Bartlett, *The Reserve Mining Controversy: Science, Technology, and Environmental Quality* (Bloomington: Indiana University Press, 1980); Gray Brechin, *Imperial San Francisco: Urban Power, Earthly Ruin* (Berkeley: University of California Press, 1999); Kathleen A. Brosnan, *Uniting Mountain and Plain: Cities, Law, and Environment along the Front Range* (Albuquerque: University of New Mexico Press, 2002); Nicholas A. Casner, "Toxic River: Politics and Coeur D'Alene Mining Pollution in the 1930's," *Idaho Yesterdays* (Fall 1991): 2–19; Christopher J. Juggard, "Mining and the Environment: The Clean Air Issue in New Mexico," *New Mexico Historical Review* 69:44 (Fall 1994): 369–88; Andrew Isenberg, *Mining California: An Ecological History* (New York: Hill and Wang, 2005); Keith R. Long, "Tailings under the Bridge: Causes and Consequences of River Disposal of Tailings, Coeur D'Alene Mining Region, 1886 to 1968," *Mining History Journal* 8 (2001): 83–101; David Stiller, *Wounding the West: Montana, Mining, and the Environment* (Lincoln: University of Nebraska Press, 2000).

5. Timothy LeCain, "The Limits of 'Eco-Efficiency': Arsenic Pollution and the Cottrell Electrical Precipitator in the U.S. Copper Smelting Industry," *Environmental History* 5:3 (July 2000): 336–51; Timothy LeCain, *Mass Destruction: The Men and Giant Mines That Wired America and Scarred the Planet* (New Brunswick, NJ: Rutgers University Press, 2009). On attitudes toward the environment in the mining industry generally, see Duane A. Smith, *Mining America: The Industry and the Environment, 1800–1980* (Lawrence: University Press of Kansas, 1987).

6. Timothy J. LeCain, "When Everybody Wins Does the Environment Lose? The Environmental Techno-fix in Twentieth-Century American Mining," in *The Technological Fix: How People Use Technology to Create and Solve Problems*, ed. Lisa Rosner (New York: Routledge, 2004), 117–32. On questions of materiality and pollutants, see Gavin Bridge, "The Social Regulation of Resource Access and Environmental Impact: Production, Nature, and Contradiction in the US Copper Industry," *Geoforum* 31 (2000): 237–56; Gavin Bridge, "Material Worlds: Natural Resources, Resource Geography, and the Material Economy," *Geography Compass* 3 (2009): 1217–44; Jennifer Gabrys, "Sink: The Dirt of Systems," *Environment and Planning D: Society and Space* 27 (2009): 666–81.

7. Jody A. Roberts and Nancy Langston, "Toxic Bodies/Toxic Environments: An Interdisciplinary Forum," *Environmental History* 13 (2008): 629–35; Gregg Mitman, Michelle Murphy, and Christopher Sellers, eds., "Landscapes

of Exposure: Knowledge and Illness in Modern Environments," *Osiris* 19 (2004); Linda Nash, *Inescapable Ecologies: A History of Environment, Disease, and Knowledge* (Berkeley: University of California Press, 2006); J. Samuel Walker, *Permissible Dose: A History of Radiation Protection in the Twentieth Century* (Berkeley: University of California Press, 2000).

8. Nancy Langston, *Toxic Bodies: Hormone Disruptors and the Legacy of DES* (New Haven: Yale University Press, 2010); Nash, *Inescapable Ecologies*; Linda Nash, "Purity and Danger: Historical Reflections on the Regulation of Environmental Pollutants," *Environmental History* 13 (October 2008): 651–58; Sarah A. Vogel, "From 'the Dose Makes the Poison' to 'the Timing Makes the Poison': Conceptualizing Risk in the Synthetic Age," *Environmental History* 13 (October 2008): 667–73.

9. Brett L. Walker, *Toxic Archipelago: A History of Industrial Disease in Japan* (Seattle: University of Washington Press, 2010); Joy Parr, *Sensing Changes: Technologies, Environments, and the Everyday, 1953–2003* (Vancouver: University of British Columbia Press, 2010).

10. Saleem H. Ali. *Mining, the Environment, and Indigenous Development Conflicts* (Tucson: University of Arizona Press, 2003); Subhabrata Bobby Banerjee, "Whose Land Is It Anyway? National Interest, Indigenous Stakeholders, and Colonial Discourses," *Organization and Environment* 13 (March 2000): 3–38; Al Gedicks, *Resource Rebels: Native Challenges to Mining and Oil Companies.* (Cambridge, MA: South End Press, 2001); Robert Wesley Heber, "Indigenous Knowledge, Resources Use, and the Dene of Northern Saskatchewan," *Canadian Journal of Development Studies* 26 (2005): 247–56; Richard Howitt, *Rethinking Resource Management: Justice, Sustainability, and Indigenous Peoples* (London: Routledge, 2001); Stuart Kirsch, *Reverse Anthropology: Indigenous Analysis of Social and Environmental Relations in New Guinea* (Stanford, CA: Stanford University Press, 2006); Marcus B. Lane and E. Rickson Roy, "Resource Development and Resource Dependency of Indigenous Communities: Australia's Jawoyn Aborigines and Mining at Coronation Hill," *Society and Natural Resources* 10 (1997): 121–42; Lianne Leddy, "Cold War Colonialism: The Serpent River First Nation and Uranium Mining, 1953–1988" (PhD dissertation, Wilfrid Laurier University, 2011); Nicholas Low and Brendan Gleeson, "Situating Justice in the Environment: The Case of BHP at the Ok Tedi Copper Mine," *Antipode* 30 (1998): 201–26; Joan Martinez-Alier, "Mining Conflicts, Environmental Justice, and Valuation," *Journal of Hazardous Materials* 86 (2001): 153–70.

11. John Sandlos and Arn Keeling, "Claiming the New North: Development and Colonialism at the Pine Point Mine, Northwest Territories, Canada," *Environment and History* 18 (2012), 5–34; Arn Keeling and John Sandlos, "Environmental Justice Goes Underground? Historical Notes from Canada's Northern Mining Frontier," *Environmental Justice* 2:3 (2009): 117–25.

12. For general overviews of mining development in northern Canada in the early to mid-twentieth century, see K.J. Rea, *The Political Economy of the Canadian North* (Toronto: University of Toronto Press: 1968); Morris Zaslow, *The Northward Expansion of Canada, 1914–1967* (Toronto: McClelland and Stewart, 1988); and Liza Piper, *The Industrial Transformation of Sub-Arctic Canada* (Vancouver: University of British Columbia Press, 2009). For details on

early staking in the Yellowknife Bay area, see Ryan Silke, *The Operational History of Mines in the Northwest Territories, Canada* (Yellowknife, Northwest Territories: privately printed, 2009).

13. For information on arsenic trioxide production in the roasting process, see Markus Stoeppler, *Hazardous Metals in the Environment* (Amsterdam: Elsevier, 1992), 289.

14. For early emissions data and a discussion of the unregulated nature of emissions from 1949 to 1951, see A. J. de Villiers and P. M. Baker, *An Investigation of the Health Status of Inhabitants of Yellowknife, Northwest Territories* (Ottawa, Ontario: Occupational Health Division, Environmental Health Directorate, Department of Health and Welfare, 1970), 3–5. Detailed emissions data were also contained in correspondence from Dr. O. Schaefer, Northern Medical Research Unit to the A/Regional Director, Northern Region, National Health and Welfare, 4 November 1971, Record Group (RG) 29, vol. 2977, file 851–5–2, pt. 1, Library and Archives Canada, Ottawa, Ontario (hereafter LAC). Details on the pollution control equipment at Con and Giant may be found in W. H. Frost, Senior Medical Advisor, Medical Services Branch, National Health and Welfare, "Arsenic—Yellowknife," 28 October 1970, RG 29, vol. 2977, file 851–5–2, pt. 1, LAC. The roasting process also produced emissions of sulfur dioxide, the impact of which is not well documented save for references to local respiratory problems among asthmatics in testimony from Yellowknives chief Fred Sangris and Deh Cho grand chief Jim Antoine at hearings on the Canadian Environmental Protection Act in 1995. Fred Sangris and Gerry Antoine, Evidence, Parliamentary Hearings on Canadian Environmental Protection Act, 11 May 1995, Parliament of Canada, http://www.parl.gc.ca/content/hoc/archives /committee/351/sust/evidence/122_95–05–11/sust122_blk-e.html#0.1.SUST122 .000001.AA1040.A (accessed March 13, 2012).

15. For discussion of wallpaper as a source of contamination, and also of the well-water contamination issue, see Andrew Meharg, *Venomous Earth: How Arsenic Caused the World's Worst Mass Poisoning* (New York: MacMillan, 2005). For the Washoe dispute, see LeCain, "The Limits of 'Eco-Efficiency.'"

16. Meharg, *Venomous Earth*, 7–12.

17. A comprehensive overview of all these issues, including the prevalence of tailings spills, is contained in de Villiers and Baker, *An Investigation of the Health Status of Inhabitants of Yellowknife*, and a report authored by the Canadian Public Health Association, *Task Force on Arsenic—Final Report, Yellowknife, Northwest Territories* (Ottawa: CPHA, 1977). The toxicity of these tailings overflows was documented in a federal Environmental Protection Service report, R. R. Wallace, M. J. Hardin, and R. H. Weir, "Toxic Properties and Chemical Characteristics of Mining Effluents in the Northwest Territories," EPS Report no. EPS-5-NW-75-4 (Ottawa, Ontario: Department of the Environment, February 1975).

18. De Villiers and Baker, *An Investigation of the Health Status of Inhabitants of Yellowknife*, 11.

19. These tensions are explored in Sheilagh Grant, *Sovereignty or Security? Government Policy in the Canadian North, 1936–1950* (Vancouver: University of British Columbia Press, 1988), 195–99.

20. Yellowknives Dene First Nation, *Weledeh Yellowknives Dene: A History* (Dettah, Northwest Territories: Yellowknives Dene First Nation Council, 1997), 53–54.

21. We have not used the boy's name in response to a request from a family member who participated in our research on this issue. The information on the fatality comes from a discussion of the coroner's report at a heavily minuted meeting among government officials and two Giant mine managers held in Ottawa on 1 June 1951 to discuss the fatality. The minutes are contained in the RG 29, vol. 2977, file 851–5–2, pt. 1, LAC. All information related to the fatality can be verified in this file and the documents cited below pertaining to follow-up measures.

22. Kirkby's assessment of the situation is summarized in a memo from Dr. M. Matas to Dr. H. Falconer, 16 May 1951, RG 29, vol. 2977, file 851–5–2, pt. 1, LAC.

23. See Minutes of Meeting held to Discuss the Death of Indian Boy, Latham Island, 1 June 1951, RG 29, vol. 2977, file 851–5–2, pt. 1, LAC.

24. Advertisements appeared in the following editions of the *News of the North* in 1951: 6 April, p. 6; 13 April, p. 5; 20 April, p. 5; 6 July, p. 6; 13 July, p. 6. One article, titled "Cottrell Plant Now Operates at Giant Mine," appeared in the 9 November 1951 issue, but even here no mention is made of the public health issues that had emerged by 1951. A study produced in the 1970s by the National Indian Brotherhood (a Native advocacy group) claimed that warning signs about drinking water from Back Bay were not posted in local Native languages until fall 1974. See Lloyd Tataryn, "Arsenic and Red Tape," National Indian Brotherhood, unpublished manuscript, n.d., University of Alberta Library.

25. A.K. Muir, General Manager, Giant Yellowknife Gold Mines, Ltd., to G.E.B. Sinclair, Director, Northern Administration and Lands Branch, Department of Resources and Development, n.d., RG 85, vol. 40, file 139–7, pt. 1, LAC.

26. The reference to Henry Lafferty is contained in a memo from P.E. Moore, Director of Indian Health Services, to L.I. Pugsley, Laboratory Services, 16 November 1953, RG 85, vol. 40, file 139–7, pt. 1, LAC. The specific references to skin conditions such as keratosis, hyperpigmentation, and paresthesia are found in a memo from Dr. O. Schaeffer to Dr. B. Wheatley, Environmental Contaminant Program, Medical Services Branch, Health and Welfare Canada, 1 May 1978, Prince of Wales Northern Heritage Centre, Yellowknife, Northwest Territories (hereafter cited as PWNHC), G-2008–028, box 9, file 16.

27. Yellowknives Dene First Nation, *Weledeh Yellowknives Dene: A History*, 52–53.

28. Laurie Cinnamon, Oral Interview, in Jackson, *Yellowknife, NWT*, 85.

29. Barbary Bromley, Oral Interview, in *Yellowknife Tales*, 97–98.

30. Helen Kilkenny, Oral Interview, in Jackson, *Yellowknife, NWT*, 114–15.

31. Minutes of Meeting held to Discuss the Death of Indian Boy, Latham Island. See also K. Raht to W.G. Jewitt, Manager of Mines, CM&S, 28 June 1951, RG 29, vol. 2977, file 851–5–2, pt. 1, LAC.

32. Minutes of Meeting held to Discuss the Death of Indian Boy, Latham Island. For Dr. Stanton's approval of the underground storage method, see his

letter to G. E. B. Sinclair, Director, Northern Administration and Lands Branch, Department of Resources and Development, 27 February 1951, in Giant Mine Environmental Assessment: IR Response, Round 1: Information Request, Alternatives North no. 01, 17 June 2011, p. 67, http://www.reviewboard.ca/upload /project_document/EA0809-001_IR_round_1_responses_to_Alternatives _North_1328902602.PDF (accessed October 29, 2013).

33. Muir's report of the shipping delays is contained in Minutes of Meeting held to Discuss the Death of Indian Boy, Latham Island. A reference to the request that Giant expedite installation of the Cottrell ESP is contained in a letter from Sinclair to S. Homulos, Mining Inspector, Yellowknife, 19 June 1951, RG 29, vol. 2977, file 851–5–2, pt. 1, LAC. See also C. K. LeCapelain, Acting Director, Northern Administration and Lands Branch, to Muir, 14 August 1951, RG 29, vol. 40, file 139–7, pt. 1, LAC.

34. The Con impinger removed 85 percent of arsenic from roaster smoke, and in any case Con processed lower amounts of arsenopyrite ore (as noted above). For emissions data, see de Villiers and Baker, *An Investigation of the Health Status of Inhabitants of Yellowknife*. Reference to the second ESP and its impact on emissions is found in D. A. Gemmill, Yellowknife Environmental Survey, Summary Report (Ottawa: Environmental Protection Service, Department of the Environment 1975), RG 29, vol. 2977, file 851–5–2, pt. 4, LAC. Negus Mines was permitted to construct an impinger using Con's wet scrubber technology, though the government preferred Giant's dry storage method. See Minutes, Meeting held in Room 101 of the Norlite Building to discuss the Arsenic Problem at Yellowknife as the Result of the Proposal of Negus Mines Limited to Commence Roasting Operations, 30 July 1951, RG 85, vol. 40, file 139–7, pt. 1, LAC. Negus Mines constructed a roaster in November 1951, and the plant went into full operation in November 1952. Problems with the plant and the recovery of lower-than-expected ore grades resulted in the closure of the mine and the sale of all claims to Cominco in March 1953. See Silke, *The Operational History of Mines in the Northwest Territories, Canada*, 316.

35. Plans for the survey are laid out in a memo from Dr. Kingsley Kay, Chief, Industrial Health Laboratory, to G. E. B. Sinclair, Director, Northern Administration and Lands Branch, 25 October 1951, RG 85, vol. 40, file 139–7, pt. 1, LAC.

36. See Minutes of Meeting Held to Discuss the Death of Indian Boy, Latham Island.

37. On the toxicological approach to health and contaminants, see Nash, "Purity and Danger," 651–58.

38. The archival files contain two graphs displaying arsenic measurements from 1953 to 1960, titled, "Arsenic in Townsite Tap Water, Yellowknife, N.W.T." A third is titled "Arsenic in Giant Tap Water, Yellowknife, N.W.T.," and a fourth, "Arsenic in Con Tap Water, Yellowknife, N.W.T.," 1953–60. All in RG 29, vol. 2977, file 851–5–2, pt. 1, LAC.

39. Dr. O. Schaefer, Northern Medical Research Unity to the A/Regional Director, Northern Region, National Health and Welfare, 4 November 1971, RG 29, vol. 2977, file 851–5–2, pt. 1, LAC. For background on liquid effluent

in Back Bay, see Canadian Public Health Association, *Task Force on Arsenic—Final Report, Yellowknife Northwest Territories,* 59–62.

40. Data on contamination of vegetation are contained in "Arsenic on Grasses in Yellowknife, 1954–61," and "Mean Values of Arsenic Yellowknife Vegetation, PPM," RG 29, vol. 2977, file 851–5–2, pt. 1, LAC. Mention of the US standard of 1 PPM is made in a memo from a person or persons coded "M17" to Dr. Procter, 10 December 1965, RG 29, vol. 2977, file 851–5–2, pt. 1, LAC.

41. M17 to Dr. Procter, 10 December 1965.

42. The results of the meeting are reported in M17 to Dr. Procter, memorandum, 20 December 1965, RG 29, vol. 2977, file 851–5–2, pt. 1, LAC.

43. Mention of the anemia is made in memos from Dr. Butler, Regional Director, National Health and Welfare, to Dr. Frost, Director General, Medical Services, Ottawa, 24 August and 17 August 1967, RG 29, vol. 2977, file 851–5–2, pt. 1, LAC.

44. Canadian Public Health Association, *Task Force on Arsenic—Final Report, Yellowknife, Northwest Territories,* 49–51.

45. See the discussion of this issue in Tataryn, "Arsenic and Red Tape," 10–14. See also Michel Sikyea to Jean Chretien, 25 September 1973, Giant Mine Remediation Public Registry http://reviewboard.ca/upload/project_document /EA0809–001_Letter_-_YKDFN_to_DIAND_Minister_J__Chretien_-_Sept _1973.pdf (accessed November 28, 2012). Sikyea was a highly regarded elder in Ndilo.

46. Dr. G.C. Butler, Regional Director, Northern Region, to Director General, Medical Services, 9 June 1970, RG 29, vol. 2977, file 851–5–2, pt. 1, LAC.

47. Dr. H.A. Procter, Director, Medical Services to Dr. E.A. Watkinson, Director, Health Services, 29 September 1967, RG 29, vol. 2977, file 851–5–2, pt. 1, LAC.

48. Butler to Procter, 30 October 1967, RG 29, vol. 2977, file 851–5–2, pt. 1, LAC.

49. Butler to Director General, Medical Services, 23 September 1970, RG 29, vol. 2977, file 851–5–2, pt. 1, LAC.

50. S.M. Hodgson, Commissioner, NWT, to Jean Chrétien, Minister of Indian Affairs and Northern Development, 25 September 1970, RG 29, vol. 2977, file 851–5–2, pt. 1, LAC.

51. Chrétien to John Munro, Minister of Health and Welfare, 2 October 1970, RG 29, vol. 2977, file 851–5–2, pt. 1, LAC. Reference to the order to Butler and the reasoning behind it are contained in a memo from Munro to Chrétien, n.d., RG 29, vol. 2977, file 851–5–2, pt. 1, LAC.

52. For an overview of this third period of heightened public concern over arsenic, see F.J. Colville, Senior Advisor, NWT Region, Arsenic in Yellowknife—A Perspective, 25 September 1979, G-2008–028, Box 9, File 17, PWNHC.

53. Statement by the Honourable Marc Lalonde on Arsenic Pollution in Yellowknife, NWT, 9 January 1975, RG 29, vol. 2977, file 851–5–2, pt. 2, LAC. The de Villiers study concluded that there was evidence of skin rashes among those occupationally exposed to arsenic in Yellowknife, and also high rates of respiratory disease that could be linked to arsenic. But the report concluded that

the links between arsenic and lung cancer were "uncertain." De Villiers and Baker, *An Investigation of the Health Status of Inhabitants of Yellowknife*, 10.

54. R.D.P. Eaton, Analysis of Hair Arsenic Results, Yellowknife, 1975, G-2008–028, Box 9, file 17, PWNHC.

55. Even when the Canadian public was concerned about air and water pollution, federal regulation tended to lag behind that of the United States. See Dimitry Anatsakis, "'A War on Pollution'? Canadian Responses to the Automobile Emissions Problem, 1970–1980," *Canadian Historical Review* 90:1 (March 2009): 99–136; Doug Macdonald, *The Politics of Pollution: Why Canadians Are Failing Their Environment* (Toronto: McClelland and Stewart, 1991); and Jennifer Read, "'Let Us Heed the Voice of Youth': Laundry Detergents, Phosphates, and the Emergence of the Environmental Movement in Ontario," *Journal of the Canadian Historical Association*, New Series 7:1 (1996): 227–50. For urban pollution, see Arn Keeling, "Sink or Swim: Water Pollution and Environmental Politics in Vancouver, 1889–1975," *BC Studies* 142–143 (Summer 2004): 69–101.

56. On the linkage between southern environmentalism, northern development, and Native concerns in the 1970s, see Robert Page, *Northern Development: The Canadian Dilemma* (Toronto: McClelland and Steward, 1986). For a further analysis of complex and conflicting Dene attitudes to development and environmentalism, see Paul Sabin, "Voices from the Hydrocarbon Frontier: Canada's Mackenzie Valley Pipeline Inquiry, 1974–1977," *Environmental History Review* 18:1 (Spring 1995): 17–48. Evidence of southern environmentalist concern for northern pollution and development issues can be seen in the large number of protest letters directed at Minister Lalonde after the story broke on the CBC and in the newspapers. These letters are contained in RG 29, vol. 2978, file 851–5-1, pt. 5, LAC.

57. Statement by Prof. Robert E. Jervis, Department of Chemical Engineering and Applied Chemistry, and Institute for Environmental Studies, re: Yellowknife Arsenic Pollution Problem, 15 January 1977, University of Alberta Library. See also the report by Tataryn, "Arsenic and Red Tape," and "Document [*sic*] Released by the National Indian Brotherhood, the United Steelworkers of America, and the University of Toronto on January 15, 1977," unpublished manuscript, Canadian Circumpolar Institute Library. The federal government's acting director of the Environmental Contaminants program cited a consensus among some researchers that hair levels below 1 PPM indicated a person who was not exposed to arsenic, while others suggested anything below 5 PPM as a normal level for arsenic. See Brian Wheatley to L.M. Black, Director General, Program Management, 28 January 1977, RG 29, vol. 2977, file 851–5-2, pt. 4, LAC.

58. Editorial, "Ottawa Hides the Poison," *Toronto Globe and Mail*, 19 January 1977. See also Victor Malarek, "Yellowknife Arsenic Level 'Horrendously High,'" *Toronto Globe and Mail*, 17 January 1977; "Private Study Shows Yellowknife Arsenic Level Dangerously High," *Edmonton Journal*, 17 January 1977; all clippings found in RG 29, vol. 2977, file 851–5-2, pt. 4, LAC. The report in question is Gemmill, Yellowknife Environmental Survey, Summary Report.

59. News Release, "Task Force to Study Arsenic," 18 January 1977, G-2008–028, box 9, file 17, PWNHC.

60. Canadian Public Health Association, *Task Force on Arsenic—Final Report, Yellowknife, Northwest Territories.*

61. H. P. Blejer, Evaluation of Canadian Public Health Association Task Force on Arsenic Interim Report of May 1977, RG 29, vol. 2978, file 851–5–2, pt. B, LAC.

62. Tataryn, "Arsenic and Red Tape"; Paul Falkowski, Environmental and Occupational Health Representative, United Steelworkers of America, Presentation to the National Indian Brotherhood 8th Annual General Assembly, 14 September 1977, unpublished transcript, University of Alberta Library.

63. Associate General Director, Medical Services Branch, to A/Regional Director, NWT Region, 11 October 1984, G-2008–028, box 9, file 17, PWNHC.

64. Lorne C. James, Pollution Control Officer, Department of Renewable Resources, Government of the Northwest Territories, to Ranjit Soniassy, Northern Affairs Program, 23 June 1986, G-1993–006, box 19, file 13 408 024, PWNHC. For discussion of the complex issue of absorption and retention in soil, see S. Mahimairaja, N. S. Bolan, D. C. Adriano, and B. Robinson, "Arsenic Contamination and Its Risk Management in Complex Environmental Settings," *Advances in Agronomy* 86 (2005): 1–82.

65. Indian and Northern Affairs Canada, *Giant Mine Remediation Plan* (prepared by Giant Mine Remediation Team, SRK Consulting and SENES Consulting, July 2007), http://reviewboard.ca/upload/project_document/EA0809 –001_Giant_Mine_Remediation_Plan_1328900464.pdf (accessed March 13, 2012), 14–15.

66. Chris O'Brien and Kevin O'Reilly to Titus Allooloo, Minister of Renewable Resources, Government of the Northwest Territories, 22 April 1991. The results of the study are summarized in Allooloo to O'Reilly, 5 July 1993. For media coverage, see Editorial, "Report Fails to Clear the Air," *Yellowknifer*, 9 July 1993, p. 7. All correspondence and clippings in the private papers of Kevin O'Reilly.

67. Muir to G. E. B Sinclair, Director, Northern Administration and Lands Branch, 24 February 1951, in Giant Mine Environmental Assessment: IR Response, Round 1: Information Request, Alternatives North no. 01, 17 June 2011, pp. 68–71, http://www.reviewboard.ca/upload/project_document /EA0809–001_IR_round_1_responses_to_Alternatives_North_1328902602 .pdf (accessed October 29, 2013).

68. These memos from mining engineers and Environment Canada officials are reproduced in "Document Released by the National Indian Brotherhood, the United Steelworkers of America, and the University of Toronto on January 15, 1977."

69. Office of the Auditor General of Canada, *Report of the Commissioner of the Environment and Sustainable Development, 2002* (Ottawa: Minister of Public Works and Services, 2002), chapters 2 and 3. Giant featured again in the commissioner's most recent report on toxic sites in Canada; see Office of the Auditor General of Canada, Chapter 3, "Federal Contaminated Sites and Their Impacts," in *Spring Report of the Commissioner of the Environment and*

Sustainable Development (Ottawa: Office of the Auditor General of Canada, 2012).

70. Cooperation Agreement Respecting the Giant Mine Remediation Project between Canada and the Government of the NWT, 15 March 2005, Northwest Territories Municipal and Community Affairs, http://www.maca.gov.nt.ca /resources/Cooperation_Agreement_Giantmine_remediation.pdf (accessed March 13, 2012).

71. Indian and Northern Affairs Canada, *Giant Mine Remediation Plan*.

72. Alternatives North, *From Despair to Wisdom: Perpetual Care and the Future of Giant Mine*, A Report on a Community Workshop, September 26–27, 2011, https://anotheralt.files.wordpress.com/2016/02/2011-09-26-giant-perpetual-care-workshop-report.pdf (accessed 14 January 2017); Carol Raffensperger, "Principles of Perpetual Care: The Giant Mine, Yellowknife, Northwest Territories" (prepared for Alternatives North as a submission to the Mackenzie Valley Impact Review Board, December 2011), http://www.reviewboard.ca/upload/ project_document/EA0809–001_Principles_of_Perpetual_Care-_Report_from_ Alt_North_1329867038.pdf (accessed March 16, 2012). For more on the alternatives for remediating the Giant mine site, see SRK Consulting, *Study of Management Alternatives: Giant Mine Arsenic Trioxide Dust* (Yellowknife, May 2001).

73. Michel Paper, Evidence, Parliamentary Hearings on Canadian Environmental Protection Act, 11 May 1995, http://www.parl.gc.ca/content/hoc/archives /committee/351/sust/evidence/122_95–05–11/sust122_blk-e.html#0.1.SUST 122.000001.AA1040.A (accessed March 13, 2012).

74. Yellowknives Dene First Nation, *Weledeh Yellowknives Dene: A History*, 50.

75. Fred Sangris, Evidence, Parliamentary Hearings on Canadian Environmental Protection Act, 11 May 1995, http://www.parl.gc.ca/content/hoc/archives /committee/351/sust/evidence/122_95–05–11/sust122_blk-e.html#0.1.SUST122 .000001.AA1040.A (accessed March 13, 2012).

76. Yellowknives Dene First Nation, *Weledeh Yellowknives Dene: A History*, 52.

77. Stuart Kirsch, "Lost Worlds: Environmental Disaster, 'Culture Loss,' and the Law," *Current Anthropology* 42 (April 2001): 167–98.

78. Yellowknives Dene First Nation presentation to Mackenzie Valley Environmental Impact Review Board Scoping Session, 23 July 2008, http://www .reviewboard.ca/upload/project_document/1219099111_15606YKDFNUndert aking10.pdf (accessed May 20, 2011).

79. Ibid.

80. Isadore Tsetta, Evidence, Parliamentary Hearings on Canadian Environmental Protection Act, 11 May 1995, http://www.parl.gc.ca/content/hoc/archives /committee/351/sust/evidence/122_95–05–11/sust122_blk-e.html#0.1.SUST 122.000001.AA1040.A (accessed March 13, 2012).

81. Alternatives North, *From Despair to Wisdom*.

82. For studies of long-term impacts of low-dose exposure to arsenic, see Chung-Min Liao, Huan-Hsiang, Chi-Leng, Ling-I Hsu, Tzu-Ling Lin, Szu-Chieh Cheh, and Chien-Jen Chen, "Risk Assessment of Arsenic-Induced Internal

Cancer at Long-Term Low Dose Exposure," *Journal of Hazardous Materials* 165 (2009): 652–63; Badal Kumar Mandal and Kazuo T. Suzuki, "Arsenic around the World: A Review," *Talanta* 58 (2002): 201–35; Jack C. Ng, Jianping Wang, and Amjad Shraim, "A Global Health Problem Caused by Arsenic from Natural Sources," *Chemosphere* 52 (2003): 1353–59; Lincoln Polissar, Kim Lowry-Coble, David A. Kalman, James P. Hughes, Gerald van Belle, David S. Covert, Thomas M. Burbacher, Douglas Bolgiano, and N. Karle Mottet, "Pathways of Human Exposure to Arsenic in a Community Surrounding a Copper Smelter," *Environmental Research* 53 (1990): 29–47.

83. For discussion of the links between landscapes of exposure and inequality, see Nash, *Inescapable Ecologies,* and Mitman, Murphy, and Sellers, "Landscapes of Exposure."

84. For the similar experience of the Yonggom people and the Ok Tedi mine, see Kirsch, *Reverse Anthropology,* chapter 7. Questions of indigenous trust in mining companies and scientific authorities are discussed in Leah S. Horowitz, "'Twenty Years Is Yesterday': Science, Multinational Mining, and the Political Ecology of Trust in New Caledonia," *Geoforum* 41 (2010): 617–26.

CHAPTER 11

Iron Mines, Toxicity, and Indigenous Communities in the Lake Superior Basin

NANCY LANGSTON

This chapter explores toxics mobilized by iron mining in the Lake Superior basin, a region shared between the United States and Canada that is currently witnessing a major mining boom (figure 11.1). Toxics from iron mines have moved from their sites of production and consumption into much broader and dispersed spaces. As they flow into water, they bioaccumulate in fish and eventually make their way into the people who eat those fish. These contaminants have permeated global ecosystems, crossing international boundaries to contaminate people far from initial sources of production and consumption. Their toxic residues not only complicate political boundaries but also confuse temporal distinctions, for their legacies persist long after they have been banned. Moreover, the risks of exposure to these chemicals is rarely distributed in a human population equitably.

In 2011, the iron ore–mining company Gogebic Taconite (GTAC) proposed the world's largest open-pit mine in the world just upstream from the reservation of the Bad River Band of the Lake Superior Tribe of Chippewa. While the mine would have been sited on one side of a legal boundary, the waters would have flowed across those boundaries to contaminate water, fish, and indigenous communities downstream (figure 11.2).[1]

If permits had been approved, the GTAC mine would have been located within ceded territories where the tribes retained hunting, fishing, gathering, and co-management rights when they signed treaties in

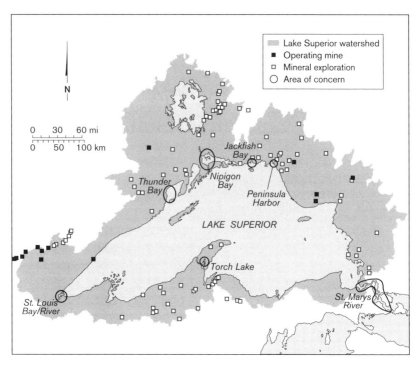

FIGURE 11.1. Extent of mining exploration and activity around Lake Superior, 2016. Map by Bill Nelson, modified from a base map created by Great Lakes Indian Fish and Wildlife Commission.

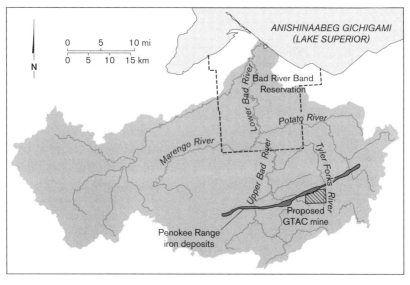

FIGURE 11.2. Location of the proposed Gogebic Taconite mine in relation to the Bad River Reservation, Lake Superior watershed. Map by Bill Nelson.

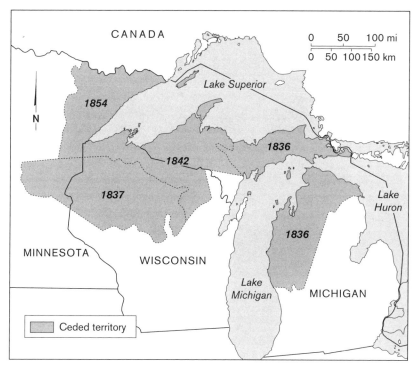

FIGURE 11.3. Map of the ceded territories of the Lake Superior Ojibwe. Map by Bill Nelson, modified from a base map created by Great Lakes Indian Fish and Wildlife Commission.

the nineteenth century enabling white settlement (figure 11.3). Bad River Band members point out that it is no accident that the proposed mine site was within their watershed. As environmental justice scholars argue, environmental exposures vary in their spatial distribution, and social inequalities influence vulnerability. Tribal members have often borne the greatest burden from the toxic wastes fostered by past mining projects, but they have rarely had much decision-making power in the planning process.[2]

The Bad River runs through the potential mine site before entering the 16,000-acre Kakagon–Bad River Sloughs—the largest undeveloped wetland complex in the upper Great Lakes. In 2012, this was designated as a Ramsar Site, recognizing it as a wetland of international importance. The Ramsar Convention on Wetlands noted that "as the only remaining extensive coastal wild rice bed in the Great Lakes region, it is critical to ensuring the genetic diversity of Lake Superior wild rice."[3]

FIGURE 11.4. Taconite launder. The conveyor discharges
taconite tailing residue into Lake Superior at Reserve Mining
Company's taconite plant in Silver Bay, Minnesota, June
1973. Photograph by Donald Emmerich, U.S. National
Archives and Records Administration (Record 3045077).

The sloughs make up 40 percent of the remaining wetlands on Lake
Superior's coast, and they contain the largest natural wild rice beds in
the entire world (figure 11.4).

For members of the Bad River Band, these wild rice beds are central
to their identity and important to their economy. When the Anishinaabe migrated westward from the St. Lawrence River valley, the ances-
tors of the Bad River Band chose to make their homes along the Kakagon Sloughs because the wild rice beds they found there had figured
significantly in their visions. In the sloughs, they found wild rice, or

manoomin, "the food that grows on water," which they continue to see as a "sacred gift from the Creator."[4] The wild rice became a major portion of their subsistence, and fisheries supported by the sloughs became equally important for subsistence and for economic development. Wild rice is extremely sensitive to sulfates in the watershed that may become mobilized by taconite mines. Band members argued that stopping the mine was essential for their survival, not just to ensure thriving wild rice beds and fisheries but also to sustain the connection with the past that is at the core of their cultural identity.

In contrast, many Euro-American residents of nearby communities such as Hurley, Wisconsin, argued that the proposed mine was the only thing that could rescue them from the economic devastation that followed the closure of local hematite iron mines in the 1960s. Hurley lies just outside the subwatershed that would have been affected by the proposed mine, so the town would not have been directly influenced by any water quality issues presented by the mine. Denying permits for the proposed mine, Hurley residents argued, would have amounted to a form of economic suicide.

As John Sandlos and Arn Keeling have argued for Pine Point Canada, "Complex and contested meanings or place and community" are common at mining sites. "While regarded by 'outsiders' as brutal, degraded or even toxic," the authors observe, "former mining landscapes may be touchstones of community identity and memory and provide both material and cultural resources for economic recovery or even political resistance."[5] This is certainly true for iron mines in the Lake Superior basin. Different communities within the basin have quite different interpretations of that mining past, and these views about the past help to shape their perspectives on current mining proposals.

IRON-MINING HISTORIES IN THE BASIN

While many Euro-American locals see iron mining as part of their past, the differences between types of iron ore are typically obscured in the retellings of history. Yet these differences are critical for both environmental and social consequences. Briefly, iron ore in the Lake Superior basin falls into two broad types: direct-shipping ore (generally hematite) that contains quite high concentrations of iron, and low-grade ore (generally taconite) that needs to be concentrated or processed before shipping. Historically, hematite, at 60–70 percent iron, needed little processing before shipping, required much labor to produce, and in

Wisconsin was mined underground. Taconite, in contrast, was more abundant in the Lake Superior basin but much less concentrated (typically about 25% iron in an ore body). Mining taconite requires enormous open pits, extensive technological and financial investments, vast quantities of water, and minimal human labor.

The Penokee Range lies in what is now northern Wisconsin; across the border in the Upper Peninsula of Michigan, the same range is named the Gogebic Range; when referring to the entire range, this essay uses the term *Penokee-Gogebic Range*. Once the center of a thriving—but short-lived—hematite-mining economy followed by clear-cutting, the region's forests have now grown back, enough so that one environmental group can write that the Penokees "form the heart of one of the most isolated, pristine and scenic portions of the state." The first phase of the proposed GTAC taconite mine would have created a pit 5 miles long and 1,000 feet deep. Eventually, plans called for 22 miles along the ridge of the Penokees to be carved off. The proposed techniques bore little resemblance to the deep-shaft mining for hematite that was the basis of the region's mining history. Rather, this would have become the first "mountaintop removal" mine in the upper Great Lakes region. As one anti-mining activist writes: "Gogebic Taconite (GTac) has grandiose plans for the Penokee mountains: To blow them to smithereens with a series of 5.5 million–pound explosives—each similar to the impact of the bomb dropped on Hiroshima—in order to extract low-grade iron known as taconite. Waste rock with the potential to leech billions of gallons of sulfuric acid from what would be the largest open pit iron mine in the world could be dumped."[6] While the rhetoric here may seem exaggerated, the details are not.

The proposed mine would have targeted a band of iron-rich ore in the Penokee-Gogebic Range known as the Ironwood Formation that is a significant ore deposit in the national and indeed global context. Best estimates suggest that the Ironwood Formation contains at least 3.7 billion tons of taconite ore that could be mined economically, or 20 percent of known iron ore deposits in the United States. This translates into 1 billion tons of steel, or sixty-six years of domestic supply.[7]

This is such an enormous deposit, GTAC argued, that mining it was inevitable. The language of inevitability figured heavily in the pro-mining discussions. But, as U.S. Steel decided three decades earlier when it did bulk sampling and then decided not to mine taconite within the Ironwood Formation, its geological context made it an extremely difficult ore deposit to exploit without losing money.[8] The rock is extremely hard (meaning one needs enormous blasting capabilities), and the deposit is tilted at a 65 degree angle, overlaid with 200–300 meters of overburden, and banded with quartzite and shale—all details that require extensive energy and infrastructure for extracting ore economically.

Environmental regulations for mines were much looser in the 1980s, and GTAC has never mined taconite. How, then, could GTAC propose to mine such a deposit, when U.S. Steel had decided it was impossible three decades before?[9] The simplest way to mine taconite cheaply is to reduce labor and environmental costs. And that is what GTAC began to do in 2011, when rising steel prices in Asia made formerly uneconomic ore bodies seem plausible again. New, enormous mining machines have now been developed that can extract up to 200 tons of rock out of the mine in a single load, thus lowering labor costs.[10] Reducing environmental costs would be possible if the industry could block the implementation of new federal standards that limit mercury emissions from the facilities needed to process taconite ore. And, if a company could persuade a state to exempt the industry from water quality regulations, tailings could be dumped cheaply into streams and wetlands.

Some environmental groups opposed to the GTAC mine argued that the Lake Superior basin and the Penokee-Gogebic Range in particular are essentially pristine and should never be mined. Residents of former iron-mining communities argued just the opposite: they said this has long been an iron-mining area, so it should continue to be one. Why did different groups have such different interpretations of the past, and how did these contested stories affect policy disputes?

Mining in the Lake Superior basin is not new, but the technologies for extensive extraction are recent. The first records of ore extraction date from thousands of years ago, with indigenous mining of copper ores on the Keweenaw Peninsula. Indians had exploited the mineral resources of the Keweenaw Peninsula on Lake Superior for at least 7,000 years, soon after the last glacier retreated. In the 1840s, word of copper deposits on the Keweenaw Peninsula spread to the East Coast and Europe.

Thousands of miners poured into the basin from Cornwall (where mines were laying off workers), eastern Europe, Italy, Finland, and elsewhere in the Americas. The federal government negotiated the Treaty of La Pointe in 1842 with the Anishinaabe nations, which required them to cede northern Wisconsin and the western half of the Upper Peninsula to the United States. Mining companies immediately moved into the area, exploiting first copper and then the rich iron ore deposits.

An 1848 report by A. Randall described the presence in the Penokee-Gogebic range of hematite iron ore, and extraction began in 1886. To the west in Minnesota, on the Mesabi Range, iron mining began with discovery of hematite iron ore in 1865; production followed in 1885 and rapidly expanded through the 1890s. In both places, mining efforts targeted the high-grade hematite ores that were concentrated and did not require extensive processing before being shipped across the Great Lakes to steel mills.[11]

By the early twentieth century, fully "85% of domestic ore production" came from soft, high-grade hematite ore mines on the U.S. side of the Lake Superior basin, in Michigan, Minnesota, and Wisconsin. Forty underground mines worked Wisconsin's Ironwood Formation between 1877 and 1967, extracting 325 million tons that were shipped to blast furnaces in the lower Great Lakes.[12] Mining towns such as Hurley, Ironwood, Iron River, and Montreal boomed briefly during the hematite era.

To reach the high-grade hematite deposits in the Penokee-Gogebic Range, miners dug deep shaft mines propped up with timbers from local forests. Waste rock was dumped near the mines, and some of these piles remain visible today. These mines did have environmental consequences: the shafts had to be dewatered to keep them from filling up with groundwater, and pumping the water from shafts into local streams presumably had some effect on watershed ecology. Some heavy metals were exposed to air, oxidizing and increasing the mobility of toxic metals into ecosystems and human bodies. Yet these consequences were much less extensive than open-pit taconite mining. As soon as the boom collapsed, first alder, then maple trees grew back, hiding the shaft holes and cloaking some of the slag piles, allowing people to imagine the forests as pristine and untouched.

The nineteenth-century mining boom had significant effects on Native communities. When the Anishinaabe were forced onto reservations in 1842 to make the rest of the region available to miners, disease, poverty, and despair often resulted. Yet the Anishinaabe successfully defended their right to remain on their homeland and its waters. They retained usufruct rights to the ceded territories, making certain that their members could continue to hunt, fish, and gather in perpetuity. These treaty rights have become central to the governance of mining conflicts in recent decades.

DEPLETION AND THE SHIFT TO TACONITE

As Jeffrey Manuel argues in this volume (see chapter 7), the shift to mining lower-grade taconite ore in the Lake Superior basin led to significant changes in environmental and political relationships. As early as the 1890s, mining engineers had argued that while surface deposits of the concentrated hematite appeared limited, beneath them lay extensive deposits of taconite. Hematite, in fact, is closely related to taconite, for it represents "the oxidized and purified surface weathering product of the much more extensive but lower grade taconite ore beneath."[13] Taconite was much more extensive, but because it lay buried under deep rock, and because it was such a low-grade ore, most engineers in the nineteenth century assumed it would be difficult to mine cost-effectively.

By the end of World War II, fears of iron ore depletion were becoming common. The war effort had demanded significant quantities of iron, and during the war, the range had supplied two-thirds of the ore for the U.S. military. After the war, a chorus of journalistic and engineering voices warned that the United States would soon run out of vital natural resources. Resource depletion, of course, is a cultural construct mediated by technology, rather than an absolute measure of a quantity of a resource. If all one has is a spoon, an ore body will be depleted for you as soon as the loose surface ore is scooped up. But if one has the 5.5 million–pound explosives that taconite operations now use, one can access entire ore bodies that were essentially invisible in earlier accounts of minerals because they were deemed impossible to mine.

As Manuel argues, depletion concerns in the Lake Superior basin were embedded in specific political contexts intended to generate pressure for government funds and new tax policies that would benefit taconite mines over hematite ore mines.[14] While most iron production within the Lake Superior basin was sited on U.S. land, the waste from mining made its way into waters that were co-managed by the United States and Canada through the International Joint Commission. In other words, transnational contaminants complicated national boundaries. The pressures to develop new mines also reflecting growing global political interconnections. Depletion fears were framed within Cold War political concerns that helped mobilize state power to promote the shift to taconite mining. During the war, U.S. iron and steel companies had developed extensive networks of international iron sources.[15] The economist Peter Kakela approvingly noted in 1978: "Rich ores were discovered on

Canada's Labrador-Quebec border and exploited by the Iron Ore Company of Canada, created in 1949 by Hanna Mining Company of Cleveland, Ohio. With the subsequent opening of the St. Lawrence Seaway in 1959, Canadian ore could be shipped more economically to steel mills bordering the Great Lakes."[16]

By 1950, engineers were insisting that hematite depletion on the iron range demanded new funds, new tax policies, and relaxed environmental standards in order to ensure U.S. national security. For example, in a 1950 issue of the *Science Newsletter,* journalist Ann Ewing reminded readers that two world wars had depleted American hematite ore and, therefore, it was essential that taconite be developed, lest the nation become vulnerable to foreign manipulation.[17] Ewing admitted that taconite development had significant environmental costs—particularly given that 48 tons of water would be needed for every ton of iron produced. Moreover, "the disposal of tailings presents a problem. They could be dumped on the ground, but it would take vast areas to accommodate them."[18] Recognizing that fishermen might object to dumping tailings in the lake that sustained the fisheries they relied upon, Ewing suggested that tailings in water might not seem ideal, but perhaps "the sand thus added to the lake will be helpful for the spawning of fish."[19]

As Manuel argues, the most important mining engineer on Minnesota's Iron Range, Edward Davis, was a longtime booster of taconite. He worked for decades to persuade a skeptical public that taconite ores could be processed cost-effectively, thus replacing hematite supplies. Davis initially borrowed processes that had been developed to work the low-grade copper deposits of Utah, recognizing that taconite was more like the western low-grade copper ores than eastern U.S. iron ore deposits.[20]

With taconite, mining for iron ore was to become less a simple matter of extracting some valuable ore from the ground (dangerous for miners, certainly, but less traumatic for the environment) than a matter of manufacturing production. As historian Timothy LeCain argues, the development of open-pit mining involved an increasingly mechanized system for extracting and refining large quantities of ore, often resulting in significant environmental destruction. Mining for copper ore—and eventually for taconite—became dominated by large corporations using enormous machines to extract low-grade deposits from open pits. LeCain describes the very visible environmental deterioration that accompanied open-pit copper mines in the U.S. West, including sulfur killing trees and clouds of poisonous gases contributing to lung diseases. In contrast, the toxicities from taconite were much subtler than those from copper, making them hard to unravel from more ordinary and easily detected forms of environmental degradation.[21]

TACONITE TOXICITIES: MERCURY AND ACID DRAINAGE

In the past three years, GTAC has done its best to present taconite as a safe ore to mine with minimal toxicity concerns. It is so safe and so pure, GTAC insisted before the Wisconsin legislature, that the state should quickly pass a new law exempting taconite mining from Wisconsin's environmental regulations. The argument had two parts: first, because taconite itself rarely contains iron sulfides or pyrites—which can produce acid mine drainage—it's pure. Second, the extraction process uses magnets and clean water, not hydrosulfuric acid.[22]

Technically, both of these details may be correct, but the conclusion that follows (therefore taconite mining has no toxicity concerns) ignores the larger geological and biological contexts of taconite within a watershed. First, taconite is a low-concentration iron ore, and to extract the valuable part, the rest of the rock (the tailings) must be crushed to a fine dust, mixed with water, then dewatered and stacked in piles or dumped into a body of water. These tailings particles are quite easily eroded by wind and water, and in this way they become mobilized into the watershed. At least 70 percent of the volume of the Ironwood Formation would be turned into waste and fine tailings, and the waste for the initial four-mile stage of mining alone would create a pile "600 million cubic yards" in size, stretching one mile square, 600 feet high.[23]

Second, there's the problem of the 200–300 meters of so-called "overburden"; that is, the rock on top of the taconite itself, and the world on top of that rock. The very term *overburden* suggests that a living ecosystem—forests, forbs, birds, habitat, streams, the many different biological communities within the soil, the layers of rock that lie under the soil—is nothing but a burden that blocks the true resource, taconite. Anishinaabe peoples have increasingly resisted the terminology of mining, arguing that it renders invisible the biological, geological, and hydrological connections that sustain them.

While taconite may contain no pyrites or iron sulfide, the rock on top of it contains quite a bit of both. And there's no way to get to that taconite without transforming the landscape into something that can cause acid mine drainage. GTAC denies the presence of pyritic ores in the formation (and the company has refused to allow the state, the tribes, or local citizens to view the samples that GTAC obtained from U.S. Steel). As long ago as 1929, however, the Wisconsin Geological Survey reported that pyrite is associated with local ore and waste rock, and a USGS report concluded the same thing in 2009. When ground to a fine dust (as required for taconite extraction), then exposed to oxygen and water, pyrites create sulfuric acid, which leaches harmful metals such as lead, arsenic, and mercury that make their way into groundwater and surface water.

Acid mine drainage represents an interesting mixture of natural and constructed toxicities. Many rock formations contain heavy metals that would be toxic if they were mobilized into biological systems. Typically, they're bound in stable formations, where they don't move into the atmosphere or the water supply on time scales that matter for biological life. (Over millions of years, of course, they do mobilize, so we're talking about a matter of scale—spatial and temporal.) But when acid conditions are present, those chemicals and heavy metals do become mobilized into biological systems.

Wild rice is particularly sensitive to even extremely low levels of acidic drainage, creating enormous concerns for the tribes. Older taconite mines in Minnesota have continued to leach sulfates into wild rice beds decades after closure. Ecological history studies have shown that wild rice was abundant in the upper St. Louis River watershed (upstream from Duluth) before the 1950s, when taconite mining boomed. Currently, sulfate levels are high in the St. Louis River, and wild rice stands are few and stunted. The tailings basin once owned by LTV Taconite still leaches sulfates and other contaminates into the St. Louis River, and from there into Lake Superior. Elsewhere on the north shore, the Minntac taconite tailings basin is leaching 3 million gallons per day of sulfates and related pollutants into two watersheds. For decades, regulations in Minnesota have required very low sulfate levels to protect wild rice (10 milligrams per liter), yet not a single taconite mine is in compliance. Why has the Minnesota Pollution Control Agency (MPCA) been reluctant to enforce its own standards? Preliminary research suggests that because those standards are based on historical data, they have been vulnerable to criticism by legislators. One legislator in Minnesota has proposed to increase the state's sulfate limit to 250 milligrams per liter, the level that models suggest is safe for adult drinking water. An unwillingness to interpret the evidence of historical change suggests that the boundaries between regulation, ecological history, and the mining industry remain contested.

What can these histories of mine problems tell communities that are trying to decide about new mine proposals? Christopher Dundas, chair of Duluth Metals Limited, argues that historical problems have no bearing on or relevance to the future proposed mines. "This is a completely different era than what happened in the '60s," Dundas has said. "Our operation will be state of the art and will be totally planned and designed to absolutely minimize every environmental issue." History is irrelevant, in other words. But to advocates for the Boundary Waters, history matters. One opponent fears that problems with the Dunka taconite mine, near Babbitt, Minnesota, are "an indicator of problems to come" from proposed mines. Bruce Johnson, a former employee of a state agency that regulated Dunka and other local mining issues, told reporters that "he fears that state agencies will shortcut environmental rules because of the intense political pressure to approve mines and put people to work. 'I want to have good jobs, too, but I want to do it right,' Johnson said. 'These guys are going to make multi-millions of dollars. We don't want to be left with a bunch of mining pits full of polluted water that even ducks won't land on.'"[24]

The Dunka mine was covered with sulfide rock similar to the overburden present in the Penokees. Its history suggests some of the difficulties of containing pyritic materials. LTV Steel operated the Dunka mine from 1964 to 1994, piling up more than 20 million tons of waste rock 80 to 100 feet high for almost a mile in length. Almost immediately, these waste piles began leaching copper, nickel, and other metals into wetlands and streams. Decades later, an average of 300,000–500,000 gallons continues to run off the piles every month, according to MPCA documents. Between 2005 and 2010, the runoff violated state water standards on nearly three hundred occasions. Rather than force LTV to spend the money to stop the toxic runoff, the MPCA fined the company that now owns the site (Duluth Metals Limited) $58,000—a cost of doing business that is far cheaper for the company than cleaning up the tailings. Yet this tainted water flows into the Boundary Waters Canoe Area, a federally designated wilderness area supposed to be protected from toxic discharges.

Mercury offers an excellent example of the complex relationships between ecological change, industrial development, and contamination data. In 2009, the U.S. Geological Survey reported that mercury contamination was found in every fish tested in nearly three hundred streams across the country.[25] The highest levels of mercury were detected in some of the places most remote from industrial activity. Remoteness offers no protection, and the very richness of the remote wetlands increases their vulnerability to toxic conversions. Methylmercury finds its way into fish and eventually into the people eating that fish. Eating fish is of great cultural significance, particularly for indigenous communities in the basin. But its potential contamination forces communities to make trade-offs between their beliefs and possible harm to themselves. How much fish do you eat when it's culturally important? How much do you eat when you're pregnant? These are difficult dilemmas posed by changes in watershed health. Contaminants transform more than the health of lakes, fish, and forests; they also transform cultural identities. Interpreting the historic evidence of fish and human contamination has become a politically and culturally complex exercise.

Taconite processing is now the primary source of mercury *produced* within the Lake Superior basin, surpassing production from coal power plants. Minnesota's taconite plants add 800 pounds per year of mercury to the basin, and a recent study found that 10 percent of newborn babies in the Lake Superior basin have mercury levels exceeding EPA standards.[26] However, much of the mercury that actually *accumulates* within the basin comes not from local sources of production, but from global sources such as coal burning that have been transported via the atmosphere into the basin. This creates a tension: will controls on mercury emitted from local taconite processing actually reduce exposure to contaminants? Or will those controls indirectly increase exposures, if they shut down the local taconite industry, displacing production to China, which might then release greater levels of mercury emissions that return to Lake Superior waters, fish, and people? Unraveling local versus global sources and exposures presents enormous challenges for regulatory communities.

Mining, microbial ecology, and mercury interrelate in complex ways in the watershed. When mining exposes natural metal sulfides in ore bodies to air and water, oxidation results, leading to acid drainage. Microbes exist in many rocks, but usually in low numbers because lack of water and oxygen keeps them from reproducing. However, during the disturbance from mining, those microbes are exposed to water and oxygen, so their numbers multiply and they form colonies that can greatly accelerate the acidification processes. These sulfates also encourage conversion of elemental mercury (not particularly toxic) to methylmercury (extremely lethal), which then bioaccumulates in fish tissue and, from there, makes its way into wildlife and people.

ASBESTOS AND THE RESERVE MINING COMPANY

Between 1896 and 1900, the U.S. steel industry experienced a radical transformation. Small companies gave way to large, vertically integrated steel corporations that controlled not just steel mills but also the iron mines that supplied those mills.[27] Nearly three-quarters of all Lake Superior iron ore in 1900 came from mines that were "either owned by or under long-term lease" to the largest American steel companies: Carnegie Steel, Federal Steel, National Steel, and several others—all to be absorbed into U.S. Steel in 1905.[28] In 1947, the Reserve Mining Company—under the ownership of Armco Steel Corporation and Republic Steel Corporation—applied for permits to process taconite on the shores of Lake Superior near the small community of Silver Bay. Lake Superior would supply both the abundant water needed for taconite processing and a convenient location for tailings disposal. The company initially requested permits to deposit about 67,000 tons of tailings each day into the lake—eventually totaling 400 million tons of tailings before the permits expired. Once the mine reached full capacity, it produced 11 percent of the country's total iron production and revitalized the faltering mining economy of the Lake Superior district.

Two decades later, conflicts over the toxic by-products of Reserve's taconite mine—particularly asbestos fibers that made their way into Duluth's drinking water and citizens—would lead to the most inflammatory environmental lawsuit in American history.[29] Before granting permits, planners and regulators held a series of hearings to decide whether the tailings would be safe. The meanings of *safety*, however, were contested among the various stakeholders. Fishermen from communities along Minnesota's north shore expressed concerns that tailings and water withdrawals might devastate fish habitat and ruin their economic base, while other citizens testified about their fears that silica in the tailings might lead to silicosis. Nevertheless, the state granted permits for the mine to dispose tailings directly into Lake Superior.

In 1947, Reserve Mining's tailings disposal permits were made subject to three key conditions: first, that the tailings would not discolor the water outside narrowly defined areas; second, that the tailings would not harm fish in the lake; and third, that Reserve would be liable for any harm to water quality. The state reserved the right to revoke the permits if Reserve violated any of its conditions, including an important condition that discharge was not to include "material amounts of wastes other than taconite."[30] By the late 1950s, local environmental organizations, commercial fishermen, and sport-fishing groups were complaining to the MPCA that taconite tailings were killing fish and clouding the waters. The state refused to use its powers to intervene despite the conditions clearly written into the permits in 1947.[31]

Only after evidence emerged that asbestos had been mobilized from taconite disposal into the drinking water and bodies of urban residents distant from the disposal site did the federal and state governments officially begin to question risks from taconite. The plant's exhaust stacks were also emitting asbestos-form fibers into the air. On behalf of the federal Environmental Protection Agency, the Department of Justice filed a lawsuit against Reserve in 1973, beginning a trial and appeals processes that would last for a decade. In early June 1973, Judge Miles Lord heard testimony from a specialist in asbestos exposure, Dr. Irving Seikoff, who confirmed that the city's drinking water contained asbestos from the tailings. The concentration was surprisingly high: 100 billion fibers per liter of water, which was at least "'1000 times higher' than any asbestos level previously found in any water sample."[32]

Reserve Mining Company disputed evidence about toxic mobilization, arguing in complex ways that asbestos could not possibly move from tailings into Lake Superior, from Lake Superior into drinking water, and from drinking water into lungs. Moreover, data suggesting toxicity did not always make it out of industry labs into regulatory spaces. In evidence presented at the trial on the ecological disruption caused by the tailings, Dr. Donald Mount of the EPA showed that Reserve was distorting the scientific process. The company designed studies with false controls in order to contaminate any possible findings, and it took measurements of tailings in places that it knew could not possibly show significant difference from controls.[33]

Discussions of new taconite mines continue to occur in the shadow of Reserve Mining Company's release of tailings into Lake Superior and the resultant concern about asbestos exposure. When GTAC argues that taconite is perfectly safe, it interprets the Reserve Mining history as an example of what happens when environmentalists and regulators over-reach and shut down a region's economy for an unproven threat. For many environmentalists in the basin, in contrast, the Reserve Mining history suggests how treacherous taconite mining can be to water quality. Mine proponents counter by arguing that because new mines will stack tailings on land rather than dumping them into Lake Superior, one cannot draw a connection between an historical episode and a technologically advanced future. People who want the mines back point to a time when miners had good jobs, rarely mentioning the lung diseases that haunted the Iron Range or the collapse of economies when the companies pulled out. Mining opponents argue, in contrast, that historical economic benefits were outweighed by the costs of past toxic exposures. For these stakeholders, while the economic benefits did not remain, the health costs have lingered.

THE ANISHINAABE AND MINING
ON CEDED TERRITORIES

The Reserve Mining Company historical precedent suggests how easy it is for scientific and legal discourses to miss what for the Anishinaabe community is the real point: clean water has values that go beyond what can be measured scientifically. And on ceded territories, members argue, the tribes have rights to comanagement of land and water resources, which means that their values must play as significant a role in policy making as other communities' values. The federal government made three major land cession treaties with the Anishinaabe in the Lake Superior basin that established reservations to be under the exclusive control of the tribes (figure 11.3). Equally important, the tribes were careful to retain the right to hunt, fish, and gather on ceded territories, which also meant the right to participate in management of natural resources.[34]

State governments rarely recognized these ceded territory rights, until a series of brutal assaults took place in the late twentieth century. For decades, Wisconsin had arrested Anishinaabe who fished and hunted on ceded territories without state licenses, which the tribes insisted were unnecessary. In 1974, two members of the Lac Courte Oreilles Band were arrested for spearing fish on ceded territories. The Lac Courte Oreilles sued the state for treaty rights violation, and in 1983, the federal district court upheld off-reservation treaty rights in a landmark judgment called the Voight Decision. The state of Wisconsin appealed (and eventually lost). Meanwhile, white supremacist vigilantes (including members of the Hurley chapter of the Ku Klux Klan) attacked Anishinaabe spear fishers who were exercising their treaty rights, and a series of violent protests at fish landings marked the late 1980s. For all the violence, as geographer Zoltán Grossman argues, in the process of fighting each other, the tribes and the whites created a series of connections that the tribes were able to build upon a few years later, when Exxon proposed to build a copper and zinc mine (named Crandon) in sulfide-containing ore bodies within the watershed of the Wolf River (a National Wild and Scenic River). Like the GTAC mine, the Exxon mine would also have been located just upstream from an Anishinaabe reservation (the Mole Lake Sokaogon Band Reservation), within ceded territories.[35] The state of Wisconsin had pushed for the mine, believing it would bring jobs to the north and money to the state. The Wisconsin Department of Natural Resources (DNR) decided to set standards for water leaving the mine at industrial water quality levels, allowing for 40 million tons of tailings and acidic mining waste that would have destroyed wild rice beds on the reservation. To stop the Crandon mine, the Mole Lake Sokaogon worked with the EPA to win the right to set its clean water standards.

Tribal lawyer Glenn Reynolds wrote regarding the case:

> Wisconsin challenged the tribe's authority to enact tribal water quality standards on the grounds that the federal government had already given Wisconsin primary authority over the state's water resources and could not rescind that authority and pass it on to tribal governments. Ironically, Wisconsin argued that the Public Trust Doctrine granted the state the exclusive right to regulate, and potentially degrade, the water quality of Rice Lake on behalf of Wisconsin citizens. Unsurprisingly, the mining company supported Wisconsin's stance. Three downstream towns and a village, however, filed a brief in support of the Sokaogon standards. After six years of litigation, the U.S. Supreme Court declined to review a federal appeals court decision that upheld the authority of the Sokaogon to set water standards necessary to protect reservation waters.[36]

Tribal efforts resulted in Wisconsin Act 171, passed in 1997, which became known as the "mining moratorium." It required a moratorium on issuance of permits for mining of sulfide ore bodies until companies provided historical information proving that they had successfully controlled mining waste from their other mines for at least ten years.[37]

The other key issue that motivated mining activism among the Wisconsin tribes was the Bad River Band's 1996 blockage of railroad tracks that would have brought sulfuric acid to a copper mine in White Pine, Michigan. On July 22, 1996, tribal members blockaded the railroad tracks that crossed their reservation, stopping a train headed for the White Pine copper mine in Upper Michigan. The train was carrying sulfuric acid that the mining company planned to experiment with in "solution-mining," which would involve injecting 550 million U.S. gallons (2.1 million cubic meters) of acid into the mine to bring out the remaining copper ore. Environmentalists and tribal members raised concerns that the acid would contaminate groundwater and Lake Superior, but the EPA granted permission without first requiring a hearing or environmental impact statement. Anishinaabe activists insisted that the project was illegal because the EPA had failed to consult with affected Indian tribes, as required by law. In the face of legal battles over treaty rights, the company withdrew its mining permit application, and the White Pine copper mine and smelter shut down.[38]

After January 2011, when news of the GTAC's proposed taconite mine in the Penokee Range first spread, the issue became extremely polarized. Because the proposed mine lay in ceded territories, the Anishinaabe tribes in the basin vowed to bring the fight against the mine out of a purely local and state discussion to a federal level, claiming violation of their tribal sovereignty. Local residents responded with death threats against tribal members, and a swirl of local, state, and federal lawsuits, hearings, and threats of violence marked the next four years. Debates over the mine became intense enough to swing a key election in Wisconsin. After the legislature defeated a pro-mining bill by one vote in March 2012, pro-mining groups donated $15.6 million dollars in campaign contributions and lobbying fees to candidates who might support a mine.[39] The result was a change in control of the Wisconsin Senate (by one vote) after the November 2012 elections, giving the far right the power to rewrite Wisconsin laws.

In February 2013, a new mining bill, written with the help of GTAC lobbyists and the American Legislative Executive Council, was passed in Wisconsin that exempts taconite mining from many of the state's water quality and wetlands standards.[40] It formally establishes that the expansion of the iron-mining industry is a policy of the state. This means that if there is a conflict between a provision of the iron-mining laws and a provision in another state environmental law, the iron-mining law overrides other state laws.[41]

With enactment of the new law, the voice of the state became the only voice allowed in negotiations about permits. Local communities and the public lost the right to challenge state scientific findings and state permits. Contested case hearings—where the state had to face expert witnesses who could challenge their versions of the evidence—were outlawed. Citizen suits against a corporate or state employee alleged to be in violation of the metallic mining laws were also outlawed, even if the mine operator or state staff knowingly violated the law. In other words, the democratic processes by which outside voices could be heard to challenge state or corporate versions of scientific claims were outlawed, so that now an echo chamber exists. Senator Fred Risser (D-Madison), the longest-serving state legislator in U.S. history, expressed the anger of many when he thundered in the senate, "This bill is the biggest giveaway of resources since the days of the railroad barons."[42]

The state of Wisconsin, however, cannot outlaw legal challenges from the Anishinaabe. The battle over the Mole Lake mine led to a U.S. Supreme Court decision in 2002 affirming the right of Indian nations to set and enforce their own clean air and water standards (working with the federal EPA). In other words, the state cannot set more relaxed standards.[43] Even though a series of treaties between the U.S. government and sovereign Indian nations makes it clear that formal consultation is required before environmental permits are issued, the Wisconsin governor and legislature decided to ignore those requirements when drafting the new law. Senate majority leader Scott Fitzgerald (R-Juneau) said that he had no plans to consult with the Bad River Band during the drafting process. "It's going to [be] difficult to ever get them on board," he added.[44]

CONCLUSION

Contested interpretations of the past continue to shape current conflicts. People who want the mines back point to a time when miners had good jobs, rarely mentioning the lung diseases that haunted Iron Range miners, the bitter battles to win the few rights they had during the brief boom, or the collapse of economies when the companies pulled out. Pro-mining individuals in Hurley stated in interviews that they feel that they can trust the mining companies, so there's no need for regulation or oversight. They remember a time when local-impact funds created good schools, decent hospitals, and well-maintained roads. But they forget that these benefits weren't just handed to them by the company. They were won, bitterly, by political fights led by unions that have since lost much of their power. Companies left to themselves never gave us anything, one resident of Minnesota's iron range said, concerned that new laws removed the protections that once marked the range.[45]

Events across scales shape the most local processes within the basin. When the Asian building boom in 2011 forced global steel prices to new highs, what had been just a pile of useless rock to U.S. Steel became reframed as the nation's most important source of iron ore. Mining advocates insist the GTAC mine is inevitable in a global economy. "Only a primitive, backward people would stand in the way of our prosperity," one white woman in Hurley said, complaining bitterly about the Bad River Band.[46] But from the band's perspective, how can we destroy the water, the wild rice, the rivers, the slough, for a few jobs and a billionaire's profit? Water isn't a resource to be commodified; it's the blood at the heart of band members' place and life.

Advocates point out that forbidding taconite mines in the Lake Superior basin will only shift more mining to other places, where environmental and labor protections may be even weaker than in today's Wisconsin. And it's true that the proportion of the world's iron ore mined in Canada and the United States has dropped to only 1.3 percent of global production in 2010. (China, in contrast, mined 41.6% in 2015 and was still the world's largest importer.)[47] Yet who gets to decide which places, if any, are unsuitable for a mine? Who decides how to measure benefit versus harm? Are there going to be places where communities decide that the local, particular harms far outweigh the benefits on the regional or national scale? In a talk at Yale in March 2012, western historian Richard White spoke of "incommensurate measures."[48] What is gained in resource development by one group cannot simply be measured against what is lost by another group, he warned. At one mining hearing, Richard White's incommensurate measures were in full force when pregnant women from the band spoke of their fears when they had to drink water poisoned by taconite mining.

There is nothing natural or inevitable about resource development. Resources are contingent and they change over time. Calling something a resource pulls it out of its intricate social and ecological relationships, isolates it in our gaze. Yet those isolations are illusions. We still live in intimate relationships with those elements, even if we think we don't. The language of inevitability masks the fact that government actions promote one vision of resources over another. So treaty rights and environmental quality must bend to the march of progress. What's hidden is the texture of the wild rice beds, the lake trout that swim through the waters of Lake Superior, the children of women poisoned by mercury, the asbestos released into the watershed by the processing of certain kinds of taconite deposits.

Selenium, mercury, arsenic—perfectly natural chemicals—lie bound and buried in rocks until miners release them while digging for something else that has become defined as a resource. Then as waters move through mining sites, these chemicals move the chemicals into fish bodies, and from there into human bodies. When minerals are dug from the ground, when trees are cut in the forest, when floodwaters are diverted, when rivers are dammed, when animals are changed from fellow creatures to livestock resources, we set into motion subtle processes of toxic transformation that have legacies far into the future.

Decisions about mine permitting are not purely scientific or technical decisions; they are also social decisions based, in part, on conflicting interpretations of historical and spatial dimensions of toxicity. Some, but not all, iron mines release sulfides into the water, which create acidic drainage that mobilizes toxics into the watershed. Some, but not all, iron mines include processing facilities that release mercury into the atmosphere, where it then mobilizes into the larger environment as methylmercury and bioaccumulates in fish. Some, but not all, iron mines release sulfates that mobilize heavy metals that can harm local wetlands and aquatic habitat. Whether or not a particular iron mine has toxic consequences depends partly on the geologic context, in particular the composition of the overburden (the rocks that cover the ore deposit). However, it is rarely the geology alone that leads to toxic exposures; social and political decisions about how to deploy contested scientific knowledge in the regulatory process ultimately shape toxic outcomes. Given that Lake Superior is the world's largest lake (by surface area), the quality of its water has significance for the entire globe, particularly given the stresses that climate change is likely to place on freshwater.

NOTES

1. While this essay was in process, on February 27, 2015, the mining company announced that it was suspending operations at the site, citing concerns that the U.S. Environmental Protection Agency might have tried to block construction of the mine under provisions of the federal Clean Water Act. Susan Hedman, regional EPA administrator, denied any such intention.

2. The literature on Native American communities and environmental justice includes Jamie Donatuto and Darren Ranco, "Environmental Justice, American Indians, and the Cultural Dilemma: Developing Environmental Management for Tribal Health and Well-Being," *Environmental Justice* 4:4 (December 2011): 221–30; James M. Grijalva, "Self-Determining Environmental Justice for Native America," *Environmental Justice* 4:4 (December 2011): 187–92; Barbara Harper and Stuart Harris, "Tribal Environmental Justice: Vulnerability, Trusteeship, and Equity under NEPA," *Environmental Justice* 4:4 (December 2011): 193–97; Jamie Vickery and Lori M. Hunter, *Native Americans: Where in Environmental Justice Theory and Research?* Institute of Behavior Science Population Program Working Paper (Boulder: University of Colorado, 2014); Kyle Powys Whyte, "Environmental Justice in Native America," *Environmental Justice* 4:4 (December 2011): 185–86.

3. The Ramsar Convention on Wetlands, "USA Names Lake Superior Bog Complex," September 3, 2012, http://www.ramsar.org/news/usa-names-lake-superior-bog-complex (last accessed February 7, 2017)..

4. Glenn C. Reynolds, "A Native American Water Ethic," *Transactions of the Wisconsin Academy of Sciences, Arts and Letters* 90 (2003): 146.

5. John Sandlos and Arn Keeling, "Claiming the New North: Development and Colonialism at the Pine Point Mine, Northwest Territories, Canada," *Environment and History* 18:1 (2012): 5–34.

6. Rebecca Kemble, "The Walker Regime Pushes for Controversial Mining Law," *The Progressive*, October 21, 2011.

7. Tom Fitz, "The Ironwood Iron Formation of the Penokee Range," *Wisconsin People and Ideas* (Spring 2012): 33–39.

8. U.S. Steel had negotiated with The Nature Conservancy (TNC) to sell the company the mineral rights under the condition that the land be maintained for logging but that mining would not be allowed (thus reducing potential competition for U.S. Steel if steel prices rose enough to make mining the deposit profitable). But in the final stages of negotiation, for reasons that the TNC negotiator doesn't understand, U.S. Steel pulled out and sold the mineral rights instead to RGGS Land and Minerals, Ltd., of Houston, Texas, and LaPointe Mining Co. in Minnesota. Matt Dallman, Director of Conservation, The Nature Conservancy, interview with the author, Wisconsin, May 2011.

9. The CEO of GTAC is Chris Cline, who has promoted longwall coal mining in Illinois, where his operations were cited fifty-three times over three years for violating water quality standards. Mary Annette Pember, "Chris Cline, the New King Coal, Likes to Swing His Big Iron Fist," *Indian Country Today*, August 22, 2013. For a discussion of Clean Water Act violations, see Al Gedicks, "The Fight against Wisconsin's Iron Mine," Wisconsin Resources Protection Council, April 16, 2013, http://www.wrpc.net/articles/the-fight-against-wisconsins-iron-mine/ (accessed January 15, 2017).

10. Gedicks, "The Fight against Wisconsin's Iron Mine."

11. Fred W. Kohlmeyer, "Pioneering with Taconite," *Minnesota History* 39:4 (1964): 163–64.

12. H. S. Harrison, "Where Is the Iron Ore Coming from?" *Analysts' Journal* 9:33 (June 1953): 98–101.

13. Fitz, "The Ironwood Iron Formation of the Penokee Range."

14. Jeffrey T. Manuel, "Mr. Taconite: Edward W. Davis and the Promotion of Low-Grade Iron Ore, 1913–1955," *Technology and Culture* 54:2 (2013): 317–45.

15. John Thistle and Nancy Langston, "Entangled histories: Iron Ore Mining in Canada and the United States," The Extractive Industries and Society, 2015, 14 July 2015, ISSN 2214–790X, http://dx.doi.org/10.1016/j.exis.2015.06.003. Bethlehem Steel Iron Company, for example, imported ore from Chile, and when the war ended, the company began developing large, high-grade concessions in Venezuela. U.S. Steel also began mining a hematite deposit in Venezuela, while Republic Steel Company developed hematite mines in Liberia.

16. Peter Kakela, "Iron Ore: From Depletion to Abundance," *Science* 212:4491 (April 10, 1981): 132–36.

17. "The Venezuela iron ore deposits are rich and extensive," Ewing admitted, but she then noted that "shipment of these or other high-grade foreign ores to the blast furnaces and steel mills in this country during an emergency might leave the ore-laden boats open to submarine attack." Ann Ewing, "Low-Grade Ore Yields Iron," *Science News-Letter* 57:20 (1950): 315.

18. Ibid., 314.

19. Ibid., 315.

20. Manuel, "Mr. Taconite."

21. Timothy LeCain, *Mass Destruction: The Men and Giant Mines That Wired America and Scarred the Planet* (New Brunswick, NJ: Rutgers University Press, 2009).

22. Al Gedicks and Dave Blouin, "Science and Facts Show a Need for Tight Regulation of Taconite Mining," *Duluth News Tribune*, February 13, 2013.

23. Ibid.

24. Dundas and Johnson quoted in Tom Meersman, "Runoff from Old Mines Raises Fears," *Minneapolis Star Tribune*, October 1, 2010.

25. U.S. Geological Survey, Press release: "Mercury in Fish Widespread," 2009, reported in Cornelia Dean, "Mercury Found in Every Fish Tested, Scientists Say," *New York Times*, August 19, 2009, A19, http://www.nytimes.com/2009/08/20/science/earth/20brfs-MERCURYFOUND_BRF.html?_r=1&em.

26. Minnesota Department of Health Fish Consumption Advisory Program and MDH Public Health Laboratory, "Final Report to EPA: Mercury in Newborns in the Lake Superior Basin," GLNPO ID 2007–942, November 30, 2011.

27. Terry S. Reynolds and Virginia P. Dawson. *Iron Will: Cleveland-Cliffs and the Mining of Iron Ore, 1847–2006* (Detroit, MI: Wayne State University Press, 2011.

28. Richard B. Mancke, "Iron Ore and Steel: A Case Study of the Economic Causes and Consequences of Vertical Integration," *Journal of Industrial Economics* 20:3 (1972): 222.

29. Michael E. Berndt and William C Brice, "The Origins of Public Concern with Taconite and Human Health: Reserve Mining and the Asbestos Case," *Regulatory Toxicology and Pharmacology* 52:1, supplement (October 2008): S31.

30. Ibid., 33.

31. Ibid.

32. Thomas R. Huffman, "Enemies of the People: Asbestos and the Reserve Mining Trial," *Minnesota History* 59:7 (2005): 296.

33. Office of Enforcement and General Counsel, *Studies Regarding the Effect of the Reserve Mining Company Discharge on Lake Superior* EPA, Washington, DC: U.S. Environmental Protection Agency, 1973). "The Reserve Mining Company performed [. . . an] in-situ study during 1972, to test algal stimulation by tailings. Unfortunately, they used as their control, water taken from a point very close to the discharge and the probabilities of contamination by tailings of this control water negates what might have been a useful study" (p. 14). In

another section of the report, Mount described research by a federal scientist that gave the "first indication that tailings, through some mechanism, stimulate growth or prolong life of bacteria in Lake water. Reserve Mining Company has frequently said that such an effect could be due just to a 'platform' effect and not due to chemical stimulation. Their implication has been that the possible physical nature of this effect makes it unimportant. This is nonsense" (p. 10). Elsewhere, the testimony states: "A significant breakthrough was achieved in 1969 when cummingtonite, a mineral composing about 40% of the tailings, was recognized as a tracer for the discharge. [. . .] This method was unjustifiably challenged by Reserve Mining Company, because they measured stream sediment just downstream from bridges on highways on which they knew tailings had been used for ice control and construction, and then contended there was much more cummingtonite in the tributaries" (pp. 4–5).

34. Patty Loew, *Indian Nations of Wisconsin: Histories of Endurance and Renewal,* 2nd edition (Madison: Wisconsin Historical Society Press, 2013).

35. Zoltán Grossman, "Unlikely Alliances: Treaty Conflicts and Environmental Cooperation between Native American and Rural White Communities" (PhD dissertation, Department of Geography, University of Wisconsin–Madison, 2002).

36. Reynolds, "A Native American Water Ethic," 154.

37. State of Wisconsin, 1997 Senate Bill 3, 1997 Wisconsin Act 171, enacted April 22, 1998, http://docs.legis.wisconsin.gov/1997/related/acts/171.pdf (accessed January 15, 2017).

38. Zoltán Grossman, "Chippewa Block Acid Shipments," *The Progressive,* October 1, 1996.

39. "Mine Backers Drill with Big Cash to Ease Regulations," Wisconsin Democracy Campaign, January 28, 2013, http://www.wisdc.org/pro12813.php (accessed August 24, 2014).

40. Senate Bill 1, Wisconsin Legislature, 2013–14, http://docs.legis.wisconsin .gov/2013/proposals/sb1 (accessed January 15, 2017). According to *The Progressive,* at least $1 million was donated to mining committee members' campaign funds, and $15.6 million was given by "pro-mining interests to Governor Walker and other state legislators, outspending groups opposed to the measure 610 to 1." R. Kemble, "Bad River Chippewa Take a Stand against Walker and Mining," *The Progressive,* January 28, 2013. In August 2014, when evidence emerged that GTAC had secretly donated $700,000 to a political group helping Governor Scott Walker win election, the *New York Times* argued that Governor Walker had likely violated campaign finance laws. "How to Buy a Mine in Wisconsin: Did Gov. Scott Walker Violate Campaign Laws?" *New York Times,* August 31, 2014. Evidence of the donation is on p. 10 of the court document "Defendant Francis Schmitz's Supplemental Opposition to Plaintiffs' Motion for Preliminary Injunction," *Eric O'Keefe and Wisconsin Club for Growth, Inc. v. Francis Schmitz et al.* (U.S. District Court for the Eastern District of Wisconsin, Milwaukee Division), Case No. 2:14-cv-00139-RTR, Case 14–2585, Document 44–6, http://www.jsonline.com/news/statepolitics/documents-governor-scott-walker-encouraged-donations-to-wisconsin-club-for-growth-john-doe-272369371.html (accessed January 15, 2017).

41. The new law diminishes water quality regulations, removing protections for streams and lake beds. Previously, the DNR was required to deny a mining permit if "irreparable damage to the environment" could not be prevented. Activities expected to cause substantial deposition in stream or lake beds, or the destruction or filling in of a lake bed, constituted grounds for denial. The 2013 law has removed these as bases for permit denial. It eliminates many of the state's existing wetlands and watershed protections, reclassifying them as sacrifice zones. It includes a legislative finding that "because of the fixed location of ferrous mineral deposits in the state, it is probable that mining those deposits will result in adverse impacts to areas of special natural resource interest and to wetlands, including wetlands located within areas of special natural resource interest and that, therefore, the use of wetlands for bulk sampling and mining activities, including the disposal or storage of mining wastes or materials, or the use of other lands for mining activities that would have a significant adverse impact on wetlands, is presumed to be necessary." State of Wisconsin, 2011–2012 Legislature, 2011 Bill Legislative Reference Bureau, 3520/1, "Analysis by the Legislative Reference Bureau," p. 41, http://www.co.iron.wi.gov/docview .asp?docid=10246&locid=180 (accessed January 15, 2017).

42. Rebecca Kemble, "Walker's Colossal Giveaway," *The Progressive,* March 5, 2013.

43. Lee Bergquist, "Decision Puts Water Quality in Tribe's Hands: Sokaogon Can Set Standard near Mine," *Milwaukee Journal Sentinel,* June 4, 2002, 1A.

44. "Walker Pushes Mining Co.'s Bill, Despite Tribe's Objections, *The Progressive,* January 8, 2013.

45. Hurley residents, interviews with author, December 31, 2011.

46. Ibid.

47. Wikipedia, s.v. "List of Countries by Iron Ore Production," last modified November 30, 2016, http://en.wikipedia.org/wiki/List_of_countries_by_iron _ore_production.

48. Richard White, "Incommensurate Measures: Nature, History, and Economics," Keynote Lecture, Yale University conference "Resources: Endowment or Curse, Better or Worse?" New Haven, CT, February 24, 2012.

CHAPTER 12

If the Rivers Ran South

*Tar Sands and the State of the
Canadian Nation*

STEVEN M. HOFFMAN

On Friday, January 27, 2012, Chief Jackie Thomas of the Yinka Dene Alliance met with Alberta and Northwest Territory First Nations to state the alliance's opposition to the Northern Gateway pipeline. As envisioned by the Enbridge Company, the almost 1,200-kilometer-long pipeline would carry tar sands crude oil from its source in Alberta to a proposed marine terminal facility at Kitimat, British Columbia. The signing of the document was timed to coincide with the end of the first week of public hearings in Edmonton, a process forecast to bring some 4,300 individuals to the witness table, including representatives from many of the First Nation communities likely to be affected by the pipeline.

The Northern Gateway pipeline and the Kitimat terminal are only two pieces of an infrastructure puzzle required for the exploitation of Alberta's abundant supplies of tar sands oil, most of which lie beneath the 140,000 square kilometers of Alberta's boreal forest.[1] The amount and value of new infrastructure required to move this oil to market is staggering by any measure. In addition to the Northern Gateway and Kitimat projects, proposals include a vast new array of additional pipeline capacity, including the politically volatile Keystone XL and numerous other pipelines that are being reconfigured to transport crude east to Ontario and U.S. markets. Significant new and retrofitted refining capacity is also under development with projects ranging from the Great Lakes to the Gulf Coast.[2]

This chapter reviews the ongoing and potential impacts of tar sands oil development for the tar sands region itself, for the lands and waters through which new pipelines go, and for the globe's climate. While each stage of the developmental process required to bring tar sands oil to market imposes actual or potentially severe environmental harms, a full rendering of the story must also encompass the political and social changes attributable to this latest stage in the development of the global petroleum economy. Indeed, the goal of those behind the project is *starkly* political; namely, to move Canada from its current position as an important but nonetheless regional player in the oil market to a world leader capable of rivaling Saudi Arabia in its ability to offer a steady stream of oil to the United States, Asia, and Europe. To achieve this end, proponents understand that Canada's social order must be reconfigured, a process that includes the marginalization of the nation's environmental community, a recalibration of the relations between Aboriginal citizens and the Canadian majority, and a rewriting of roughly forty years of Canadian environmental policy.[3]

FROM SAND TO OIL: AN OVERVIEW

Tar sands, also referred to as oil sands, is composed of a mixture of 80 to 85 percent sand and clay, 10 to 15 percent raw bitumen, and 5 percent water, with small amounts of minerals such as titanium, zirconium, tourmaline, and pyrite.[4] While almost six hundred tar sands deposits have been identified in twenty-two countries,[5] the largest economically recoverable deposits are located in Canada and Venezuela, though the latter is often considered at the heavy end of conventional crude.[6] Canadian tar sands deposits occur primarily in the western provinces of Alberta and Saskatchewan with the most significant deposits found in the former. As seen in Figure 12.1, Alberta's tar sands "plays," or fields, are primarily located in the northern portions of the province at Cold Lake and in the Peace and Athabasca River regions. The area amounts to some 140,000 square kilometers, a territory equivalent in size to New York or North Carolina and twice the size of New Brunswick.[7]

Canada's First Nations mixed bitumen with tree gum to patch their canoes long before the Europeans landed in America. While European explorers noted the presence of the material as early as the late 1700s, the resource was left idle until the 1930s, when it started to be used as largely unrefined tar to coat roads.[8] For most of the twentieth century, however, exploitation remained quite modest, due to the high cost and

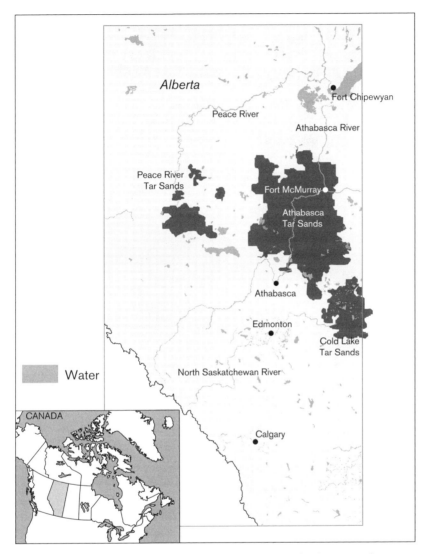

FIGURE 12.1. Alberta's tar sands region, 1960s to today. Map by the Geography Department, University of St. Thomas.

technological difficulties associated with both extraction and processing and due to relatively low world oil prices. Instead, the Canadian oil industry remained dominated by eastern producers at least until the discovery of Alberta's Leduc No. 1 oil well in 1947. While the find marked the beginning of a new era in petroleum exploration and production for western Canada, petroleum producers were forced to deal with a number of relatively unfavorable geographic and geologic factors. Chief among these was the high cost of transporting crude to West Coast refineries or finished products over the Rocky Mountains to the potentially lucrative markets that stretch from Vancouver to Washington State and Southern California. While producers did succeed in transporting limited volumes of crude westward, most of their attention was directed toward eastern Canadian markets and, more important, the American Upper Midwest. So successful were their efforts that by the early 1960s not only was the West responsible for supplying virtually all of Canada's crude oil, but the amount of oil the region was bringing to market was rising dramatically.[9] This achievement was made possible by the incremental development of an extensive system of pipelines and tar sands–capable refineries that reached outward from Alberta and into the northern tier of midwestern American states.[10]

While so-called conventional crude dominated Canadian production for most of the twentieth century, several companies were making significant investments in new extraction technologies suitable for unconventional tar sands oil,[11] beginning with the Great Canadian Oil Sands company (now called Suncor Energy) in the 1960s. This was followed by Syncrude and Imperial Oil Limited in the late 1970s and Shell Canada Limited in 1980. Despite these investments, the recession that hit Alberta in the mid-1980s nearly brought tar sands exploitation to a halt.[12] A decade later, however, increasing world oil prices and declining supplies of conventional oil in Alberta and elsewhere in Canada stimulated a significant level of renewed interest in the resource. The combination of these factors has meant a profound and continuing change in the structure of Canada's petroleum economy: as of 2010, of the 2.8 million barrels per day of Canadian production, oil sands production accounted for 1.6 million barrels, the remainder being supplied by conventional oil from the western provinces and an ever-dwindling share of Atlantic offshore production. The Canadian Association of Petroleum Producers (CAPP) projects that by 2025 no more than 10 percent of the country's crude production will be derived from conventional sources.[13]

Driving all of the investments in tar sands technology was the pursuit of a process that could economically break the bonds that bind oil and sand, one that begins with the removal of oil from the ground in one of two ways, depending on the depth of the deposit. When deposits are located near the surface, generally less than 80 meters or so below ground level, deposits are mined through a process similar to that used to access large, shallow coal deposits. Immense shovels are used to strip the surface clean, after which equally large trucks transport the material to crushers, in which the material is filtered through screens and mixed with hot water and caustic soda. The mixture is then delivered through a pipeline to a processing plant where it is placed in separation vessels. Air is injected into the mixture to separate the bitumen, or tar, from the sand, which sinks to the bottom of the pit while the air bubbles float the bitumen to the surface where it forms a thick froth.

In Alberta, virtually all mined bitumen is upgraded on site into so-called synthetic crude oil (SCO). Upgrading requires two steps, the first being a distillation process that separates diluents from bitumen. The bitumen is then heated to extremely high temperatures to break up the large bitumen hydrocarbon molecules into smaller hydrocarbon chains and ultimately into different products, including fuel gas, diesel, gas oil, naphtha, and coke. The coke is stored for future utilization, while the fuel gas is used on site to power the upgrader's furnaces. The remaining hydrocarbons are put through a secondary upgrading process in which hydrogen is added to stabilize the hydrocarbon molecules while the remaining sulfur, nitrogen, and other impurities are removed. The remaining products, including gas oil, naphtha, and diesel are blended into the SCO, most of which is shipped via pipeline to American refineries in the Upper Midwest that further refine it into a variety of petroleum products, including gasoline and aviation fuel.[14]

While surface deposits accessible through mining have historically constituted the bulk of Alberta's production, the future exploitation of the vast majority of the province's reserves will require so-called in situ extraction, a process used when the deposit is 80 meters or more underground. While a variety of in situ techniques are available, each relies in some way upon the injection of steam into the reservoir at temperatures high enough to liquefy the oil, allowing it to flow to the surface.[15]

In contrast to the bulk of the oil obtained from surface mining, the majority of in situ–extracted oil is being transported via pipelines as so-called dilbit, or diluted bitumen, due to the extreme viscosity (i.e., heaviness, thickness, like tar) of raw bitumen and its inability to flow in the confined space of a pipeline. While avoiding the cost of upgrading at the extraction site, dilbit requires other costly investments throughout the supply chain. Refiners, for instance, must separate the oil from the diluents prior to processing, as well as invest in upgrades required to refine this more unwieldy form of crude.[16] The cost of the diluent also means that producers are reluctant to use it on a one-off basis. As a result, new dilbit-sourced pipelines are constructed in parallel, with one line bringing dilbit to the refinery and a second line transporting diluent back to Alberta. Despite these additional costs, the "narrow price differential of bitumen relative to light crude," the extremely high costs of preprocessing, and the declining global availability of sweet, conventional crude relative to rising demand mean that the bulk of the oil obtained though in situ methods will continue to be shipped as dilbit, though Alberta's Energy Resources Conservation Board (ERCB) predicts that the "portion of in situ production upgraded in the province will increase from 11 per cent in 2010 to 13 per cent" by 2020.[17]

A REGION AT RISK

The extraction, transportation, refining, and, of course, consumption of tar sands oils is freighted with environmental consequences that far exceed even the most egregious instances of environmental destruction occasioned by conventional crude. Indeed, the mass exploitation of tar sands crude is routinely referred to as the most environmentally destructive industrial project ever executed by humankind.[18]

One of the more striking facts about tar sands oil is the amount of energy required to obtain useful work from the resource. While a somewhat contested notion, the energy balance of any particular energy source is most commonly stated as the EROI ratio; that is, the energy return on (energy) investment, or "the ratio of how much energy is gained from an energy production process compared to how much of that energy (or its equivalent from some other source) is required to extract, grow, etc., a new unit of the energy in question."[19] The fuel with the highest EROI ratio is hydropower, with a measured return of some 100:1. Conventional oil and gas returns some 11 to 18 units of energy, while shale oil generates a much lower return of 5:1. At the bottom of the list is bitumen from tar sands, which returns, at best, some 4 units of energy output for every unit of energy input and, in some cases, only 2.[20]

At present, most of the required energy inputs are met by natural gas, in quantities sufficient to power some 5 million Canadian homes.[21] The intensity of energy demands, and the preferred uses to which natural gas may be put, has led to a number of controversial proposals. For instance, a variety of organizations have proposed the construction of nuclear power plants dedicated solely to tar sands extraction and processing.[22]

However the required energy is secured, damages rapidly accumulate once production commences. In the case of surface mining, the process begins with what Environment Canada blithely refers to as the removal of "landscape features"—rocks, soil, trees, and water bodies,[23] or, in the language of the miners, "overburden"—some four tons of which must be removed for every barrel of oil produced. Given that tar sands production has grown from 1.065 million barrels per day to its current rate of 1.6 million barrels per day with an expectation of 4.1 million barrels of daily production by 2025,[24] the amount of overburden generated during the extraction process is staggering. Even more ominous are the projected environmental costs. Alberta's ERCB estimates that the province also holds over 18 billion barrels of recoverable tar sands oil,[25] a number the agency believes might be low based on "recent reserve growth in low permeability oil plays."[26]

Given such projections, there is scant comfort in the Alberta government's declaration that while "the oil sands underlie a 142,200 square kilometer area [. . .] the surface mining area is limited to a 4,800 square kilometer region near Fort McMurray, 715 square kilometers of which has been disturbed by oil sands mining."[27] Indeed, in concert with the overall increase in tar sands mining, the rate of disturbance has been rapidly increasing since about 2005. For instance, between 1977 and 2005, the single largest additional disturbance was 2005's 4,562 hectares (45.6 square kilometers).[28] Between 2005 and 2008, however, an *average* of 6,122 hectares of disturbance was recorded, the high point being 7,303 hectares in 2008.[29] Whether a future built on in situ extraction is more or less disruptive is a matter of substantial debate. For instance, while the process requires significantly less continuous disturbance, it nonetheless creates extremely disruptive fragmentation effects, a feature highly problematic for all species that require large expanses of continuous habitat for successful foraging and breeding. In any event, the fact that surface mining will continue to be a feature of Alberta's operations for many years to come leaves little doubt that the region will see a vast increase in a type of landscape where "most of the native biota, composed of thousands of species and millions of individuals, has been extirpated."[30]

In addition to requiring extensive surface disturbance, tar sands processing also requires prodigious amounts of water. While industry asserts that in situ extraction will decrease future water demands, studies comparing the projected versus the actual use of water in operational in situ mines have indicated that "water use is much higher than was predicted in project designs."[31] Even an optimistic reading of future water demands will, however, be extremely problematic for both surface water and aquifers in the region, given the vast increase in projected in situ projects. According to the Council of Canadian Academies, for instance,

"Knowledge is lacking as to whether the aquifers in the oil sands Atha-basca region can sustain [projected] groundwater demands and losses."[32] A federal study obtained under the Access to Information Act came to the same conclusion: "If all proposed projects are approved, water use for oil sands development is expected to more than double and may exceed the threshold for instream flow needs as defined by the Department of Fisheries and Oceans. [...] Presently, very little water that is removed from the Athabasca River for [tar] sands development is returned, and its flow is decreasing along with the flow of other rivers that feed the Mackenzie River."[33]

Both surface disturbance and high rates of freshwater consumption also create significant waste disposal problems. At present, the preferred method for dealing with wastewater is also the simplest: the use of so-called tailings ponds, or large, uncovered, and untreated surface repositories filled with extremely toxic liquid that can be neither reused nor recycled. Between 1992 and 2008 tailings ponds grew by some 422 percent, and by 2011 covered more than 50 square kilometers, all of which is held back by some of the largest dams on earth.[34] Since an additional 480 million gallons of toxic tailings are being generated daily, with Syncrude alone dumping 250,000 tons of liquid waste every year, there is little doubt that the region will be pockmarked with toxic ponds for many years to come.[35]

Despite the virtually complete destruction of the landscape and the serious impacts on the region's water bodies, both government and industry maintain that restoration or reclamation will eventually return the affected areas to a pre-mining state. Provincial officials, for instance, have claimed that the industry will achieve "one hundred percent long-term restoration on the lands it makes use of," re-creating landscapes capable of supporting "scenes reminiscent of the way things once were: human, land and beast coexisting in harmonious simplicity."[36]

Unfortunately, there is little evidence that reclamation projects generally succeed even on a limited scale, much less regionwide or at a pace anywhere near what is required. For instance, through the first forty years of widespread tar sands mining, the Royal Society of Canada estimates, the cumulative disturbance for oil sands mine development was 47,832 hectares. As of 2006, even using the criteria preferred by industry, only 13.6 percent (6,498 hectares) of this disturbed area was considered reclaimed. As of 2010, the one area eligible for a government reclamation certificate was a 104 hectare parcel known as Gateway Hill.[37]

The moonscape that results from surface mining and the damage to local and regional water supplies pose enormous risks for many species in the region, the best studied being the caribou herds. According to a 2011 assessment conducted by Environment Canada, virtually all of the herds in Alberta were considered "very unlikely" to be self-sustaining, a ranking to be considered the most dire of all possible threats and one tied directly to the extraordinarily high rate of territorial disturbance imposed on the various herds by tar sands–related activities.[38] Thus, while Canada as a whole has an 8.4 percent range disturbance rate, almost 71,000 of the 132,000 square kilometers utilized by Alberta's herds have been significantly disturbed.

Within existing oil and gas developments, caribou herds are considered particularly vulnerable due again to extraordinarily high levels of disturbance within their respective ranges. As reported by Environment Canada, 69 percent of the range used by the West Side Athabasca River herd is considered significantly disturbed, as is 78 percent of the range used by the Cold Lake herd.[39] Even harder hit is the East Side Athabasca River herd, which has seen more than 82 percent of its range seriously degraded.[40] To the extent that recovery does occur, the old-growth forests removed during mining operations will be replaced with early successional forests. This type of "habitat alteration" will do little to assist in the recovery of caribou, since it provides very favorable conditions for prey species (i.e., wolves) to further reduce or even eliminate whatever caribou populations survive landscape removal.[41]

In addition to endangering Alberta's caribou herds, tar sands development is also destroying habitat required for the survival of numerous other species. For instance, waterfowl and songbirds that come from all over the Americas to nest in the Canadian boreal forest suffer from both surface and in situ mining operations:

> Each year between 22 and 170 million birds breed in the 35 million acres of boreal forest that could eventually be developed for tar sands oil. Faced with tar sands development, migrating birds don't just move elsewhere since they depend on a certain type of habitat. Not only do many adult birds die when faced with lost and fragmented habitat and ponds of mining waste, but future generations of birds will have lost their chance to exist.[42]

Paul Wells estimates that 166 million birds could be lost over the next five decades due to habitat loss, water withdrawals, and the contaminated tailings ponds associated with tar sands development.[43]

IF THE RIVERS RAN SOUTH: TAR SANDS AND
ABORIGINAL COMMUNITIES

While the environmental impacts associated with tar sands development are on display throughout the province, they are particularly evident in those areas of northern Alberta inhabited by the province's
numerous First Nation communities.[44] As figure 12.1 shows, the bulk of
the tar sands plays are significantly north of Alberta's primary population centers of Calgary and Edmonton, in areas largely inhabited by
First Nation communities. Many of these communities sit astride the
waters most directly affected by tar sands development, namely, the
Peace River Delta and the Athabasca River and its watershed. Since
these rivers flow northward, eventually ending up in Hudson Bay, they
avoid the southern communities inhabited mainly by non-Aboriginal
populations. As a result, northern communities suffer a disproportionate share of the harms while enjoying few of the benefits associated with
the developmental process, harms greatly exacerbated by a social and
economic structure closely allied with the land and its capacity to sustain a functioning economy as well as a distinct cultural heritage.

The decline in the region's caribou herds is emblematic of the disproportionate impacts inflicted upon First Nation communities. In 2010, a
petition from various Cree and Chipewyan First Nations calling themselves the First Nations Petitioners requested the immediate protection
of the remaining woodland caribou in northeastern Alberta, generally
citing the essential link between caribou and the traditional way of life
that defines northern First Nation communities. Chief Walter Janvier of
the Chipewyan Prairie Dene First Nation argued: "The extinction of
caribou would mean the extinction of our people. The caribou is our
sacred animal; it is a measure of our way of life. When the caribou are
dying, the land is dying. We see no respect from government for the
caribou or for us as humans. The way Alberta is operating, profit for
the oil industry is number one, and everything else can be sacrificed."[45]

Chief Al Lameman of the Beaver Lake Cree Nation was equally
explicit in his denunciation of the impact of tar sands development,
again citing the essential link between the animal and the cultural heritage of the region's First Nations:

> We are calling on government to immediately halt the destruction of our
> lands, lands that sustain our caribou and our people. It is difficult for me to
> express the anger I feel at the loss of this noble animal in our territory. Our

traditional land is dwindling. We need habitat for our animals to ensure there is a healthy surplus. These animals sustain us and, as they die, our future becomes uncertain. We must act now to take care of Mother Earth.[46]

According to the chiefs, the obligation of the Canadian government to protect the species is clear, since under the "terms of Treaty 6 or Treaty 8, as modified by the Natural Resources Transfer Agreement, the First Nations Petitioners have (at a minimum) the constitutionally-protected right to hunt woodland caribou for sustenance on all unoccupied Crown lands, and on all other lands to which they have a right of access, within their respective traditional territories."[47] These treaties also recognize the essential cultural role of the woodland caribou "as a preferred species to hunt (when available), as part of a traditional diet that helps maintain the health and welfare of indigenous peoples, and for cultural and spiritual purposes."[48] The chiefs also argued that federal law, in particular the Species at Risk Act, dictates that protection measures be implemented, up to and including limiting tar sands–related activity in order to protect the habitat required for species survival.

The urgent need for protection was illustrated by the Beaver Lake Cree Nation. According to tribal chief Lameman, "Tar sands production within Beaver Lake Cree traditional territories stands at 546,000 barrels of oil per day (bpd), accounting for around thirty percent of Canada's total tar sands production," with current plans calling for more than a three-fold increase in "production within or near the caribou herd ranges." The potential impact on the caribou's range in the region is dramatic:

Within the Beaver Lake Cree's traditional territories there are 34,773 oil and gas well sites and of those 11,111 are within the caribou herd ranges. Each well site accounts for roughly 1 hectare of habitat loss. Within these ranges there are also 21,700 km of seismic lines, 4,028 km of pipelines and 948 km of roads. [Since] caribou are known to avoid habitat within 250 metres of these linear features, a 5–8 metre wide seismic line therefore degrades the habitat value of the forest for 500 metres.[49]

Despite these conditions, neither the provincial nor the Canadian government has "implemented any conservation-based land use plans in [. . .] traditional territories."[50] Instead, both federal and provincial authorities have accepted harms such as the eventual extinction of the Alberta herds as a necessary sacrifice, one well worth the benefits of a tar sands petroleum future. Thus, notwithstanding staff warnings delivered to his office in September 2011, Environment Minister Peter Kent decided that the "elevated risk of regional extirpation due to industrial," that is, tar sands–related, activity did not warrant action under any existing federal law.[51]

In addition to the loss or decline of caribou, there are also widespread concerns about adverse health impacts in the region's First Nation communities. George Poitras, former chief of the Mikisew Cree First Nation, points to the "terrifyingly high rate of cancer in Fort Chipewyan where I live. We are convinced that these cancers are linked to the Tar Sands development on our doorstep. It is shortening our lives. That's why we no longer call it 'dirty oil' but 'bloody oil.' The blood of Fort Chipewyan people is on these companies' hands."[52]

At the root of the health concerns is, of course, a reliance on the land and its resources that is endemic to these northern Aboriginal communities. Local fisheries, for instance, have been found to contain levels of mercury and arsenic elevated to the point that Health Canada recommended that consumption of large predatory fish should not exceed one meal per week for adults, while pregnant women, women of childbearing age, and children are told to consume no more than one fish meal per month.[53] Such limits represent a particular hardship on Aboriginal residents both because of the nutritional value of fish and because of the resource's cultural significance.

Proving a specific link between particular tar sands operations and adverse community health, however, is a difficult task. There is, for instance, ample evidence showing that leukemias and lymphomas are linked to petroleum products, including volatile organic compounds, dioxin-like chemicals, and other hydrocarbons. Cancers have also been tied to petroleum and to polycyclic aromatic hydrocarbons (PAHs), chemicals commonly found in tar and soot. At this point, however, relatively few studies have demonstrated specific cause-effect relationships between tar sands operations and Aboriginal communities. A study conducted by Suncor in 2005, for instance, concluded that at least in the Buffalo Wood region, a lifetime exposure to arsenic, a substance associated with tar sands operations, could result in 312–453 additional cases of cancer per 100,000 people, depending on levels of local fish and vegetable consumption. In 2006 a local community physician also asserted that he found an elevated number of cholangiocarcinomas, a rare form of bile duct cancer.[54]

Such findings prompted the Alberta Cancer Board to conduct a study in response to community demands. While the board's report determined that the number of cases of cholangiocarcinoma was within the expected range, the report also found that the *overall* cancer rate was approximately 30 percent higher than expected.[55] Nonetheless, Alberta Health and Wellness concluded that indigenous population was at little, if any, risk of developing cancer as a result of exposure to "existing (or) [. . .] future anthropogenic activity in the region," that is, tar sands–related activity.[56]

In confirming the findings of the provincial authorities, the Royal Society of Canada also found that "there is currently no credible evidence of environmental contaminant exposures from oil sands reaching [Aboriginal communities] at levels expected to cause elevated human cancer rates."[57] The society concluded that while "the overall cancer rate was higher than expected [and while] higher than expected cancers of the blood and lymphatic systems, biliary tract cancers as a group, and soft tissue cancers were found, these findings were based on a small number of case studies, and could be due to chance increased detection or increased risk in the community."[58] The Royal Society also emphasized the fact that a number of affected individuals had worked in the region's uranium mines, thus increasing their risk of occupational exposures to cancer-causing dust and other contaminants. In this regard, as in many contexts in North America (and beyond) where oil or mineral extraction has dominated small communities, health issues have found their way onto the public agenda in the tar sands region, but so far have produced many more questions than answers.

EXPANDING THE FOOTPRINT

The environmental damages associated with Alberta's all-out effort to exploit tar sands oil are pervasive and likely to endure for many generations. Unfortunately, should the growth strategy currently pursued by the Canadian government and its oil industry partners be realized, both the extent and scope of damages will increase greatly in the future, reaching far beyond the borders of this one province.

A critical part of this expanding geography of harm is the rapid expansion of the pipeline system currently radiating outward from Alberta.[59] In addition to proposals for new pipelines such as the Keystone XL, companies such as Enbridge and TransCanada have proposed repurposing existing natural gas pipelines to carry oil, "reversing" lines that now flow east to west, or pursuing a combination of all three approaches. In addition to being essential to the securing of new markets, all of these projects share another attribute, namely, the likelihood that they will expand the range of environmental harms far beyond those currently on display in Alberta. The Northern Gateway project illustrates this fact. Despite the widespread attention given to the Keystone XL project and its passage across Nebraska's sand dunes and aquifers, Northern Gateway appears to be the riskier of the two. As proposed, the line would carry some 525,000 barrels of dilbit per day 1,177 kilometers to the marine terminal at Kitimat, British Columbia; a twin line would return 193,000 barrels of diluent per day from Kitimat to Edmonton. The pipelines would cross 773 watercourses, 669 of which are fish bearing, including an extremely sensitive salmon-spawning habitat in the upper Fraser, Skeena, and Kitimat watersheds.[60] In addition, two tunnels would have to be built through the extraordinarily rugged Coast Mountains, along with a large number of pumping stations, the number and location of which are yet to be determined.[61]

The use of the Kitimat marine terminal to accommodate the raft of supertankers required to receive and then ship tar sands crude to China and other markets also poses enormous risks. The facility is located in the midst of the Great Bear Rain Forest, an area that holds some of the most rugged, wild, and unspoiled country to be found in the whole of North America. As proposed, the terminal would have to be significantly upgraded if it is to serve as the loading site for an estimated 220 mammoth oil tankers per year, each longer than the Empire State Building is tall. While the area around Kitimat has routinely accommodated standard petroleum product tankers for at least thirty-five years, no "very large crude carrier" of the type that will be used for tar sands oil has ever navigated Douglas Channel, the primary route between the open sea and Kitimat. The channel is an inlet laced with extremely narrow passages and filled with what Bruce Barcott describes as "a jigsaw of islands." Tankers would be forced to navigate their way through this maze in the midst of rough and unpredictable waters occasioned by the area's often turbulent weather.[62]

In addition to Keystone XL and Northern Gateway, numerous other tar sands–driven projects are being pursued. The Alberta Clipper and Southern Lights projects, for instance, will accelerate the flow of tar sands crude into Minnesota and Wisconsin, along with a new, thirty-five-mile long, twenty-inch-diameter pipeline that will supply the Marathon Refinery in Detroit.[63] Energy East is a 4,500-kilometer pipeline proposed to carry 1.1 million barrels of crude oil per day from Alberta and Saskatchewan to refineries in eastern Canada, and the Flanagan South pipeline project is a nearly six-hundred-mile, thirty-six-inch-diameter interstate crude oil pipeline that will originate in Flanagan, Illinois, and terminate in Cushing, Oklahoma, crossing Illinois, Missouri, Kansas, and Oklahoma. The majority of the pipeline will parallel Enbridge's existing Spearhead crude oil pipeline right-of-way and will have an initial capacity of 600,000 barrels per day.[64] Like many domestic pipelines, the $2.6 billion Flanagan South pipeline does not cross an international border and therefore does not require a U.S. State Department permit, a fact that contributes to the relative lack of public attention the project has received.

When completed, these and numerous other related pipelines will both greatly increase tar sands consumption in the midwestern and eastern portions of the United States and Canada and provide access to refineries or transshipment points capable of supplying markets beyond the North American continent.[65] While each proposed segment is portrayed as discreet in its purpose and impact, when taken together, they constitute a system in service to a growth agenda that rivals the most ambitious of industrial dreams.

The projects also demonstrate another important dimension of the development process: the willingness of proponents to change their approaches and modify their rhetoric in the face of opposition while continuing to pursue the ultimate goal of large-scale expansion. For example, in 2009, Enbridge halted the development of its so-called Trailbreaker project, a scheme designed to transport oil from Alberta to the U.S. East Coast that faced serious local opposition. In May 2012, however, the company "announced expansion plans that confirmed a long-range goal to ship tar sands to Montreal." According to local opponents, "The overall goal of moving tar sands oil out of Quebec and through New Hampshire and Vermont, eventually reaching the Maine coast appears to be the larger goal of the project," one completely consistent with the earlier Trailbreaker project—an assertion roundly denied by the company.[66] At the same time, while the Northern Gateway pipeline is explicitly intended for European and Asian markets, the ultimate destination for the product shipped via the Keystone XL project remains subject to debate. Thus, while both the Harper government and its American supporters argue that accelerating the pace of tar sands development is consistent with the larger goal of U.S. energy security, opponents point to actions by Gulf Coast refiners as evidence of an intent to ship tar sands–sourced products to Europe.[67]

While both the Northern Gateway and the Keystone XL projects are being built with dilbit in mind, all of the pipelines coming out of Alberta will eventually carry ever-increasing quantities of this highly toxic material. The result of this transition will be a substantially more brittle and spill-prone system. Several factors explain the increased risk, including dilbit's acidic content, which is up to twenty times higher than that of conventional crude oil; the presence of abrasive materials such as quartz and pyrite sand particles; and the higher operating pressures required to push the viscous dilbit though the pipeline. These factors are particularly important given the older pipelines that Enbridge is proposing to use to carry bitumen from Alberta to the East Coast, including the sixty-two-year-old pipeline on the Portland-Montreal route and the forty-year-old Line Nine pipeline that extends from Montreal, Quebec, to Sarnia, Ontario.[68]

When a spill does occur, dilbit is also more problematic than conventional oil. According to Anthony Swift and his colleagues:

> The low flash point and high vapor pressure of the natural gas liquid used to dilute the DilBit increase the risk of the leaked material exploding. DilBit can form an ignitable and explosive mixture in the air at temperatures from o degrees Fahrenheit. [. . . O]ne of the potential toxic products of a DilBit explosion is hydrogen sulfide, a gas which can cause suffocation in concentrations over 100 parts per million and is identified by producers as a potential hazard associated with a DilBit spill. [. . .] DilBit contains benzene, polycyclic aromatic hydrocarbons, and n-hexane, toxins that can affect the human central nervous systems. DilBit also contains vanadium, nickel, arsenic, and other heavy metals in significantly larger quantities than occur in conventional crude. These heavy metals have a variety of toxic effects, are not biodegradable, and can accumulate in the environment to become health hazards to wildlife and people.[69]

Finally, dilbit poses particular problems for spills that occur in or near a water body, as demonstrated by the 2010 rupture of a pipeline that spilled some 819,000 gallons of dilbit into a tributary of the Kalamazoo River. Because the diluents can quickly evaporate, the heavy bitumen sinks to the bottom, a fact that greatly exacerbates the cleanup problem. Indeed, according to the EPA incident commander at the Kalamazoo spill site, "This was the first time the Environmental Protection Agency or anyone has done a submerged cleanup of this magnitude. I would never have expected [. . .] that we would have spent two or three times longer working on the submerged oil than [on] surface oil."[70] An August 2013 regulatory filing by Enbridge estimated the cost of the Kalamazoo cleanup at $1.039 billion dollars.[71]

The new and upgraded refineries to which the increased volumes of both synthetic crude oil and dilbit will be shipped also pose significant risks and hazards. According to a 2012 study by Aaron Sanger, communities downstream from those refineries that rely on tar sands–based crude oil as their primary feedstock "have elevated levels of cancer. [. . .] The sulfur dioxide pollution from tar sands refineries also puts more pressure on those who already suffer from disease-weakened hearing and blood circulation problems."[72]

While significant, all of the impacts and risks discussed to this point can be discreetly located—in a particular water body, in a specific estuary, or across a particular stretch of sensitive grassland. The effect of tar sand development on climate change is, on the other hand, profoundly global. It is an undeniable fact that the use of fossil fuels is the major factor in anthropogenically induced climate change. In the United States, for instance, the EPA estimates that some 43 percent of all greenhouse gas emissions derive from various petroleum products, primarily gasoline and other transportation-related fuels.[73] Yet, according to critics such as NASA's James Hansen, tar sand development on the scale proposed by Canada will mean "game over for the climate":

> Canada's tar sands, deposits of sand saturated with bitumen, contain twice the amount of carbon dioxide emitted by global oil use in our entire history. If we were to fully exploit this new oil source, and continue to burn our conventional oil, gas and coal supplies, concentrations of carbon dioxide in the atmosphere eventually would reach levels higher than in the Pliocene era, more than 2.5 million years ago, when sea level was at least 50 feet higher than it is now.[74]

Hansen's conclusions derive from any number of recent studies assessing the life cycle impact of tar sands versus other petroleum sources. For example, the U.S. Department of Energy estimates that on a well-to-wheels emissions basis the variation between the lowest–carbon intensity crude oil source, such as that produced domestically in the United States, and highest–carbon intensity crude oil source (i.e., an average of all Canadian tar sands–sourced oil) is 130 percent.[75] A 2012 Congressional Research Service study that surveyed the primary published literature came to the same conclusion:

> Canadian oil sands crudes are on average somewhat more GHG emission–intensive than the crudes they would displace in U.S. refineries, as Well-to-Wheel GHG emissions are, on average, 14%–20% higher for Canadian oil sands crude than for the weighted average of transportation fuels sold or distributed in the United States [and that] discounting the final consumption phase of the life-cycle assessment (which can contribute up to 70%–80% of Well-to-Wheel emissions), Well-to-Tank (i.e., "production") GHG emissions

are, on average, 72%–111% higher for Canadian oil sands crude than for the weighted average of transportation fuels sold or distributed in the United States.[76]

Canadian tar sands crude is also higher in emissions than any type of imported oil, being "9% to 19% more emission-intensive than Middle Eastern Sour, 5% to 13% more emission-intensive than Mexican Maya, and 2% to 18% more emission-intensive than various Venezuelan crudes, on a Well-to-Wheel basis."[77]

Increases of this magnitude no doubt explain the Canadian government's decision to withdraw from the Kyoto Protocol, since, under the enhanced level of tar sands production desired by both the government and industry, Canada will fall well short of its obligations. As Marc Hout, Lindsay Fischer, and Nathan Lemphers observe:

> In the last two decades, [tar sands] emissions have more than doubled. Government forecasts predict annual GHG emissions from [tar sands] will double again from 2009 to 2020 with current policies and practices. [. . .] The Canadian Association of Petroleum Producers forecasts that GHG emissions from [tar sands] are likely to continue climbing well into the 2030s. [As a result] based on Canada's projected emissions in 2020, the combined effect of all currently announced federal and provincial climate policies would accomplish just one quarter of the emissions reductions required by the 2020 deadline.[78]

TAR SANDS AND THE STATE OF THE NATION

The drive to exploit tar sands is hardly a unique chapter in Canadian history; indeed, it is fair to say that resource development has always defined and, to a great extent, divided Canada. Indeed, one of the most important features of the country's constitutional structure is the considerable power granted to provincial authorities for resource development.[79] While the federal government can claim jurisdiction over off-shore resources, trade and commerce, and international resources, the environmental regulation of fossil fuels occurs primarily at the provincial level with direct federal environmental regulation of oil and gas activities within provinces being limited to interprovincial and export pipelines.[80]

One factor that makes this "center-periphery cleavage" particularly important is the relative resource endowments of the various provinces.[81] While the western provinces are rich in many types of exploitable natural resources, the conventional energy resources of the East are mainly confined to the hydro resources of Manitoba and Quebec and the off-shore oil wells of the maritime provinces.[82] This disparity has resulted in periodic regional clashes over the nature and direction of national energy policy. In the early 1970s, for instance, a major objective of energy policy was to foster a strong petroleum and gas industry by, in part, enhancing the

> constitutional responsibilities of the provinces over natural resources [and] allowing Alberta, Ontario, and Quebec to put forward their respective energy programs. [. . .] Alberta chose to adapt the main features of systems that were developed in energy-producing states such as Texas. Simple licensing regulations provided the province with the means to maintain surveillance and control over general exploration activity without intervening in corporate strategy. [This] was bound to clash with a federal energy policy essentially devised, in the Trudeau years, by politicians from Central Canada.[83]

These tensions came into full public view with the 1980s-era National Energy Program (NEP), which, according to Michel Duquette, "was resented by corporate and provincial opinion makers altogether and denounced as a unilateral policy towards forced unification at the expense of the actual balance of power within the Canadian constitutional tradition."[84] Elections in 1980 brought an end to NEP-type centralization as the "nation-accentuating" policies were trumped by the "periphery [which] succeeded in reversing a powerful trend towards economic nationalism."[85]

Tar sands development is now taking its place in this long history of intense regional conflict, with the political center again advocating aggressive development. In this case, however, rather than a unified provincial response, a clear regional split is evident among both provincial leaders and the public at large. To some extent, the split is being driven by a number of larger economic issues, chief among which is the so-called Dutch disease, a condition often associated with resource-based economies of the sort represented by Canada. This term, first noted in reference to the 1950s-era natural gas developments in the Netherlands, refers to the case in which a resource boom in an economy leads to a real exchange rate appreciation and to the crowding out of the tradable manufacturing sector. In the case of Canada,

> between 2002 and 2008, the spectacular rise in commodity and oil prices has led to an important development of the Canadian energy and commodity sectors. [. . .] While the extraction from oil sand is difficult and costly, the rise of the oil price to new high levels has triggered an important expansion of the extraction and export of the oil. This has in turn led to a big increase

[in] Alberta nominal GDP per head, to a spectacular improvement of the provincial public finance and to important spillovers for the other provinces. During the same period, the Canadian exchange rate appreciated and the manufacturing sector has contracted.[86]

By one reckoning, "fifty-four percent of the manufacturing employment loss (in Eastern provinces is) due to exchange rate development between 2002 and 2007 are related to a Dutch disease phenomenon."[87]

In addition to the macroeconomic regional consequences of a resource-dominated economy, there are also questions of just how tax and royalty revenues generated by tar sands operations will be allocated between Alberta and Ottawa. A 2011 report authored for the Canadian Energy Research Institute illustrates the significance of this issue and its potential for regional conflict. According to the institute, the Canadian government will collect some $311 billion in total taxes between 2010 and 2035; Alberta's share of taxes over the same time period is projected at $105 billion. However, the Alberta government will also receive many billions in additional royalties starting with $3.2 billion in 2010, with annual receipts forecast to grow to $47 billion by 2035. The cumulative total of royalties collected by the Alberta government will exceed $623 billion over the following twenty-five years.[88]

Given all of this, it is not surprising that debate over tar sands development often surfaces in the nation's political discourse. In May 2012, for instance, Ontario premier Dalton McGuinty stated that given a choice over "a rapidly growing oil and gas sector in the West or a lower dollar benefiting Ontario, I stand with the lower dollar." Alberta's premier, Alison Redford, called McGuinty's argument a "false paradigm. [. . .] We know how the value of the dollar works. It's in relation to an overall national economy. The reason the Canadian dollar is high is partly because the United States has been going through some economic difficulties."[89] The split between provincial leaders is mirrored in public opinion, with polls conducted since at least 2008 consistently showing marked differences between the residents of the petro-provinces versus the rest of the country regarding the benefits of tar sands production.[90]

The fracturing of Canadian society is also evident in the treatment of those who object to tar sands development, in particular those in the environmental community. For instance, in response to the thousands of potential interveners who wished to speak in opposition to the Northern Gateway project, many of whom were identified with or had been encouraged to participate by various environmental charities, Natural Resource Minister Joe Oliver stridently proclaimed that such charities were threatening to "hijack our regulatory system to achieve their radical ideological agenda. They seek to exploit any loophole they can find, stacking public hearings with bodies to ensure that delays kill good projects. They use funding from foreign special interest groups to undermine Canada's national economic interest."[91]

Joining in the condemnation, Environment Minister Peter Kent argued that environmental charities were "laundering foreign funds, claiming that there are allegations—and we have very strong suspicions—that some funds have come into the country improperly to obstruct, not to assist, in the environmental assessment process."[92] Conservative Senator Nicole Eaton also referred to green charities as "master manipulators who are operating under the guise of charitable organizations in an effort to manipulate our policies for their own gain."[93]

Indeed, according to a former government employee and self-proclaimed whistle-blower, during internal cabinet deliberations the prime minister's office characterized ForestEthics, a prominent Canadian environmental charity leading the opposition to the Northern Gateway project, as an "Enemy of the Government of Canada," and an "Enemy of the people of Canada."[94] While such a claim may be seen as overheated rhetoric by a disgruntled civil servant, members of the Harper administration employed similar language in characterizing a visit to the United States by members of the political opposition as a "treacherous course" inimical to the country's long-term health and welfare.[95]

Tar sands development is also driving ever-deeper wedges between First Nations and nonaboriginal Canada. In August 2010, for instance, the indigenous hereditary chiefs of the Wet'suwet'en First Nation issued a final notice of trespassing to Enbridge, stating that the company was no longer welcome on their territory, and members subsequently constructed a traditional longhouse directly in the path of the proposed pipeline.[96] Also in 2010, sixty-one British Columbia First Nations signed a resolution stating their opposition to the Northern Gateway project. Building on this resolution, in February 2011 the Yinka Dene Alliance rejected an offer from Enbridge for "revenue sharing" benefits of more than $1.5 billion in cash, jobs, and business opportunities over the next thirty years, as well as a 10 percent stake in the project. Water, land, and cultural heritage, the alliance explained, were more important than short-term financial gain. A similar point was made in the 2012 Save the Fraser Declaration, signed by the Yinka Dene Alliance along with Alberta and Northwest Territory First Nations, which declared that the federal processes granting approval to the Northern Gateway project "violate our laws, traditions, values, and our inherent rights as Indigenous Peoples under international law."[97]

Finally, Canada's environmental policy regime is also being undone in response to the tar sands industry's demands. In 2012, the Omnibus Budget Bill, otherwise known as C-38 or the Jobs, Growth, and Long Term Prosperity Act, came into force in order to expedite development of and reduce delays for so-called major energy projects such as tar sands–related pipelines and terminals. The act repealed the Canadian Environmental Assessment Act and diminishes the Fisheries Act, the Species at Risk Act, and the National Energy Board Act. The bill also imposed significant restrictions on the activities of environmental charities and limited the scope of intervention to parties with a direct vested interest in major projects, an element of the bill that, as of late 2012, was already being used to limit participation in tar sands–related hearings.[98]

CONCLUSION

The environmental degradation of the Canadian North; the hazards posed by oil extraction, pipelines, and terminals; the further degradation of the planet's atmosphere, the relentless attacks on Canada's environmental and Aboriginal communities; and the undermining of long-observed policy practices are all issues of tremendous significance. Yet the development of Canada's tar sands also underscores a more fundamental fact about the political economy of contemporary resource development.

In its quest to make Canada a dominant player in international commodity markets, the Harper administration is actively courting, and has succeeded in attracting, significant amounts of international capital. A 2012 report by Nikke Skuce, for instance, found that 71 percent of all companies operating in the Fort McMurray, Alberta, area are not Canadian owned, including a number of Chinese state-owned enterprises. While almost all of the companies operating in the area have roughly 50 percent foreign ownership, others have a much higher share. For instance, 91 percent of Husky Energy, nominally a Calgary-based company, is owned by non-Canadians. At the same time, some 51 percent of all oil and gas operating revenue in Canada goes to foreign entities; this compares with the national average of 28 percent. Canada's oil and gas industry also has twice the amount of foreign-controlled operating profit compared to the national average.[99] In this respect, the tar sands development process is understood by its proponents as an exercise that properly brings together international private capital and a regulatory system conducive to full-scale and unhindered resource exploitation.

This fact stands in sharp contrast to the rigidly defined localism that, at least in the eyes of the Harper administration, is the only legitimate form of opposition to the exploitation of tar sands. Thus, while industry and the state are allowed to operate as a united global enterprise, opposition groups are being told that they must fragment their efforts, focusing, one at a time, on the many discreet projects that make up the tar sands system. Further, they are being told to limit their opposition to projects by which they are immediately affected. Operating within this logic, opposition forces in eastern Canada cannot legitimately question the need for or impacts of the Northern Gateway project, First Nations in Manitoba cannot join with First Nations in British Columbia in opposing the Kitimat terminal, residents of BC are expected to remain silent in their opposition to the eastward-flowing pipelines, and anyone questioning the wisdom of tar sands development in light of its impact on global climate change is regarded as voicing an opinion bordering on treason.

Such a logic is, of course, designed to place the resource development process safely out of the reach of those who might object to its pace or impact. This goal was clearly enunciated by Natural Resource Minister Oliver in announcing the administration's intent to secure passage of C-38: "We believe reviews for major projects can be accomplished in a quicker and more streamlined fashion. We do not want projects that are safe, generate thousands of new jobs and open up new export markets, to die in the approval phase due to unnecessary delays."[100] Just how much Canada and its citizens will allow themselves to be reshaped by the rush to oil is, of course, uncertain. Given events to date, however, it might well be the case that First Nations organizer and activist Clayton Thomas-Muller was sadly prescient when he declared that "in our country everything revolves around Tar Sands, even human rights."[101]

NOTES

1. Environment Canada, *Environment Canada and the Oil Sands* (Ottawa, Ontario, May 2011).

2. Anthony Andrews, Robert Pirog, and Molly F. Sherlock, *The U.S. Oil Refining Industry: Background in Changing Markets and Fuel Policies* (Washington, DC: Congressional Research Service, 2011); Benjamin Wakefield and Matt Price, *Tar Sands: Feeding U.S. Refinery Expansion with Dirty Fuel* (Washington,

DC: Environmental Integrity Project, 2008); David Israelson, *How the Oil Sands Got to the Great Lakes Basin: Pipelines, Refineries and Emissions to Air and Water* (Toronto: Munk Centre for International Studies, 2008).

3. The proponents of this ambitious transformation of Alberta and Canada are led by Stephen Harper, the Alberta-based politician who has served as prime minister of Canada since 2006. PM Harper moved to Edmonton, Alberta, in the late 1970s, where he found work in the mail room at Imperial Oil. Later, he advanced to work on the company's computer systems. He took up post-secondary studies again at the University of Calgary, where he completed a bachelor's degree in economics, later returning to earn a master's degree in economics in 1993. Harper became involved in politics as a member of his high school's Young Liberals Club. He later changed his political allegiance because he disliked the NEP. Given this professional background and political history, his strong affinity with oil and gas development and the full-bore development of tar sand is not unexpected.

4. Dictionnaire de l'environnement, *Sable bitumineux, définition* (2012), http://www.dictionnaire-environnment.com/sables_bitumineux_ID5758.html (accessed January 30, 2012); Office national de l'énergie, *Les sables bitumineux du Canada: Perspectives de l'offre et du marché jusqu'en 2015* (Ottawa, Ontario, 2000).

5. Claude Bandelier, *Problématique environnementale de l'exploitation des sables bitumineux en Alberta (Canada)* (Bruxelles: Université Libre de Bruxelles, Institut de Gestion de l'Environnement et d'Aménagement du Territoire, Faculté des Sciences, Master en Sciences et Gestion de l'Environnement, 2010), 11.

6. Countries such as the Democratic Republic of Congo, Madagascar, and Russia also have deposits, they these are of less importance and the extreme conditions of eastern Siberia have not permitted the exploitation of its deposits.

7. Greenpeace Canada, *What If It Was In My Home?* (2012), http://www .ifitweremyhome.ca/ (accessed June 17, 2012).

8. Andrew Nikiforuk, *Tar Sands: Dirty Oil and the Future of a Continent* (Vancouver: Greystone Books, 2008), 25.

9. Canadian Association of Petroleum Producers (CAPP), *Crude Oil Production, 1971–2010* (Calgary, Alberta, 2012), http://membernet.capp.ca/SHB /Sheet.asp?SectionID=3&SheetID=76 (accessed April 9, 2012); CAPP, *Crude Oil Production, 1947–1970* (Calgary, Alberta, 2012), http://membernet.capp .ca/SHB/Sheet.asp?SectionID=3&SheetID=75 (accessed April 9, 2012).

10. Steven M. Hoffman, "A Legacy of Dependence: Alberta, Minnesota and the Coming Tar Sands Future" (paper presented at Minnesota's Environmental History, sponsored by the Minnesota History Center, Carleton College, and Macalester College, St. Paul, Minnesota, 2012).

11. Oil is divided between conventional and unconventional crude. The former is further divided into light/medium and heavy crude, defined by the specific gravity or viscosity of the oil. Unconventional oil is defined as such both by the method of extraction and its viscosity or its ability to flow, or, in this case, its lack of ability to flow in its undiluted state. Some commentators use the term *oil sands* on the assumption that it conveys a more suitable affinity with other

forms of crude oil. In this paper, "tar sands" will be used except when in quoted material.

12. P. Biays, *Le Canada: environnement naturel, économie, regions* (Paris: SEDES-DIEM, 1987).

13. CAPP, *Crude Oil Production, 1971–2010*.

14. CAPP, *Crude Oil: Forecasts, Markets and Pipelines* (Calgary, Alberta, 2012); Energy Resources Conservation Board (ERCB), *Alberta's Energy Reserves 2010 and Supply/Demand Outlook 2011–2010* (Calgary, Alberta, June 2011).

15. Petroleum Technology Research Center, *What Is VAPEX (Vapour Extraction) and How Does It Work?* http://www.ptrc.ca/faqs.php?f_action=news_detail&news_id=8734 (accessed March 3, 2012); Graham Chandler, "Excelsior to Test COGD Bitumen Production," *Schlumberger* (2009): http://www.slb.com/services/industry_challenges/heavy_oil/heavyoilinfo.aspx (accessed March 3, 2012); James G. Speight, *The Chemistry and Technology of Petroleum* (Boca Raton, FL: CRC Press, 2007); Kurt Cobb, "Will Toe-to-Heel Air Injection Extend the Oil Age?" *Scitizen* (2002), http://scitizen.com/future-energies/will-toe-to-heel-air-injection-extend-the-oil-age-_a-14-3449.html (accessed March 3, 2012); Roger M. Butler, *Thermal Recovery of Oil and Bitumen* (Englewood Cliffs, NJ: Prentice Hall, 1997).

16. Aaron Sanger, *Tar Sands Refineries: U.S. Communities at Risk* (San Francisco: Forest Ethics, 2012).

17. ERCB, *Alberta's Energy Reserves 2010*, 6.

18. Marc Hout, Lindsay Fischer, and Nathan Lemphers, *Oil Sands and Climate Change: How Canada's Oilsands Are Standing in the Way of Effective Climate Action* (Calgary, Alberta: Pembina Institute, 2011); Keith Stewart and Melina Laboucan-Massimo, *Deep Trouble: The Reality of In Situ Tar Sands Operations* (Toronto: Greenpeace Canada, 2011).

19. David J. Murphy and Charles A.S. Hall, "Year in Review—EROI or Energy Return on (Energy) Invested," *Annals of the New York Academy of Sciences*, 1185 (2010): 102.

20. Ibid., 109.

21. Greenpeace Canada, *What If It Was In My Home?*

22. World Nuclear Association, *Alberta Tar Sands: Nuclear Power in Canada Appendix 2* (2010), http://www.world-nuclear.org/info/inf49a_Alberta_Tar_Sands.html (accessed June 19, 2012).

23. Environment Canada, *Environment Canada and the Oil Sands*.

24. CAPP, *Crude Oil: Forecasts, Markets and Pipelines* (Calgary, Alberta, 2011), appendix B1.

25. ERCB, *Alberta's Energy Reserves 2010*, 3–2 and table 3.1.

26. Ibid., 6 and figure 4.

27. Province of Alberta, *Oil Sands Reclamation* (Edmonton, Alberta: Government of Alberta, 2011).

28. There are 100 hectares (ha) per square kilometer; thus, 4,562 ha = 45.6 square kilometers of disturbance.

29. Royal Society of Canada, *The Environmental and Health Impacts of Canada's Oil Sands Industry* (Ottawa, Ontario, 2010), 166, table 9–2.

30. Kevin P. Timoney and Peter Lee, "Does the Alberta Tar Sands Industry Pollute? The Scientific Evidence," *Open Conservation Biology Journal* 3 (2009): 71, http://www.deadducklake.com/wp-content/uploads/2009/11/timoney-and-lee_toconbj.pdf (accessed December 12, 2011).

31. Environment Canada, *Environment Canada and the Oil Sands*; Stewart and Laboucan-Massimo, *Deep Trouble*.

32. Quoted in Stewart and Laboucan-Massimo, *Deep Trouble*, 8.

33. *A Federal Perspective on Water Quality Issues* (December 2007), 10–11.

34. Timoney and Lee, "Does the Alberta Tar Sands Industry Pollute?" 71.

35. The lethality of these waters was vividly on display in April 2008 when 1,600 migrating ducks died when they landed on one of Syncrude's tailings ponds—an incident that may only be the tip of the iceberg in terms of harm to wildlife. The company paid a $3 million fine and installed loudspeakers and scarecrows to scare off any waterfowl with the bad sense to land on the ponds. See the Ecojustice website, www.ecojustice.ca.

36. Nikiforuk, *Tar Sands*, 94, 96.

37. Jennifer Grant, Simon Dyer, and Dan Woynillowicz, *Fact or Fiction: Oil Sands Reclamation* (Edmonton, Alberta: Pembina Institute, 2008); Royal Society of Canada, *Environmental and Health Impacts*, 166.

38. Environment Canada, *Recovery Strategy for the Woodland Caribou (*Rangifer tarandus caribou*), Boreal Population, in Canada [PROPOSED]* (Ottawa, Canada, 2011), accessed July 7, 2012 at: http://www.sararegistry.gc.ca/default.asp?lang=En&n=B82CFC52-1.

39. Ibid.

40. Peter G. Lee, *Canada's woodland caribou: Industrial disturbances in their ranges and implications for their survival* (Edmonton, Alberta: Global Forest Watch, 2011), 4–6, http://globalforestwatch.ca/publications/20120110A (accessed January 16, 2017).

41. Environment Canada, *Recovery Strategy for the Woodland Caribou*, 4.2.

42. Paul Wells, *Danger in the Nursery: Impact on Birds of Tar Sands Oil Development in Canada's Boreal Forest* (Washington, DC: Natural Resources Defense Council, 2008), 1.

43. Ibid., iv.

44. The disproportionate impact suffered by Aboriginal populations is hardly unique in the history of Canada's resource development policies and practices, a fact illustrated by enormous hydroelectric projects in various parts of Canada. See Thibault Martin and Steven M. Hoffman, eds., *Power Struggles: Hydro Development and First Nations in Manitoba and Quebec* (Winnipeg: University of Manitoba Press, 2008).

45. Demand Letter, *From First Nations Petitioners to Minister Prentice* (July 15, 2010), accessed June 30, 2012 at: http://www.ienearth.org/news/Demand_letter_to_Minister_Prentice_-_2010–07–15_FINAL.pdf.

46. Ibid.

47. Ibid.

48. Ibid., 2.

49. The Co-operative, Raven and Beaver Lake Cree Nation, *Save the Caribou: Stop the Tar Sands* (2010), 3, 8, http://www.coop.co.uk/upload/ToxicFuels/docs /caribou-report.pdf (accessed January 16, 2017).

50. Mel Evans et al., *Cashing in on Tar Sands: RBS, UK banks and Canada's "blood oil"* (London, UK: Platform, n.d.), archived at http://www.banktrack .org/download/cashing_in_on_tar_sands_rbs_uk_banks_and_canada_s_blood _oil_/100308_cashing_in_on_tar_sands.pdf (accessed January 16, 2017).

51. Mike De Souza, "Woodland Caribou Threatened by Industry, Memo Warns," *Postmedia News/Vancouver Sun* (July 7, 2012), http://www.vancouversun .com/business/Woodland+caribou+threatened+industry+memo+warns/6899076 /story.html#ixzz208rGLbvX (accessed July 10, 2012).

52. Poitras quoted in Evans et al., *Cashing in on Tar Sands,* 44.

53. H.R. Guo, "The Lack of a Specific Association between Arsenic in Drinking Water and Hepatocellular Carcinoma," *Journal of Hepatology* 39 (2003): 383–88.

54. Kevin P. Timoney, on behalf of the Nunee Health Board Society, *A Study of Water and Sedimentary Quality as Related to Public Health Issues, Fort Chipewyan, Alberta.* (Fort Chipewyan, Alberta, 2007), 7.

55. Pembina Institute and Environmental Defence, *Les sables bitumineux du Canada: les obligations du gouvernement fédéral* (Calgary, Alberta, 2010), archived at http://www.equiterre.org/sites/fichiers/rapport_sables_bitumineux _oct2010_fr.pdf (accessed January 17, 2017).

56. Timoney, *A Study of Water and Sedimentary Quality,* 7.

57. Royal Society of Canada, *Environmental and Health Impacts,* 5.

58. Ibid., 230.

59. A number of analysts have argued that trucks or railcars, or both, can serve as a sufficient substitute for pipelines, a claim much in dispute. See U.S. State Department, *Keystone XL Pipeline Project: Draft Supplementary Environmental Impact Statement,* ES-15 (Washington, DC: March 1, 2013), at http://keystonepipeline-xl.state.gov/documents/organization/205719.pdf (accessed June 23, 2013).

60. Candice Bernd, *BREAKING: Obama Rejects Keystone XL Pipeline,* posted Wednesday, January 18, 2012 (Center for American Progress/Campus Progress, 2012), http://campusprogress.org/articles/breaking_obama_rejects _keystone_xl_pipeline/ (accessed January 26, 2012).

61. Enbridge Northern Gateway Pipeline, *Pipeline Information and Plan* (2012), http://www.northerngateway.ca/project-details/pipeline-information-and- plan/ (accessed February 1, 2012).

62. Bruce Barcott, "Pipeline through Paradise," *National Geographic* 220:2 (August 2011): 54–65.

63. Minnesota Public Utilities Commission (MNPUC), *In the Matter of the Application of Enbridge Energy, Limited Partnership, and Enbridge Pipelines (Southern Lights) LLC for a Certificate of Need for the Alberta Clipper Pipeline Project and the Southern Lights Diluent Project,* Issue Date: December 29, 2008, Docket No. PL-9/CN-07–465; Order Granting Certificate of Need (St. Paul, MN: MNPUC, 2008); Enbridge Pipelines (Toledo) Inc., *Enbridge Application Line 79,* MPSC Case No. U-16937 (2012).

64. See Enbridge Energy, LLC, *Flanagan South Pipeline Project,* http://www.enbridge.com/FlanaganSouthPipeline.aspx (accessed August 26, 2013).

65. Lauren Krugel, "Enbridge First-Quarter Net Earnings Beat Street: Eyeing Eastern Oil Pipelines," *Canadian Business,* May 9, 2012, http://www.canadianbusiness.com/article/83449-enbridge-first-quarter-net-earnings-beat-street-eyeing-eastern-oil-pipelines (accessed May 11, 2012).

66. Natural Resources Defense Council (NRDC), *Going in Reverse: The Tar Sands Threat to Central Canada and New England.* (Washington, DC, June 2012), 4–5, http://www.nrdc.org/energy/files/Going-in-Reverse-report.pdf (accessed June 19, 2012).

67. Oil Change International, *Exporting Energy Security: Keystone XL Exposed* (Washington, DC, 2011).

68. NRDC, *Going in Reverse.*

69. Anthony Swift, Susan Casey-Lefkowitz, and Elizabeth Shope, *Tar Sands Pipelines Safety Risks* (Washington, DC: National Resources Defense Council, 2011), 7.

70. NRDC, *Going in Reverse,* 14. For a report highly critical of the operation of Enbridge pipelines, see National Transportation Safety Board, *Enbridge Incorporated Hazardous Liquid Pipeline Rupture and Release,* NTSB No. PAR-12–01, NTIS No. PB2012–916501, Adopted July 10, 2012, http://www.ntsb.gov/news/events/2012/marshall_mi/index.html (accessed October 30, 2012).

71. Enbridge Energy, Limited Partnership, Before the Minnesota Public Utilities Commission, Docket No. PL-9/CN-13–153, Application for a Certificate of Need for a Crude Oil Pipeline, table 7853.0270–3, p. 13. https://www.edockets.state.mn.us/EFiling/edockets/searchDocuments.do?method=showPoup&documentId={F1B13575–3D71–4CAA-A86A-05CE1EBBCA38}&documentTitle=20138–90363–03 (accessed August 26, 2013). See also Jeffrey Tomich, "Oil Sands Pipeline Avoids Keystone XL Scrutiny," *St. Louis Post Dispatch,* August 25, 2013, at http://www.stltoday.com/business/local/oil-sands-pipeline-avoids-keystone-xl-scrutiny/article_5a80dcef-44aa-5272–8451-fb4511a33522.html (accessed August 26, 2013).

72. Sanger, *Tar Sands Refineries,* 3, 5.

73. Energy Information Administration, *What Are Greenhouse Gases and How Much Are Emitted by the United States?* (Washington: U.S. Department of Energy, 2012), http://www.eia.gov/energy_in_brief/greenhouse_gas.cfm (accessed June 20, 2012).

74. James Hansen, "Game Over for Climate," *New York Times* (May 9, 2012), http://www.nytimes.com/2012/05/10/opinion/game-over-for-the-climate.html?_r=2&emc=eta1 (accessed May 12, 2012).

75. Natural Resources Defense Council (NRDC), *Setting the Record Straight: Lifecycle Emissions of Tar Sands* (Washington, DC, 2010), http://docs.nrdc.org/energy/files/ene_10110501a.pdf (accessed June 20, 2012).

76. Richard K. Lattanzio, *Canadian Oil Sands: Life-Cycle Assessments of Greenhouse Gas Emissions* (Washington, DC: Congressional Research Service, June 18, 2012).

77. Ibid. See also Adam R. Brandt, *Upstream Greenhouse Gas (GHG) Emission from Canadian Oil Sands as a Feedstock for European Refineries* (Stanford,

CA: Department of Energy Resources Engineering, Stanford University, January 18, 2011); James T. Bartis, Frank Camm, and David S. Ortiz, *Producing Liquid Fuels from Coal: Prospects and Policy Issues* (Berkeley, CA: RAND Corporation, 2008). For an extensive survey of the available literature comparing relative emission levels, see NRDC, *Setting the Record Straight*.

78. Hout, Fischer, and Lemphers, *Oil Sands and Climate Change*, 1, 4.

79. Hendrik Spruyt, *Juggling the New Triad: Energy, Environment, and Security; A Case Study of the Canadian Oil Sands* (Montreal: Université de Montreal and McGill University, Center for International Peace and Security Studies, 2010), 17.

80. Paul Muldoon, Alastair Lucas, Robert B. Gibson, and Peter Pickfield, *Environmental Law and Policy in Canada* (Toronto: Emond Montgomery, 2009), 94.

81. Michel Duquette, "From Nationalism to Continentalism: Twenty Years of Energy Policy in Canada," *Journal of Socio-Economics* 24:1 (1995): 229–52.

82. Thibault Martin and Steven M. Hoffman, *Power Struggles: Hydro Development and First Nations in Manitoba and Quebec* (Winnipeg: University of Manitoba Press, 2008).

83. Duquette, "From Nationalism to Continentalism," 234.

84. Ibid., 237.

85. Ibid., 239.

86. Michel Beine, Charles Bos, and Serge Coulombe, *Does the Canadian Economy Suffer from Dutch Disease?* (Amsterdam: Tinbergen Institute, 2009), 2.

87. Ibid., 1.

88. Afshin Honarvar, Jon Rozhon, Dinara Millington, Thorn Walden, Carlos A. Murillo, and Zoey Walden, *Economic Impacts of New Oil Sands Projects in Alberta (2010–2035)* (Calgary, Alberta: Canadian Energy Research Institute, 2011), xi, 7.

89. Daniel Tencer, "Alberta Oil Sands Royalties to Bring In $1.2 Trillion over 35 Years: CERI," *Huffington Post,* March 27, 2012, http://www.huffingtonpost.ca/2012/03/27/alberta-oil-sands-royalties-ceri_n_1382640.html (accessed January 17, 2017).

90. Innovative Research Group, *Alberta Government: July 2008 Provincial Image and Issue Survey,* www.alberta.ca/home/286.cfm (accessed February 2012); Harris Decima, *Government of Alberta: Quarterly Issues Study* (May 2010), www.alberta.ca/home/286.cfm (accessed February 2012); Forum Research Inc., News Release: "Majority Think Alberta Oils Sands Bad for the Environment" (Toronto, February 8, 2012); Darren Campbell and Jeff Lewis, "Mood Swing," poll conducted by Leger Marketing on behalf of Alberta Oil (2012).

91. Natural Resources Canada, "An Open Letter from the Honourable Joe Oliver, Minister of Natural Resources, on Canada's Commitment to Diversify Our Energy Markets and the Need to Further Streamline the Regulatory Process in Order to Advance Canada's National Economic Interest," The Media Room, January 9, 2012, http://www.nrcan.gc.ca/media-room/news-release/2012/1/1909 (accessed January 17, 2017).

92. "Environmental Charities 'Laundering' Foreign Funds," CBC News, last updated May 2, 2012, http://www.cbc.ca/news/politics/environmental-charities-laundering-foreign-funds-kent-says-1.1165691 (accessed January 17, 2017).

93. Eaton quoted in Max Paris, "Oilsands Critics Put Spotlight on Foreign Ownership," CBC News, last updated May 10, 2012, http://www.cbc.ca/news /politics/oilsands-critics-put-spotlight-on-foreign-ownership-1.1247414 (accessed January 18, 2017).

94. Andrew Frank, "A Whistleblower's Open Letter to the Citizens of Canada," Scribd (2012), http://www.scribd.com/doc/79228736/Whistleblower-s-Open-Letter-to-Canadians (accessed June 12, 2012).

95. Tim Harper, "For Conservatives, Contrary Positions Are Treasonous," *The Star.com* (November 17, 2011), https://www.thestar.com/news/canada/2011/ 11/17/tim_harper_for_conservatives_contrary_positions_are_treasonous.html (accessed January 18, 2017).

96. Dave Vasey, *A Risky Business: Tar Sands, Indigenous Rights and RBS* (Bemidji, MN: Indigenous Environmental Network, 2011), 6, http://www.ienearth .org/docs/NTSN_Brief-RBS_11.pdf (accessed June 30, 2012).

97. Indigenous Environmental Network (IEN), *Oil Development and Its Threat to Our Native Way of Life* (Bemidji, MN, n.d.), http://www.ienearth .org/docs/factsheet_human_health.pdf (accessed June 30, 2012).

98. Ecojustice, *EnviroLaw Watch,* http://www.ecojustice.ca/envirolaw-watch (accessed June 20, 2012); Bob Weber, "Shell Canada Tries New Rules to Block Greenpeace from Jackpine Oilsands Hearings," *Financial Post,* October 17, 2012, http://business.financialpost.com/news/energy/shell-canada-tries-new-rules-to-block-greenpeace-from-jackpine-oilsands-hearings (accessed January 18, 2017). In late 2012 a similar measure, C-45, was introduced in the Canadian parliament.

99. Nikke Skuce, *Who Benefits? An Investigation of Foreign Investment in Tar Sands* (Vancouver: ForestEthics, 2012), http://forestethics.org//sites /forestethics.huang.radicaldesigns.org/files/FEA_TarSands_funding_briefing .pdf (accessed June 12, 2012).

100. Natural Resources Canada, "An Open Letter."

101. Bernd, *BREAKING: Obama Rejects Keystone XL Pipeline.*

CHAPTER 13

Quebec Asbestos

Triumph and Collapse, 1879–1983

JESSICA VAN HORSSEN

For much of the twentieth century, the cities and industries of the world relied on fireproof materials made from asbestos. As the use of asbestos became increasingly pervasive, its harmful effects on human health became apparent to medical researchers. Industry leaders, however, hid the dangers surrounding the mineral for decades until the industry collapsed in the 1980s. In Quebec, the source of most of the asbestos mined in the twentieth century, the industry's cycles of boom and bust, and certain social and cultural features pushed asbestos-mining communities to accept extraordinary environmental risks.

This chapter examines the global asbestos industry from a local perspective, showing how the miners of the aptly named community of Asbestos, Quebec, negotiated changes in the environment and their own health through the work they did and the industry they fed. It focuses on how asbestos was mined and processed, how it impacted human health, and how community members interpreted the massive environmental and cultural changes that took place between 1879 and 1983. People and place collided in Asbestos, creating an intensely local understanding of environmental risk.

The town of Asbestos is located midway between Montreal and Quebec City, roughly a two-hour drive from each. Asbestos, a name the Royal Mail gave the mining camp in 1884, is the site of the Jeffrey mine, the largest chrysotile asbestos mine in the world, which was founded in 1879 and owned by the American company Johns-Manville (JM) from

1918 to 1983. JM was the largest asbestos producer in the world during the twentieth century and connected the Jeffrey mine and its workers to a vast global industrial network. The residents of Asbestos were unilingual francophone and Catholic. They were part of a growing trend in Quebec of young men and women leaving family farms and entering new industries that exploited the rich and diverse natural resources of the province.[1] The opencast Jeffrey mine is located in the center of the community and is the source of the local population's pride and sorrow, success and collapse.

The fact that asbestos causes cancer led to the industry's collapse. For most of the twentieth century, however, the mineral was considered indispensable to modern life because of its fireproof qualities, and it is important to remember this fact when considering the history of Asbestos and the global asbestos trade. The mineral was added to an exponential number of goods that would not burn, rust, or decay with age. Quebec chrysotile at one point made up 95 percent of the global trade in the fireproof mineral, and the Jeffrey mine produced the majority of this supply.[2] Asbestos promised safety for those who used it and profits for those who sold it.

Foreign—especially American—ownership of Canada's natural resources is not unusual. The fact that the majority of people in Asbestos were French Canadian was a benefit to JM when it came to shielding workers from information concerning the specific occupational health risks present throughout the community. Medical reports and pamphlets on asbestos-related disease were rarely written in French, and for much of their history, townspeople were isolated from the communication of risk due to the language they spoke and their relatively rural location. This isolation allowed JM to do things in Asbestos it could not do elsewhere.

Between 1879 and 1983, the people of Asbestos experienced both triumph and tragedy as the mineral moved from being synonymous with safety to something that invoked widespread fears of cancer. Throughout these experiences, the local population remained committed to the Jeffrey mine and organized their lives around the massive opencast pit located in the center of town. The town of Asbestos complicates the history of mining communities by demonstrating an agency within the local working-class population that reveals an awareness of disease and a sacrifice of health for long-term community survival.

ENVIRONMENTAL CHANGE

The Jeffrey mine is the largest of its kind thanks to a local geological quirk. Most asbestos deposits in the world form along a linear plane, which results in a number of open-pit mines along the vein to access an entire deposit. Historically, multiple pits have been dug to reach a single vein, many of them because of competing land ownership claims, and also to help prevent landslides that are prevalent in especially large opencast mines. Contrary to the norm, the deposit at Asbestos runs in a circular pattern and forms a rounded mineral-based knoll. The Jeffrey mine was the only pit needed to access the deposit in Asbestos. Unaware of the dimensions or shape of the deposit, early community members constructed the town right on top of it. From the beginning, the land in and around Asbestos was unique.

Gentleman farmer William H. Jeffrey founded the Jeffrey mine in 1879. The circular deposit at Asbestos meant that the majority of fibers at the surface were shorter than those found in linear deposits. These short fibers could, among other things, be added to lead paint to fire-proof walls, applied to roofing shingles to retard fires, and added to cement to make it more durable. Once clear of surface rock, 90 percent of the deposit contained both long and short asbestos fibers.[3] The pit contained very little waste.

Lacking the skills needed to realize this potential wealth, Jeffrey went bankrupt in 1892, and the town went bust. Johns-Manville purchased the mine from local owners in 1916. World War I demonstrated the importance of fireproof buildings and materials, and the market boomed. As the war increased demand for both long and short fibers, the potential of the land transformed Asbestos into a place of extraor-dinary value. JM was primarily an asbestos-manufacturing company that specialized in building supplies. From 1916, the growing war and construction industries in the United States more than made up for the wartime loss of European markets, and Jeffrey mine employees worked night and day to keep up with demand.[4]

JM changed both land and people in Asbestos. The mine and the community suddenly became connected to an immense industrial net-work. JM began an American revolution at the Jeffrey mine, assembling a far more technologically advanced, economically connected, and managerially cutthroat operation than anything the region had seen before. Each person in Asbestos was in some way dependent on the industry, and this gave JM power. The company introduced multiple shifts to operations, and townspeople lived according to the rhythms of work. Everyone came to a halt each day at noon and again at 5:30 P.M., when blasting in the mine marked the turnover of each shift.[5]

JM initially relied on horses to transport large loads of rock and fiber. The company quickly replaced them with a steam-powered railroad that ran from the bottom of the pit to the nearby Grand Trunk Railway station. After workers extracted raw asbestos from the mine, they sent it to the mill, also located in Asbestos. Workers then crushed the ore and sifted it through screens so the surrounding rock was removed. Inspectors next classified the fiber according to length, and it was either packaged and shipped elsewhere or sent to the Textile Department. In the Textile Department, local women would card, spin, and weave the mineral into fabric much like textile workers wove cotton or wool.[6]

Everything and everyone in the community was becoming more and more involved in mine operations as the company continued to expand production in the 1920s. This focus on a single industry was not an experience unique to Asbestos, but rather one shared across many mining towns. As Thomas G. Andrews describes in *Killing for Coal*, similar community involvement and dependency took shape in Colorado's coal-mining industry in the early twentieth century.[7] North America in the early twentieth century featured many so-called company towns, of which Asbestos was most decidedly one.

The Jeffrey mine was a dominant part of life and labor in Asbestos. With money and new technology, JM brought the town and the mine exponential growth. By 1928, industry insiders knew the region as "the most important asbestos producing territory in the world."[8] However, the Great Depression put this growth on hold. JM controlled almost half of the Canadian asbestos industry, and it significantly reduced operations at the Jeffrey mine because of the economic crisis, closing it entirely between May 1932 and April 1933.[9] For the first time since 1916, the Jeffrey mine went silent.

The closure of the mine significantly impacted local workers and community members. Without money to pay bills or feed their families, laid-off Jeffrey mine workers applied to the town council for financial aid. At the end of 1933, the council applied to the provincially managed Secours-Direct for $800 to help feed thirty local families and clothe and shelter forty-one others.[10] That seventy-one families were in need of aid even after JM resumed operations demonstrates the extent to which a year's worth of lost wages affected the people of Asbestos. Council requests for provincial money continued throughout the Depression, even after JM reopened the Jeffrey mine, with town officials asking for $950 on 9 May 1934 and for an additional $900 one week later.[11]

Despite the suffering experienced at the local level, JM became focused on developing new markets and products during the Depression. By the end of 1933, after the company had reopened the Jeffrey mine, these innovations had helped bring a 29 percent increase in production in the North American industry and a 71 percent increase in the price of asbestos on international markets.[12] The majority of the Jeffrey mine's fiber went to the several hundred factories JM ran in the United States, the world's leading exporter of finished asbestos products.[13]

The people of Asbestos were shaken by the Depression. However, while other Canadians continued to suffer high unemployment as businesses failed left and right, JM steadily increased production—and shifts—at the Jeffrey mine after it reopened in 1933. With the global economy slowly recovering in the late 1930s, Canada's asbestos production increased.[14] The company also introduced new technologies to speed production, and the community increasingly heard, saw, and breathed the sounds and dust of progress. Other asbestos operations in the region struggled to step up production, because there was no available space in which to expand operations. This was not the case in Asbestos, where industry trumped community: JM sought the town council's permission to enlarge the mine in 1938. Without hesitation, the Asbestos town council approved the company's request for land on which to expand the mine.[15] The pit had reached its physical limits and resembled a steep inverted cone. Due to the threat of landslides, workers could not dig deeper without first expanding laterally.[16] To maintain the land's structural stability and the town's financial stability, the Jeffrey mine had to grow.

Sacrificing community land for mine growth is not a story unique to Asbestos. In *Mass Destruction*, Timothy J. LeCain examines the history of the large-scale opencast copper-mining industry in the American West.[17] LeCain highlights mining's destruction of the natural environment. The history of Asbestos shows that this pattern was not confined to the West. JM's expansion of the Jeffrey mine did destroy the local environment, but the community was built on the premise that its land was not meant for parks or swimming pools or homes. It was meant for mining, and everything else was of secondary importance.

The land in Asbestos tells a story of large-scale progress and environmental change. The original hill on which Jeffrey had established the mine had almost disappeared by 1939. The pit was 510 feet deep and 300 feet wide with spiraling benches 35 feet high and 75 feet wide to accommodate the trains emerging with loads of fiber.[18] Locomotives pulled cars carrying supplies around the pit, and three 4-yard electric shovels worked in tandem with one 8-yard shovel to load the fiber into empty cars heading back up to the surface. The increased use of large machinery took Asbestos to a new level of industrialization. Writing for the *Canadian Mining Journal* in April 1939, R.C. Rowe described the Jeffrey mine as both a natural and a technological phenomenon,[19] unique in its geology but now conventional for North American mining enterprises of the time in its reliance on gargantuan machinery.

Canada's asbestos production increased as the world economy pulled out of depression. In 1916, Canada had produced 139,751 short tons in 1916, worth $5,211,157.[20] By 1937, that number rose to 337,443 metric tons of asbestos worth $14,505,541.[21] The Jeffrey mine had become the largest pit in the region, with shifts operating twenty-four hours a day as 1,000-watt floodlights shone on the mine at night. JM's increased production in and industrialization of the pit changed the connection between the people of Asbestos and the mine, but it remained strong. The sounds, smells, and dust emerging from the pit were a constant reminder that the small town depended on a global industry, which also depended on it.

The market for asbestos increased exponentially after the outbreak of World War II. A thirty-year boom began in the community as a publication from the U.S. Department of the Interior termed the mineral "indispensable to modern life."[22] Wartime production increased sharply by 1941, and the American industrial surge more than compensated for the loss of overseas sales. Jeffrey mine workers struggled to meet demand. Burgeoning production and the departure of men to the war meant a labor shortage for JM's operations, addressed by recruiting women. By 1943, 25 percent of the workers at the manufacturing plant were female.[23]

It was an exciting era for Asbestos. American manufacturers developed several new technologies and products containing the mineral during the war and quickly adapted them for postwar society. These innovations included brake pad linings, and whole communities of prefabricated houses that relied on asbestos for insulation, wall plaster, paint, shingles, and floor tiles, as well as cement. These new applications made asbestos pervasive in Western society and led many contemporaries to believe that the "Asbestos Age" was just beginning.[24]

The Jeffrey mine was at the center of the Asbestos Age. By 1942 it covered 115 acres of surface land, and workers extracted 6,000 tons of rock and mineral daily. JM operated three shifts round the clock every day, seven days a week. In 1947 the industry in Quebec exported 10,785,189 tons of fiber worth $438,356,805.[25] Despite the end of World War II, demand continued to grow thanks to the new civilian uses of asbestos, as well as continued military production for the Cold War.

As the mine's footprint grew, so did the town's. The community of Asbestos underwent a period of immense physical growth in the post-war era. The town council purchased land from the county, local property owners, and even JM in order to create space for new housing to be constructed. The combination of open-pit mining with a new block-caving system that the company introduced led to extraction levels at the Jeffrey mine far surpassing those of its competitors. Block-caving is a form of underground mining particularly suited to land that already has an opencast pit. The benches that spiraled up the Jeffrey mine were hollowed out and mined from underneath. Block-caving was an efficient way to extract fiber, especially because the mine had reached the limits of JM's property by 1948. This new method allowed the company to continue operations without purchasing more land from the town. In addition to block-caving, JM constructed new roads and new housing for the growing community in the postwar era.

JM continued to introduce new machinery into its operations in Asbestos. In 1951 the company stopped using trains to carry fiber from the bottom of the pit and adapted the Jeffrey mine to accommodate giant 35-ton trucks.[26] These trucks grew in size over the years, eventually reaching a capacity of 200 tons in the 1970s. JM also changed how shifts were run at the mine, with no pause in operations as one group of workers replaced another.[27] The Jeffrey mine had become an efficient open-air factory: its workers were the tiny gears that kept it running, and because of them, the land was constantly changing as well.

By 1952 Canada produced 70 percent of the world's asbestos supply.[28] The global market price for the mineral continued to rise, further boosting the value of the Jeffrey mine.[29] Part of this exponential growth and production can be attributed to global market changes during the Cold War, as Western nations and corporations pulled away from business transactions with the USSR, another major producer of raw asbestos. Canada had an advantage in the Cold War, as industry leaders in the United States perceived it as the "friendlier" asbestos-extracting nation. By 1967, the *Canadian Mining Journal* had termed the Jeffrey mine

"The Free World's Largest Asbestos Producer." The magazine attributed this reputation to JM, which had transformed the community into an industrial complex that by 1967 produced over 600,000 tons of asbestos annually.[30] This reputation had long-term implications that outlasted the Cold War. Industry heads and government officials used it to cast Canadian asbestos in a positive light even while the general public increasingly became aware of the mineral's negative health effects.[31]

Locally, JM dominated the land and the community. Globally, JM reaped the commercial benefits of being the company in control of the Jeffrey mine.[32] By 1954 the giant trucks that hauled the fiber out of the Jeffrey mine via 15-foot-high spiraling benches were large diesel trucks that made twenty-two trips to the surface during each of the three daily shifts, five days a week.[33] JM developed a new form of blasting that used dynamite without wires so the mineral would remain free from foreign materials. These and other technologies replaced employees who gathered asbestos by hand at the bottom of the pit and picked out blasting debris. The relationship the workers had with the land, rock, and mineral was changing.

JM's industrialization of community land dramatically increased Quebec's asbestos production, and for a while it seemed no end was in sight. In 1955 the *Canadian Mining Journal* declared that JM had ensured "asbestos mining as a principal industry in Quebec for at least another century."[34] Wealth and stability came from the land at Asbestos. It seemed unfathomable that anything could change this state of affairs. The unfathomable came in the 1970s and 1980s. As the Western world became aware of the dangers asbestos posed to human health, the industry was threatened with collapse. This new development promised to be devastating for the town of Asbestos. Without employment at the Jeffrey mine, the community was doomed.[35]

JM management knew what was coming, and the company began to extract as much asbestos from the Jeffrey mine as it could before the industry financially collapsed. This caused the Jeffrey mine to literally collapse. Landslides occurred throughout 1975 and destroyed large portions of the pit's southeast spiral benches. Asbestos had become an increasingly dangerous place to live as rocks were blasted into neighborhoods at all hours as the company spent $77 million not on the community, but on a new factory and new equipment that would increase production levels by enabling more blasting each day.[36] Nothing would get in JM's way as the company frantically extracted as much fiber as it could while the industry was still viable.

ENVIRONMENTAL CONTAMINATION

For much of the twentieth century, asbestos was synonymous with safety. With increasing urbanization and industrialization following World War I, corporations, marketing firms, and government agencies taught consumers that asbestos was the remarkable mineral needed to keep them safe from fire. By and large, this was true: asbestos did help contain the spread of fire and, therefore, helped reduce the number fire-related casualties. From oven mitts to fire-fighting equipment to bed sheets, asbestos-based goods quickly became a part of everyday Western life. Asbestos's ability to prevent the spread of fire, however, did not negate the harmful effects it had on human health.

There are three main diseases associated with asbestos: asbestosis, lung cancer, and mesothelioma. Asbestosis develops when microscopic asbestos fibers are inhaled over an extended period of time and build up in the lining of the lungs, ultimately leading to death by suffocation. Inhaling asbestos dust and microfibers can also cause lung cancer. Mesothelioma is another asbestos-related cancer that manifests in the linings of major organs and is highly aggressive. These three diseases can occur individually or together. Asbestos also causes skin, breast, ovarian, colon, and intestinal cancer. These diseases can take up to thirty years to develop and depend on an uncertain dose-to-longevity exposure ratio.

The first asbestos-related death to be widely reported in medical journals and court records was that of Nellie Kershaw in 1924. Kershaw was a weaver of asbestos fabric in Rochdale, England, and died of asbestosis after seven years of working in the industry.[37] JM may not have been aware of the dangers the mineral posed to human health, or of Kershaw's death, in 1924 when it built a new fiber-processing factory in Asbestos. This factory offered employment opportunities to local women in the Textile Department and was similar to the plant Kershaw had worked in.

Along with the new factory, JM brought to Asbestos its own medical professionals. These doctors reported the health conditions of workers to company officials, not to the patients. Company-run medical care was common in mining communities,[38] and it contributed to paternalistic corporate control over all aspects of local life. The health of workers at the Jeffrey mine was of great interest to JM because of the large number of employees exposed to the raw mineral. Company medical professionals monitored the health of laborers in Asbestos as though they were mice in an experimental laboratory. As long as Jeffrey mine employees were shown to be healthy, the industry and the town would be safe.

JM subtly addressed the risks posed by asbestos, in part, by introducing a policy in 1930 mandating the transfer of every male Jeffrey mine worker to a new department every ten years to limit his exposure to the dangerous mineral dust.[39] Workers were not told why they were transferred, and female employees remained in the Textile Department. For decades, this was as close as the company came to acknowledging workplace risk.

In addition, in 1930, JM invited Dr. Frank G. Pedley to examine Jeffrey mine workers. At that time, Pedley was one of the only medical professionals researching asbestos-related disease in Canada. The resulting study—thoroughly edited by JM before it was published—claimed that, "If work with asbestos presented a hazard to the worker it would be reasonable to suppose that cases of disease would be reported from time to time, but so far as can be determined no cases of specific disease have been reported among asbestos workers in the Province of Quebec."[40] In his unpublished work, submitted only to industry leaders, Pedley highlighted this fact by describing four cases of asbestosis among Jeffrey mine workers, and among almost half of the workers at a neighboring mining community, Thetford Mines.[41] Disease was present among the region's asbestos workers, but a lack of education and information concerning the risks the mineral posed to them, combined with company-funded doctors withholding the results of medical examinations from patients, left Jeffrey mine employees ill-equipped for self-diagnosis or to push for better dust control measures in the early 1930s. Because those afflicted were not coughing and did not complain of ill health, Pedley concluded that the disease was not severe.[42]

Although Pedley was not alarmed by asbestosis, JM was. Not only did the company remove from his published report any mention of the cases he had discovered,[43] but it also launched a media campaign to promote confidence in the safety of the mineral throughout the 1930s. Furthermore, in a 1931 letter from Metropolitan Life to JM attorneys, a Dr. McConnell wrote that Pedley's unpublished report "will be given no publicity by us except with the consent of the firms concerned."[44] This corporate suppression of medical evidence allowed JM to market asbestos as safe and to combat the rising number of occupational health lawsuits the company was facing from its manufacturing employees in the United States.

The first civil case JM employees filed against the company for occupational health compensation was in 1929 by asbestos textile workers in New Jersey.[45] Initially passed over to industry insurers, legal action continued to grow in the United States in terms of the number of lawsuits and the amount in damages they sought. In a letter from one JM vice president to another in 1931, E.M. Voorhees wrote S.A. Williams that "ever since dust suits have been brought against us at Manville [New Jersey] we have considered, first, the possibility of installing the most modern and improved dust collecting systems."[46] Manville was not the only American asbestos textile factory conscious of the potential threat of employees filing lawsuits against the company. In 1932, the manager at JM's Waukegan, Illinois, factory wrote to Williams concerning his plant's initiatives to reduce worker exposure to asbestos dust "in case suits develop."[47]

Although Jeffrey mine workers were not complaining of ill health, they were becoming increasingly vulnerable to asbestos-related disease. The enormous escalations in production during the war years ratcheted up the risks to residents. In 1943, JM researchers discovered that asbestos caused cancer in lab mice 81.8 percent of the time.[48] This discovery had severe implications for the health of workers and the general public, but again, the company remained silent. Although JM did not inform Jeffrey mine workers of the increased risk the industry posed, the people of Asbestos were aware something was wrong.

The female workforce at the Jeffrey mine increased significantly as wartime production rose. The Textile Department was the dustiest— and therefore the deadliest—place to work in Asbestos, and the women who worked there had alarmingly high absentee rates, widely discussed in the community. JM official Joan Ross went to Asbestos in 1944 to investigate the problem and acknowledged in her report that the "situation has become a topic of conversation throughout the entire community and is a serious detriment to the reputation of the company."[49] Despite local complaints, Ross concluded that although the department was excessively dusty, the fact that female employees missed work frequently was based on the "higher absentee rate among women in general,"[50] not asbestos-related disease. The hypermasculinity of the mining industry overshadowed female employment and risk. The fact that the issue of female absenteeism was a concern to townspeople is significant, but they did not raise the issue again following Ross's visit. Few, if any, medical researchers examined the effects of asbestos on female industry workers until the 1960s.[51]

In the meantime, JM officials remained concerned over the effects of asbestos dust on male Jeffrey mine workers. Company doctors in

Asbestos rarely performed official autopsies on deceased workers. This was partly due to the staunch Roman Catholicism of the local population, and partly due to the apparent lack of need for such postmortem investigation: miners died, and families believed company doctors when they said the death was related to lifestyle choices such as smoking. Despite the lack of official autopsies, JM instigated a policy of secret dissections during the war and postwar period in Asbestos. Company doctors removed the lungs from deceased Jeffrey mine employees without family members' consent. JM lawyer Yvan Sabourin smuggled these lungs across the U.S. border to a company-funded laboratory in Saranac Lake, New York.[52] This lab had much more sophisticated equipment and specialist researchers who could trace the cause and progression of disease more efficiently than those in Asbestos.

Company officials deemed this transborder lung-smuggling necessary in part because, according to Gerrit W. H. Schepers, who interned at Saranac Laboratory during this study and later became director of the lab, "such a large number of cases in such a small and well-defined group of industrial employees suggested a significant problem."[53] JM had decided to gather as much information on this "significant problem" as it could without informing its workers—or the general public—of the risk. By 1958, Saranac Laboratory researchers had discovered seventy-eight cases of unreported asbestos-caused lung cancer in the bodies of Jeffrey mine workers.[54] At the same time, the company used independent medical professionals as pawns to prove the mineral was safe. In 1948, JM invited independent doctors from New York to visit Asbestos and assess a series of employee X-rays. Company doctor Kenneth Smith provided the investigators with slides taken from employees he had already deemed healthy. "We never have let anyone know that this company (JM) had anything to do with the scheme," Smith wrote; "we are merely co-operating with [. . .] the Board to the best of our ability. [. . .] Even the head of the union here thinks that."[55] Smith's deception was never exposed.

By 1949, however, it was clear that the head of the union in Asbestos was not convinced by Smith's report and neither were Jeffrey mine employees. In January 1949, the major newspapers of the province printed an exposé on Quebec's asbestos industry written by independent journalist Burton LeDoux. The exposé compared mining communities like Asbestos to concentration camps and explained that diseases like asbestosis acted like a spider spinning a web tightly around a worker's lungs until he died.[56] Unions distributed LeDoux's text in pamphlet form. This was the first time Jeffrey mine workers saw such information written in French, the only language most of them could read. A few weeks after LeDoux's report was released, the workers in Asbestos went on strike for five months.

For many historians of Quebec and Canada, the Asbestos Strike of 1949 sparked the province's Quiet Revolution. This was a sociopolitical movement mainly during the 1950s and 1960s in which the French Canadian majority became increasingly secular; gained control of the province's major industries and businesses, which had traditionally been run by a minority Anglophone upper class; and rallied their political strength and ambitions to effect major change within Quebec and the rest of Canada through waves of neonationalism and reform liberalism.[57] JM president Lewis H. Brown even called the demands workers made during the strike a "revolutionary doctrine," designed to seize control of managerial policy.[58] Revolutionary or not, Jeffrey mine employees were quickly joined in the strike by the workers in every asbestos-mining community in the province except one, and the slowdown crippled the North American industry. Workers raised a number of issues during the strike, including job security, but among the most "revolutionary" elements of the conflict was environmental health.[59]

JM adamantly refused to acknowledge the concern that workers had over the health effects of asbestos. Instead, the company reframed the conflict as an attempt by the union to stage an anticapitalist revolution. After five months of negotiation, the union withdrew the proposed clause to reduce employee exposure to deadly asbestos dust on the condition that the company would rehire its employees once a settlement was reached. This was the last time Jeffrey mine workers went on strike over issues of asbestos and health.

Despite the growing local and international concern over the mineral's effects on human health the strike had ignited, the Quebec government took asbestos-related disease off its list of compensable industrial diseases in the early 1950s. Industry lobbying convinced officials that the mineral was safe.[60] In fact, the idea that asbestos was not only safe but helped ensure the safety of the entire population persisted as a common mantra into the 1960s in the Western world.

JM was instrumental in the spread of this mantra. To combat the growing body of medical evidence proving asbestos was harmful to human health, the company cofunded a new study on Quebec's asbestos miners that mirrored Pedley's 1930 report. The thoroughly edited study by Daniel C. Braun and David Truan was published in 1958 and concluded that while workers were indeed developing lung cancer, cigarettes—not the mineral—caused the disease.[61] The emphasis on smoking, rather than on asbestos dust, highlights the ways in which medical professionals viewed miners: because they lived unhealthy lifestyles, they were expected to get diseases. This understanding has its roots in the Victorian era, when medical professionals and company officials began to regulate working-class culture through health reform.[62]

This insight is crucial to understanding how the people of Asbestos viewed their health. They knew they had respiratory problems because of their work at the Jeffrey mine, but they did not know about the risk of cancer. International medical studies increasingly showed that exposure to asbestos led to cancer of the pleura, stomach, colon, and rectum, as well as mesothelioma, suggesting that asbestos-related disease went beyond the respiratory system.[63] These reports remained relatively absent in Quebec, with unions and provincial officials focused on the sociopolitical changes sweeping the province during its Quiet Revolution.

As the province experienced dramatic changes, Jeffrey mine workers did as well. No longer did they agitate on issues of asbestos and health. Instead, they increasingly refused to acknowledge the dangers the mineral posed on the job and in the community. One of the most significant ways they did this was by declining to wear respirators at the Jeffrey mine. Knowing the dangers asbestos dust posed to the health of its workers and therefore its profits, JM instigated a series of policies designed to promote the use of protective devices.[64]

Despite these efforts, workers continued to refuse to wear the respirators until JM made them mandatory in 1975. The company now disciplined any Jeffrey mine employee who refused to wear the mask.[65] The reluctance of workers to wear respirators can be brought into a larger story of masculine bravado often seen in mining communities, as well as a working-class rejection of corporate attempts to regulate behavior, but it also highlights another issue. The respirators used in the asbestos industry were useless in environments like the Jeffrey mine's mill. The concentration of dust was so high that the filters clogged immediately, making it even harder to breathe. It was easier to cast them aside. A company study later proved that "respirators were not as efficient as we thought they were," with one official noting that JM had "had a dirty house and now we have to pay for it."[66]

By 1968, community members began to complain about the clouds of dust emerging from the mine and mill. Children wrote their names in the small particles as asbestos dust coated cars.[67] Workers took it home with them on their clothes. The local paper reported that the people of Asbestos "dine[d] on dust and noise."[68] The clouds of mineral dust hovering over the community meant that asbestos-related disease went beyond the borders of the Jeffrey mine. It was not until 1971 that JM launched an internal study of asbestos-related cancers among community members not directly involved in the industry. The study found that people living in Asbestos were at a heightened risk of developing diseases because of their proximity to the mine.[69]

JM knew its operations in Asbestos could not continue indefinitely. The company needed to extract as much raw asbestos as it could before widespread knowledge of the risks the mineral posed—both to workers and the general public—led to the industry's collapse.[70] If community members were adversely affected in the process, that was not the company's concern.

Once again viewing Jeffrey mine workers as test mice, in 1972 JM cofunded a study of the people of Asbestos by McGill University's Dr. J. C. McDonald, the result of which would "preserve the industry on which their business depends [. . . and] avoid any undesirable publicity or any precipitate action by the USA or Canadian Federal Government which might be detrimental to the industry."[71] McDonald did exactly what JM asked. He found that there was lung damage in Jeffrey mine workers, but concluded that this was not caused by the mineral. While he acknowledged that high levels of asbestos dust led directly to mesothelioma, he believed that cigarettes caused more damage than asbestos.[72] McDonald also concluded that while female employees worked in extremely dusty areas, few of them exceeded ten years of employment, resulting in negligible cases of disease.

Two years later, French Canadian medical researchers released a report that contrasted with McDonald's sharply. This study indicated that while mesothelioma occurrences were 1 in 10,000 for the general population, they were 1 in 10 for those working in Quebec's asbestos industry.[73] These numbers were alarming. In May 1975, JM's vice president for health, safety, and environment, Paul Kotin, sent filmmaker Walter Cooper to Asbestos to make a pro-industry documentary called "Asbestos and Health." The mill had been closed and cleaned by employees for two days, but Cooper still found it too dusty to film in. Cooper also reported to Kotin that, "the bagging operation on the main floor was shocking. There were accumulations of dust everywhere

[. . . and] I noticed an ankle-high accumulation of fiber, which was being shovelled into an open cart for disposal by a worker who was not wearing a respirator."[74]

Kotin was shocked by Cooper's experience at the Jeffrey mine. He wrote to JM officials: "If the division cannot complete the environmental clean-up of this textile operation, then serious consideration should be given to shutting down the operation. The Jeffrey Textile Plant is an embarrassment."[75] JM was rapidly losing interest in preserving its operations in Asbestos as the company became increasingly smothered by occupational health lawsuits in the United States. Before filing for bankruptcy protection in 1982, JM profiled Norman Chartier, a Jeffrey mine employee, in its shareholder magazine. Chartier had worked at the mine for four decades and explained that no job was 100 percent safe, but "if a man uses common sense on the job and follows the rules set down for his protection, he's more apt to get into trouble when he's not working."[76] Chartier's life and the risks he faced had been expected and accepted by townspeople in Asbestos for generations. While JM was prepared to file for bankruptcy and move on, community members were not.

COMMUNITY AND MINE

The town had collapsed when Jeffrey went bankrupt in 1892, but the community was well on its way to dominating the Quebec asbestos industry by the time JM took ownership of the mine in 1916.[77] The people of Asbestos saw the American company's arrival as a boon for the community, bringing the Jeffrey mine and the town that had grown around it immeasurable success. In many ways, they were correct.

JM brought Asbestos sixty years of almost uninterrupted growth and stability. The company's international asbestos manufacturing network ensured that as long as there was a market for the mineral, there would be jobs for community members at the Jeffrey mine. Although both JM and the people of Asbestos were committed to the industry's success, the cultural divide within the community occasionally caused friction. The large, predominantly French, Roman Catholic working class lived on the edge of the growing Jeffrey mine. Boys went to the local school run by Catholic priests until they were teenagers and old enough to work at the Jeffrey mine. Girls typically did the same, working in the Textile Department after they left school until they married. In a feature on Asbestos in JM's monthly employee magazine in 1950, the company claimed that the average family in the community had at least ten children, capable of consuming "half a peck of potatoes at one meal and ten loaves of bread a day."[78] In fact, the average birth rate in the province of Quebec from 1926 to 1961 remained between 3.77 and 4.39 children per household.[79] In Asbestos, these children were raised knowing their future success would depend on the Jeffrey mine and, to some extent, on the company that owned it.

The powerful managerial elite of the community, most of them American-born, lived on a hill much farther away from the mine. For the most part, they spoke only English, were Protestant, and were university educated. This divide led to some conflict in Asbestos, especially over linguistic rights and promotions. When JM arrived in Asbestos, Jeffrey mine dynamite workers welcomed them by going on strike in 1918.[80] Although the asbestos workers in the neighboring community of Thetford Mines had unionized in 1915, those at the Jeffrey mine had yet to do so. The absence of a union in Asbestos did not mean workers were docile, however, and the short 1918 strike echoed a similar labor dispute for higher wages they had instigated in 1912.[81] In 1919, Jeffrey mine workers joined l'Union ouvrière Catholique du Québec, but JM refused to recognize the union. Despite this refusal, workers remained committed to the union, which was led by Catholic Church officials, with the local priest in Asbestos heading the branch at the mine. The workers did not initially push JM to recognize the union. Instead, they focused on their work, and an excited spirit of industrial cooperation flourished in Asbestos. Record profits in the interwar period ensured the continuation of this cooperative spirit.

As asbestos production climbed and the town's population rose, as noted earlier, the pit and the community both needed more room, which brought on another sort of conflict. In March 1927, the town council wrote to the Quebec government requesting permission to extend the boundaries of Asbestos into the surrounding countryside to make room for the future prosperity of the community, which had reached a population of 3,602. This was the first major step toward the community's domination of the surrounding land, and JM immediately requested a portion of this new land.[82]

The community had been constructed on top of the mineral deposit. Because of this, JM justified its request for 55 acres of new land by explaining that the people of Asbestos were in the way of the Jeffrey mine's necessary expansion. The global price of the mineral was increasing, and with Canada providing 85 percent of the fiber worldwide, this

was the perfect opportunity for the company to increase its landhold-
ings. To combat any negative feelings toward expansion, which would
require citizens to move farther away from the mine as it grew into
existing neighborhoods, JM constructed new homes and roads for its
employees equipped with such modern conveniences as running water,
electricity, and streetlights.[83] The people of Asbestos became indebted
to the company as JM reshaped the land around them.

Pit expansion changed not only the land, but also the way townspeople related to the Jeffrey mine. As company and community negotiated
a balance between livable space and workable space, the Jeffrey mine
divided and defined the two. The increasing cultural and economic
importance of land defined the community. Under JM rule, the town
did not have a durable history preserved in buildings or roads, but
rather a future ensured by unrestricted environmental changes.

As JM changed the land, the Great Depression changed Jeffrey mine
employees and thereby the company's community relations. The Catholic Church also evolved in the interwar period as left-wing ideologies
spread throughout the institution. This was evident in Quebec's labor
movement. In 1936, Sherbrooke's abbé Aubert became the leader of the
province's Catholic union, and his social activist principles infiltrated
the local union in Asbestos. The industry's instability during the 1930s,
and the fact that JM supplied medical care to the community and owned
many of the homes Jeffrey mine workers lived in, meant that employees
had a lot to lose in pressuring the company to recognize the union.
Despite this fact, the crisis of the Depression and the urging of Catholic
union leaders that workers demand JM recognize the collective convinced Jeffrey mine employees to go on strike in 1937.

The strike lasted eight days. In total, 1,100 male and 50 female Jeffrey mine employees participated in the dispute, demanding a wage
increase of 33 percent and recognition of their union.[84] The *Toronto
Clarion* described the conflict as "one of the most important strikes in
the province" because of the financial value of the asbestos industry.[85]
Greater still, however, was the importance of the dispute to the community of Asbestos, which provided the labor and the land that JM
relied on to pull itself—and perhaps the industry—out of the Depression. Over the strike's eight days, Jeffrey mine workers received the
wage increase they demanded and recognition of their union, now
called the Confédération des travailleurs catholiques du Canada
(CTCC). In addition to these labor advances, a committee of workers
and company officials banished the head of JM's employment office in

Asbestos and the vice president of JM's Canadian operations.[86] The head of the employment office did not speak French, which made industrial relations difficult. And, despite his honorary seat and voting privileges in the Asbestos town council, the vice president was too often in the United States on business and thus had failed to fulfill his duties at the Jeffrey mine.

JM employees in Asbestos went on strike four more times between 1937 and 1949. Workers demanded greater control over how JM operated the Jeffrey mine and the community land surrounding it. Employees urged the company to slow down production and to decrease its reliance on new technologies that made their labor redundant. Furthermore, employees routinely submitted suggestions for reform at the Jeffrey mine, some ninety-two of them between January and April 1948 alone.[87] The people of Asbestos enjoyed the economic gains the industry brought them, but were growing increasingly frustrated with the company's dominance over every aspect of community land and life.

Each time JM expanded the boundaries of the pit to prevent landslides and ensure the quantity and quality of the extracted mineral, a portion of the community had to move away from the heart of Asbestos. Following the mine's 1928 expansion, JM further enlarged the pit in 1933 and in 1938.[88] Then, in the late 1940s, at the height of postwar production, the company again announced it would extend the limits of the Jeffrey mine. It also introduced new mechanized shovels, which replaced 40 percent of the workforce.[89] In the post-Depression, postwar Asbestos of 1948, these were unacceptable changes to the community. Quebec Minister of Labour Antonio Barrette sensed "a problem brewing" in Asbestos.[90] After the CTCC distributed Burton LeDoux's exposé on asbestosis to its union members, and after contract negotiations with JM broke down in February 1949, the workers at the Jeffrey mine voted to strike. Along with wage increases, prime among the workers' demands were mandatory union dues of 3 percent of the wages of all employees (even nonunionized ones), job security, and the aforementioned dust clause, which would mandate the elimination of the dangerous asbestos dust.[91] After five months without wages, and with JM compromising very little, Jeffrey mine workers voted to end the strike and agreed to a settlement lacking provisions for job security and better health and safety measures.

The pivotal 1949 strike convinced workers and the community that having a job was more important than occupational health. JM did not immediately reemploy Jeffrey mine workers who had joined the strike.

In the community, this led to panic; workers begged the company and the provincial government to be taken back. In an August 1949 letter to Quebec minister of labor Antonio Barrette, JM employee Bertrand McNeil voiced the uncertainty and desperation some felt in the aftermath of the strike: "I need to work. I've had no bad relations with the company. I would like to know if they are going to take all of us back or if we're waiting for nothing. My father has a large family and I'm the only one who can help them."[92] The strike had taught workers and their families that nothing mattered more than employment, not even health.

In Asbestos, bitterness and animosity reigned during and after the 1949 strike. The local population began to use the Chez Nous Ideal, a local cooperative home-building group, to reduce its reliance on JM. Townspeople bought shares, pledging material and five hours of labor toward a new house for every member.[93] Townspeople had become increasingly dependent on rented company-built housing and were forced to leave whenever JM took more land to expand operations. The flaws of this employer-employee, landlord-tenant relationship became clear during the 1949 strike, when the company threatened to evict strikers to house strikebreakers.[94] Although residents had heretofore enjoyed the perks of living in a company town, JM's response to the 1949 strike taught the people of Asbestos their vulnerability.

As the people of Asbestos came to more fully understand that vulnerability, JM became even more aware of the industry's power. Although the strike had frozen the global asbestos trade due to lack of raw mineral coming from the asbestos towns in the province that joined Jeffrey mine workers during the conflict, the supply shortage had actually benefited the industry. Confronted with a shortage of some quarter million tons because of the months of inactivity,[95] JM attempted to purchase as much land in Asbestos as it could. Despite the company's power at the boom's height, the town council hesitated to satisfy its need for land. Before the strike, things had been far different; the council had acquiesced to almost everything JM requested. Policy regarding the Chez Nous Ideal also changed after the strike: when the collective asked for land at a discounted price for the construction of twenty homes and new roads and sewers, to which the council agreed.[96] The strike had changed the community's land politics.

As community members and company officials negotiated these changes, JM continued to assert its authority over the town's land. When the Chez Nous Ideal attempted to build one hundred homes on new land, JM warned that the parcel was an unstable mix of sand and gravel, with several large, deep holes that the company had created while testing its value.[97] JM proved itself the expert on land in Asbestos, and the collective suspended the project. The incident showed how industrialization had scarred the town and how residents were running out of space. The Jeffrey mine was literally devouring the community.

In 1955, almost 10,000 people lived in Asbestos. Anticipating further population growth because of the industry's prosperity, the town council purchased more land for expansion, paying for it in part from money collected from JM construction permits.[98] The Chez Nous Ideal was now the largest provider of housing in Asbestos, building 124 homes in 1956 alone. Homes built by the Chez Nous Ideal were signs that local land was meant for the community, not the industry. However, in 1958, the company again presented plans to expand the mine.

The expansion would be gradual but massive. Echoing its past reasoning, the council agreed to the extension because the development of the mine was necessary for the community's continued prosperity, even though JM had no intention of hiring more workers in the future.[99] This was a dramatic change in company-community relations. It showed JM's conviction that the land in Asbestos was to be used for mining purposes, not community development.

The global demand for asbestos doubled between 1955 and 1965; by the end of this period, the industry was worth over $148 million in annual income to the province.[100] With this boom came greater danger. Giant rocks were blasted out of the pit and into local homes in 1965, but because the company had asked people to leave at-risk areas, JM believed it was not at fault.[101] The constant noise, dust, and mine expansion changed how locals saw JM, the Jeffrey mine, and themselves, and prompted the editor of the local paper to write that the community was becoming the "hellhole of Quebec."[102]

By 1967, JM's expansion of the mine had consumed 54 percent of town land and destroyed 250 buildings. Despite increased industrialization, and despite the danger of huge rocks blasting through the community, the people of Asbestos maintained their connection to the mine. During the 1967 St-Jean Baptiste Day parade, workers waved from a float while holding up a sign loaded with poignant, troubling double meanings: "Asbestos: Our Heritage."[103] While everyone in Asbestos saw the land in terms of financial security and gain, Jeffrey mine employees also looked to it for identity.

JM did not acknowledge this local identity. In the 1970s, company-community tension surrounding workable versus livable land persisted.

Housing remained a major issue, and residents forced to move because of JM's endless expansion of the Jeffrey mine had nowhere to go. Many families also lived in dangerous proximity to the mine. They lobbied for a 1,000-foot buffer zone between the town and the pit, but this was impossible to achieve without additional residential relocation. JM was turning its back on the people of Asbestos, as new technologies reduced the company's reliance on its human workforce. Officials were also beginning to understand that because of the mineral's negative health effects, the future of the community—and the industry—was limited.

As the 1970s wore on, community objections to JM's practices grew more vocal. The town council described rocks flying from the Jeffrey mine in 1977 as acts of vandalism. Local officials also complained of rising clouds of dust so thick that life in Asbestos had become unacceptable and intolerable.[104] Thick clouds of toxic dust, flying rocks, and the constant noise of new machines had transformed the community into an industrial horror. This was not how land and people were supposed to interact.

Despite the realities and risks involved and the town council's complaints, townspeople were committed to keeping the industry thriving. The local population was convinced that the community's survival depended on the industry's survival. As the negative health effects of asbestos became widely known throughout the Western world, the people of Asbestos remained silent on the issue. JM's American employees, thanks in part to differences in U.S. and Canadian law, reacted completely differently to this knowledge. By the early 1980s, JM was increasingly plagued by multimillion-dollar asbestos-related class-action workers' compensation lawsuits in the United States. The industry was collapsing, and there was little the people of Asbestos could do to stop it. In August 1982, the company filed for bankruptcy protection in the United States, and the local newspaper in Asbestos claimed that without JM, the community would collapse.[105]

Local investors purchased the Jeffrey mine from JM in 1983. The company left Asbestos after having been an important presence there for more than half a century. The pit had grown to 6,500 feet east to west, 6,000 feet north to south, and 1,000 feet deep. Its immense size made life in the community difficult, but not as difficult as a collapsed industry would. Despite JM's bankruptcy and abandonment, the local population remained committed to the industry. Some citizens left town to seek more stable, healthy employment, but many remained, hoping for a revival of the asbestos industry. Municipal officials declared that 1984 would be the "year of asbestos."[106]

CONCLUSION

The Quebec government has subsidized the province's asbestos industry since 1978. After JM sold the Jeffrey mine in 1983, both provincial and federal government officials became involved in the survival of Quebec's asbestos towns. This began a trend in which the government supported the industry by doing what JM could no longer do: denying medical evidence, overlooking the welfare of community members, refusing to adequately label shipments of asbestos to other countries, and agreeing to sell the mineral to developing nations that would not uphold strict health regulations.

Government support sustained the acceptance of risk in Asbestos until 2012, when the mine was finally closed. Generations worked at the Jeffrey mine surrounded by dust they knew to be dangerous. JM employees consistently refused to wear respirators and chose not to push the company for better workplace safety. This was a toxic environment, but neither the community nor the company could afford to admit it. JM manipulated medical evidence and the workers in Asbestos, but the community also played an active role in developing a local understanding of risk, environmental change, and collapse.

The collapse of Asbestos was unlike that of other mining communities, such as Cobalt, Ontario, or St. Clair, Pennsylvania, because the local mineral deposits have not been exhausted. It was different than the bust of uranium towns in the American West, which survive by marketing their communities to tourists in search of 1950s atomic nostalgia.[107] In fact, throughout their community's history of triumph and collapse, Jeffrey mine workers were unlike other asbestos industry employees, including those who worked elsewhere for JM. They were not the industry's only miners, and they were not its only French Canadian workers. But the community's conduct complicates and challenges the international literature on the asbestos industry and mining communities.

Everything in Asbestos has occurred in the extreme: land exploitation, profits, disease rates, global renown, and industrial collapse. The international rejection of asbestos led to the collapse of the community and the loss of JM, a dominant presence from 1916 to 1983. The town's different factions did not always live in harmony and often went through periods of great animosity, but through constant negotiation, and recognition that they shared common goals, a fierce identity was created that was rooted in the Jeffrey mine and its seemingly limitless circular veins of asbestos.

NOTES

1. For more context on this population shift, see Jean-Pierre Kesteman, Peter Southam, and Diane Saint-Pierre, *Histoire des Cantons de l'est: Les Régions du Québec* (Sainte-Foy, Québec: Les Presses de l'université Laval, 1998); and Paul-André Linteau, René Durocher, Jean-Claude Robert, and François Ricard, *Quebec: A History, 1867–1929*, trans. Robert Chodos (Toronto: James Lorimer, 1983).

2. Geoffrey Tweedale, *Magic Mineral to Killer Dust: Turner and Newall and the Asbestos Hazard* (Oxford: Oxford University Press, 2000), 2.

3. *Canadian Mining Review* (October 1896), 218.

4. W.G. Clarke, W.G. Clarke Fonds, Eastern Townships Research Centre (hereafter cited as ETRC), Sherbrooke, Quebec.

5. W.G. Clarke, W.G. Clarke Fonds, ETRC.

6. *Johns-Manville News Pictorial* 1:1 (December 1938): 1; *Johns-Manville News Pictorial* 2:1 (January–February 1939): 3; Joan Ross, "Survey of Female Employees in Canadian Textile Department," 1944, "Asbestos Chronology," Asbestos Claims Research Facility (hereafter cited as ACRF), https://www.claimsres.com/about-us/services/asbestos-claims-research-facility/, 34.

7. Thomas G. Andrews, *Killing for Coal: America's Deadliest Labor War* (Cambridge, MA: Harvard University Press, 2008).

8. *Asbestos: Its Sources, Extraction, Preparation, Manufacture, and Uses in Industry and Engineering* (Berlin: Becker and Haag, 1928), 17.

9. Elizabeth W. Gillies, "The Asbestos Industry since 1929 with Special Reference to Canada" (PhD diss., McGill University, 1941), 40.

10. *Procès-verbal*, La ville d'Asbestos (minutes of town council, village of Asbestos), 6 December 1933, p. 146.

11. *Procès-verbal*, La ville d'Asbestos, 9 May 1934, p. 187, and 16 May 1934, p. 190.

12. Oliver Bowles and B.H. Stoddard, "Asbestos," in *Minerals Yearbook, 1934*, ed. O.E. Kiessling (Washington, DC: U.S. Government Printing Office, 1934), 1014.

13. Oliver Bowles and M.A. Cornthwaite, "Asbestos," in *Minerals Yearbook, 1937*, ed. Herbert H. Hughes (Washington, DC: U.S. Government Printing Office, 1937), 1363.

14. Quebec asbestos production went from 301,287 tons worth $9,958,183 in 1937 to 389,688 tons worth $14,072,000 in 1938. *Canadian Mining Journal* (February 1938), 65.

15. *Procès-verbal*, La ville d'Asbestos, 30 September 1938, p. 158.

16. W. Gillies Ross, "Encroachment of the Jeffrey Mine on the Town of Asbestos, Quebec," *Geographic Review* 57:4 (1967): 529.

17. Timothy J. LeCain, *Mass Destruction: The Men and Giant Mines That Wired America and Scarred the Planet* (New Brunswick, NJ: Rutgers University Press, 2009).

18. R.C. Rowe, "Mining and Milling Operations of the Canadian Johns-Manville Company Ltd. at Asbestos, PQ," *Canadian Mining Journal* (April 1939), 190.

19. Ibid, 185.

20. *Canadian Mining Journal* (15 March 1917), 121.

21. *Canadian Mining Journal* (February 1938), 65.

22. Oliver Bowles and K.G. Warner, "Asbestos," in *Minerals Yearbook, 1939*, ed. Herbert H. Hughes (Washington, DC: U.S. Government Printing Office, 1939), 1309.

23. Jock McCulloch and Geoffrey Tweedale, *Defending the Indefensible: The Global Asbestos Industry and Its Fight for Survival* (Oxford: Oxford University Press, 2008), 25.

24. Gillies Ross, "Encroachment of the Jeffrey Mine," 8.

25. H.R. Rice, "The Asbestos Industry in Quebec," *Canadian Mining Journal* (October 1948), 148.

26. Marc Vallières, *Des Mines et des Hommes: Histoire de l'Industrie Minérale Québécois des Origines au Début des Années 1980* (Québec: Publications du Québec, 1989), 348.

27. *Entre Nous* (Montreal: The Canadian Johns-Manville Co., February 1951), 9.

28. This was 100,000 tons annually. *Canadian Mining Journal* (February 1952), 106.

29. *Canadian Mining Journal* (February 1953), 101.

30. *Canadian Mining Journal* (May 1967), 45.

31. For more on this widening divide in views of asbestos mining, see David Egilman, Corey Fehnel, and Susanna Rankin Bohme, "Exposing the 'Myth' of ABC: 'Anything but Chrysotile'; A Critique of the Canadian Asbestos Mining Industry and McGill University Chrysotile Studies," *American Journal of Industrial Medicine* 44 (2003); McCulloch and Tweedale, *Defending the Indefensible*, 226.

32. L.K. Walkom, "New Shaft, Unusual New Mill: Feature Expansion at World's Largest Asbestos-Producing Property," *Canadian Mining Journal* (October 1954), 57.

33. Ibid., 58.

34. *Canadian Mining Journal* (February 1955), 89–90.

35. Hugh Jackson, JM, 13 March 1981, "Asbestos Chronology," ACRF, 188–89.

36. *Canadian Mining Journal* (February 1977), 125.

37. W.E. Cooke, "Fibrosis of the Lungs Due to the Inhalation of Asbestos Dust," *British Medical Journal* (1924): 487.

38. See, for example, Larry Lankton, *Cradle to Grave: Life, Work, and Death at the Lake Superior Copper Mines* (New York: Oxford University Press, 1991).

39. R.H. Stevenson, CJM Asbestos, "Talk by Dr. Stevenson to Quebec Asbestos Producers," 23 May 1938, p. 1, Turner & Newall Archives, Manchester, England.

40. Frank G. Pedley, "Asbestosis," *Canadian Medical Association Journal* 22:2 (1930): 253.

41. The reason the workers at Thetford had a higher incidence of asbestosis was because significant asbestos mining had been in operation there since the

early 1870s, whereas it was not until JM purchased the Jeffrey mine, in 1918, that extraction and processing levels in Asbestos increased in scale and number. As asbestos-related disease is often dose-to-longevity-specific, the early start of Thetford mine employees meant that they were among the first in Quebec to develop asbestosis.

42. Frank G. Pedley, "Report of the Physical Examination and X-Ray Examination of Asbestos Workers in Asbestos and Thetford Mines, Quebec" (Montreal: McGill/Metropolitan Life, 1930), ACRF, 10.

43. At this time, Metropolitan Life provided insurance for JM. David Egilman and Candace M. Hom, "Corruption of the Medical Literature: A Second Visit," *American Journal of Industrial Medicine* 34 (1998): 402.

44. Dr. McConnell, Metropolitan Life, to JM Attorneys, 9 July 1931, "Doc 7," ACRF, 129.

45. "Asbestos Chronology," 1929, ACRF, 1.

46. E.M. Voorhees, Johns-Manville Co., to S.A. Williams, VP, Johns-Manville Co., 28 July 1931. "Asbestos Chronology," ACRF, 6.

47. J.P. Kottcamp, Waukegan Plant Manager, to S.A. Williams, VP, Johns-Manville Co., 25 November 1932, "Asbestos Chronology," ACRF, 5.

48. Leroy Gardner, Saranac Laboratory, to Hektoen, JM, 15 March 1943, "Doc. 7," 31; Dr. Leroy Gardner, "Draft Report," "Doc. 7," 29, both in ACRF.

49. Joan Ross, "Survey of Female Employees in Canadian Textile Department," 1944, "Asbestos Chronology," ACRF, 34.

50. Ibid.

51. E.E. Keal, "Asbestosis and Abdominal Neoplasms," *Lancet* (December 1960), 1211.

52. Gerrit W.H. Schepers, "Chronology of Asbestos Cancer Discoveries: Experimental Studies of the Saranac Laboratory," *American Journal of Industrial Medicine* 27 (1995): 593–606; "Asbestos Chronology," 1930s–1964, ACRF, 3.

53. Schepers, "Chronology of Asbestos Cancer Discoveries," 600.

54. Ibid., 602–3.

55. Kenneth Smith, CJM Asbestos, to Paul Cartier, Thetford, 6 July 1948, "Asbestos Chronology," ACRF, 43.

56. Burton LeDoux, *L'Amiantose à East Broughton: Un Village de Trois Mille Âmes Étouffe dans la Poussière* (n.p.: privately printed, 1949), 3, 55.

57. Historical perspectives on the Quiet Revolution vary. See, for example, Michael D. Behiels, *Prelude to Quebec's Quiet Revolution: Liberalism versus Neo-nationalism, 1945–1960* (Montreal: McGill-Queen's University Press, 1985); Linteau et al., *Quebec*; Kenneth McRoberts, *Quebec: Social Change and Political Crisis*, 3rd ed. (Toronto: McClelland and Stewart, 1988); and Pierre Vallières, *White Niggers of America: The Precocious Autobiography of a Quebec Terrorist*, trans. Joan Pinkham (Toronto: McClelland and Stewart, 1971).

58. *(Toronto) Globe and Mail*, 23 April 1949, p. 7.

59. For more information on environmental health as a strike issue, see Jessica van Horssen, "'À faire un peu de poussière': Environmental Health and the Asbestos Strike of 1949," *Labour/LeTravail* 70 (Autumn 2012).

60. Maurice Duplessis, Premier of Quebec, "Letter of Address to the Documentation catholique de Paris," May 1950, reprinted in Esther Delisle and

Pierre K. Malouf, *Le Quatuor d'Asbestos: Autour de la Grève d'Amiante* (Montreal: Les Éditions Varia, 2004), 11.

61. Daniel C. Braun and David Truan, "An Epidemiological Study of Lung Cancer in Asbestos Miners," *American Medical Association Archives of Industrial Health* 17 (June 1958): 31–33.

62. See, for example, René Dubos and Jean Dubos, *The White Plague: Tuberculosis, Man, and Society,* introductory essay by Barbara Gutman Rosenkrantz (1952; reprinted, New Brunswick, NJ: Rutgers University Press, 1987); Nadja Durbach, "'They Might as Well Brand Us': Working-Class Resistance to Compulsory Vaccination in Victorian England," *Social History of Medicine* 13:1 (2000): 45–63; and Sheila M. Rothman, *Living in the Shadow of Death: Tuberculosis and the Social Experience of Illness in American History* (Baltimore: Johns Hopkins University Press, 1995).

63. I.J. Selikoff, J. Churg, and E.C. Hammond, "Asbestos Exposure and Neoplasia," *Journal of the American Medical Association* 188:1 (April 1964): 146.

64. I.H. Sloane to H.M. Jackson, 25 April 1954, "Asbestos Chronology," ACRF, 74.

65. JM Memo, 11 August 1975, "Asbestos Chronology," ACRF, 167.

66. Dr. Kent Wise, notes on discussion with Clifford Sheckler, Manager, JM Occupational Environmental Control, "Asbestos Chronology," 1969, ACRF, 144.

67. CJM to T.H. Davidson, February 1970, "Asbestos Chronology," ACRF, 138; Reitze to Paul Kotin, JM Health, Safety and Environment VP, 11 July 1978, "Asbestos Chronology," 175.

68. "Nous en avons soupé de la poussière et du bruit," *Le Citoyen,* 23 April 1968, p. 1 (English translation by author).

69. H.M. Jackson to Drs. G.W. Wright and T.H. Davidson, 17 March 1971, "Asbestos Chronology," ACRF, 153.

70. *Canadian Mining Journal* (February 1972), 139.

71. John Beattie, QAMA Meeting Minutes, 15 December 1965, Quebec Asbestos Mining Association archives (privately held; hereafter cited as QAMA), 2.

72. John Corbett McDonald, McGill University, "Report," 1972, QAMA, 4–9.

73. J. Turiaf and J.P. Battesti, "Le rôle de l'agression asbestosique dans la provocation du mésothéliome pleural," *La vie médicale au Canada français* (June 1974), 653.

74. Walter Cooper to Paul Kotin, 29 July 1975, "Asbestos Chronology," ACRF, 166.

75. Paul Kotin to JM executives, memorandum, 20 September 1977, "Asbestos Chronology," ACRF, 173.

76. *JM Today* 2:3 (1980): 2.

77. *Canadian Mining Review* (October 1896), 218.

78. "Asbestos: Where We Live and Work," *Johns-Manville News Pictorial* (October 1959), 11.

79. Linteau et al., *Quebec,* 155.

80. "Une Grève se Déclare dans les Mines de la Manville Asbestos Co.," *La Tribune*, 29 May 1918, p. 1.

81. *Sherbrooke La Tribune*, 3 September 1912, p. 7.

82. *Procès-verbal*, La ville d'Asbestos, 27 April 1927, p. 137.

83. Réjean Lampron, Marc Cantin, and Élise Grimard, *Asbestos: Filons d'histoire, 1899–1999* (Asbestos, Québec: Centenaire de la ville d'Asbestos, 1999), 140.

84. H.K. Sherry, Strike Report, 2 February 1937, p. 8, Department of Labour, Strikes and Lockouts, Record Group (RG) 27, Vol. 330, Reel T-2713, Library and Archives Canada (hereafter cited as LAC).

85. *Toronto Clarion*, 27 January 1937, Department of Labour, Strikes and Lockouts, RG 27, Vol. 330, Reel T-2713, LAC.

86. *Montreal Gazette*, 27 January 1937; *Toronto Telegram*, 29 January 1937, both in Department of Labour, Strikes and Lockouts, RG 27, Vol. 330, Reel T-2713, LAC.

87. *L'Asbestos*, 20 April 1948, p. 1.

88. Gillies Ross, "Encroachment of the Jeffrey Mine," 529.

89. Léopold Rogers, Government of Quebec, "Rapport Final d'Intervention," Arbitration Report, 21 May 1948, P659 7C 018 05–02–008B-01, 1982–11–008\1, Bibliothèque et Archives nationales du Québec, Quebec City, Quebec (hereafter cited as BANQ).

90. Gérard Tremblay, Quebec Deputy Minister of Labour, to Paul E. Bernier, Secretary, Commission de Relations ouvrières, 21 April 1948; and Gérard Tremblay, Quebec Deputy Minister of Labour, to Cyprien Miron, Director, Service de conciliation et d'arbitrage, 21 April 1948, P659 7C 018 05–02–008B-01, 1982–11–008\1, BANQ.

91. *Le Devoir*, 15 February 1949, p. 1; *La Tribune*, 14 February 1949, p. 1, and 15 February 1949, p. 5.

92. "J'ai 20 ans, j'avais 1 ans et demi de service et j'ai besoin de travailler. Je n'ai pas de mauvais raports avec la compagnie. Je voudrais savoir s'ils vont tous nous reprende où s'ils nous font attendre pour rien. Mon père a une grosse famille et une maison à payer et je suis seul pour l'aider. Je payais pour mon frère de 16 ans qui fait des études pour devenir religieux." Bertrand McNeil to Antonio Barrette, Quebec Minister of Labour, 19 August 1949, P659 7C 018 05–02–008B-01; 1982–11–008\1, BANQ (English translation by author).

93. *Entre Nous*, 8.

94. *Le Devoir*, 22 April 1949, p. 3.

95. G.W. Josephson and F.M. Barsigian, "Asbestos," *Minerals Yearbook*, 1949, ed. Allen F. Matthews (Washington, DC: U.S. Government Printing Office, 1951), 139; *Canadian Mining Journal* (August 1949), 54.

96. *Procès-verbal*, La ville d'Asbestos, 9 September 1949, p. 73, and 7 June 1950, p. 138.

97. *Procès-verbal*, La ville d'Asbestos, 2 September 1953, p. 73.

98. *Procès-verbal*, La ville d'Asbestos, 21 July 1955, p. 236.

99. *Procès-verbal*, La ville d'Asbestos, 20 May 1958, p. 174, and 27 May 1958, p. 176.

100. This figure is in 1965 Canadian dollars. See *Canadian Mining Journal*, February 1965, p. 129.

101. Antonio Hamel to Bérubé, Government of Quebec, 10 October 1979, E78 S999, 7 A 009 03–06–004B-01, 1993–06–004\12, BANQ.

102. "Bouge au Québec," *Le Citoyen*, 14 October 1964, p. 4 (author's translation).

103. "Amiante: Notre Patrimoine," *Le Citoyen*, 28 December 1974, p. 186 (author's translation).

104. *Procès-verbal*, La ville d'Asbestos, 18 May 1977.

105. James Kelly, "Manville's Bold Maneuver," *Time*, 6 September 1982; *Le Citoyen*, 10 August 1982, p. 2.

106. "L'Année de l'Amiante," in *Procès-verbal*, La ville d'Asbestos, 19 December 1983, p. 143 (author's translation).

107. For more information on these communities, see Charlie Angus and Brit Griffin, *We Lived a Life and Then Some: The Life, Death, and Life of a Mining Town* (Toronto: Between the Lines, 1996); Anthony F.C. Wallace, *St. Clair: A Nineteenth-Century Coal Town's Experience with a Disaster-Prone Industry*, 3rd ed. (New York: Alfred A. Knopf, 1987); and Michael A. Amundson, *Yellowcake Towns: Uranium Mining Communities in the American West* (Boulder: University Press of Colorado, 2004).

Afterword

Mining, Memory, and History

ANDREW C. ISENBERG

As with most children raised just outside Chicago, one of my favorite forays into the city was to the Museum of Science and Industry. The museum, a celebration of commercial technology modeled on the Deutsches Museum in Munich, was founded in 1933 and installed, just in time for Chicago's Century of Progress World's Fair, in a cavernous edifice in Jackson Park that had been built for the World Columbian Exposition in 1893. The undisputed highlight of my childhood trips to the museum was the coal mine exhibit: one descended an elevator made to resemble an elevator in a working mine shaft, with a docent dressed as a miner; from the elevator one spilled out into dark and cramped basement rooms in which the walls and ceilings had been molded with concrete to resemble a deep mine; the mining machinery in the exhibit was not static but shown at work, appearing to tunnel into the walls while spitting ersatz lumps of coal on to conveyer belts. When the museum opened in 1933, the exhibit demonstrated the newest mining technologies; by the time I first visited the museum in the late 1960s, however, an air of musty obsolescence surrounded it. Still, the exhibit seemed real enough to me. As a very young child, I was convinced, after viewing the exhibit for the first time, that a working coal mine lay beneath Jackson Park.

Notwithstanding the archaic quality of some of the mining machinery on display, the coal mine exhibit told a celebratory story of technological ingenuity triumphing over obstacles in the natural world. That narrative was entirely consistent with the other exhibits at the museum as well as with the theme of the 1933 Century of Progress exposition, for which the museum was founded. However deeply buried or otherwise inaccessible, the docents at the coal mine exhibit told us, mining engineers developed technologies to extract this resource vital to American industry. The tour concluded (before disgorging visitors directly into the museum's basement cafeteria) with a demonstration of mining safety in which the docents engineered a tiny, controlled explosion as a way of illustrating how miners—once again employing clever technologies—protected themselves against the buildup of dangerous gases in the mines. This demonstration was the exhibit's only nod toward miners' health and safety. The exhibit made no mention of pneumonoconiosis, more commonly known as black lung disease, the deadly occupational disease caused by the repeated inhalation of coal dust. Neither did the exhibit even allude to the environmental consequences of coal mining: the pollution of rivers and groundwater from coal slag; the dumping of thick, sticky water used to wash coal into rivers; or the particulates (which contribute to respiratory illnesses), sulfur dioxide (which contributes to acid rain), and carbon dioxide (which contributes to global warming) emitted when coal is burned.[1]

The celebration of mining (and the concomitant minimization of its environmental costs) is a long-standing practice in North America. It dates to one of the earliest and, I would argue, culturally and economically most significant mining booms in North America: the California gold rush. For the first year or two following the discovery of gold in California in 1848, the gold rush was an economic windfall with minimal environmental costs. Most gold seekers merely scooped gravel from the beds of the Feather, Yuba, Bear, and American Rivers (where flakes and nuggets of gold were visible to the naked eye in 1848) in a pan and swished the gravel around to separate the heavier gold from lighter sand. Yet, by 1852, the most accessible gold was gone, and miners began to apply a new technology, with much greater power to transform the environment, to reach gold deposits more deeply buried. In that year, an engineer used pressurized water in a canvas hose to propel gold-bearing gravel into a wooden-planked sluice that, like a pan, separated heavier gold from lighter sand and gravel. By the mid-1860s, the technology had been developed to the point that water shot from cannons at speeds up to 150 miles per hour. Hydraulic mining carved enormous craters into California hillsides and flushed everything—soil, boulders, tree stumps, unfortunate small animals, and gold-bearing gravel—into sluices that were slathered with mercury (which amalgamates readily with gold) to try to trap the precious mineral. The wilderness advocate John Muir observed that in the gold country, "the hills have been cut and scalped and every gorge and gulch and broad valley have been fairly torn to pieces and disemboweled, expressing a fierce and desperate energy hard to understand."[2]

The environmental costs of hydraulic mining were not confined to the mines. Thousands of tons of debris (including the mercury, a potent neurotoxin mined on the California Coast Range, much of which escaped through the sluices) were deposited in the river valleys directly below the mines. One observer calculated that the amount of debris deposited in the ravines of the Sierra between the mid-1850s and 1885 was more than three times greater than the amount of earth moved to make a path for the Panama Canal. Debris in piedmont belts of the Bear and Yuba Rivers buried pine trees and telegraph poles. While larger pieces of debris accumulated in the higher ravines, the current carried smaller, lighter amounts of sand, soil, and mercury farther downstream, where it filled river channels, destroyed salmon habitats, rendered water unfit to drink, and transformed the swift, clear rivers flowing from the Sierra into slow, murky, broad effluents. By 1878, the Yuba River had filled with sediment, and the stream ran a mile away from its original channel. The raised riverbeds of the Yuba, Bear, Feather, and Sacramento caused spring floods to overtop the natural levees of the rivers, inundating farmlands with a watery mixture of sand and gravel. The environmental costs of hydraulic mining were so great that in 1884 a federal judge issued an injunction putting a stop to the practice—one of the earliest legal impediments to industrial pollution in North America.[3]

Barred in California, engineers exported the technology of hydraulic gold mining to other parts of North America. By the end of the nineteenth century, the largest hydraulic mine on the continent was the Bullion pit in British Columbia. Established in 1894 and spurred by American investment, the Bullion pit eventually grew to be 400 feet deep, 800 feet wide, and a mile long. Its hydraulic cannons drew from reservoirs created by damming two nearby lakes to impound a billion cubic feet of water. Thirty-three miles of ditches delivered that water to the mine, where it was used to wash away more than 200 million tons of material. At its peak, the gold mine, which operated until 1942, used more water every day than the city of Vancouver.[4]

Despite the rapid transformation of gold mining from the simple techniques of prospectors to environmentally costly industrial technologies, many North Americans' understanding of mining remains stuck in the

1840s. Even as hydraulic mining was displacing the prospectors, California artists such as August Wenderoth and Charles Christian Nahl churned out romantic canvases depicting prospectors using simple tools—pans, picks, and shovels—to coax wealth out of the soil.[5] Their sanitized, ruralized, and romanticized understanding of gold mining endures.

In 1893, the influential historian Frederick Jackson Turner smoothly incorporated the icon of the prospector into his national narrative of progress, the frontier thesis, arguing that mining was part of what he viewed as the central narrative of United States history: western settlers' transformation of "wilderness" into "civilization." Ignoring the ways that miners, mining technology, and capital investment in mining readily transcended national boundaries (tens of thousands of people arrived in California from China, Latin America, and Europe, and tens of thousands of Americans left California for the gold fields of Australia and British Columbia), Turner, and generations of popular historians who followed him, argued that among other benefits, subduing nature made the United States prosperous. The frontier, according to Turner, spurred Americans to a technological genius that remained, in Turner's analysis, purely abstract, with no environmental context or consequences. "To the frontier the American intellect owes its striking characteristics: inquisitiveness, that practical, inventive turn of mind, quick to find expedients, that masterful grasp of material things, lacking in the artistic but powerful to great effects, that restless, nervous energy," Turner wrote.[6] (Canadian intellectuals such as Charles Mair and Nicholas Flood Davin embraced Turner's thesis with one caveat: Turner had erred in attributing the blessings of the frontier to the United States rather than to Canada.[7])

North Americans continue to rehearse that narrative of progress at nostalgia sites. A visitor to the Empire Mine in California, once the largest below-ground gold mine in the state, will learn of the amount of gold extracted from the mine, but little of the human or environmental costs of mining gold. The docents at the Empire Mine State Historic Park deemphasize the industrial techniques of mining, and instead treat visitors to a tour of a blacksmith's shop. Likewise, among the distractions at the Knott's Berry Farm amusement park in suburban Los Angeles, visitors are invited to pan for gold in a trough filled with sand in a reenactment of the techniques of Forty-Niners. A promotional video tells potential visitors to the park that they can pan for gold not only in the same manner of Forty-Niners, but in the same way that miners pan for gold "today."[8] At neither Knott's Berry Farm nor the Empire Mine park do docents hand you a copy of Turner's 1893 thesis, but they might as well.

A history of mining in North America that reckons with the environmental context and consequences of the industry must confront these romantic understandings of mining. The romantic idea, promulgated in Chicago and California and countless places in between, holds that mining is benign for the environment and a boon to the economy. That notion has become ingrained in what the French historian Pierre Nora called collective memory. According to Nora, memory is "a perpetually active phenomenon" in which commemoration ties the past to the present.[9] The coal mine exhibit in Chicago and the gold-panning amusement at Knott's Berry Farm in California are important parts of that perpetually active commemoration. Through participants' reenactments, the exhibits root mining in North Americans' collective memory. Yet as Nora suggested, memory vitiates context, one of the central analytical tools of the historian. Connecting past events to us with seeming directness and immediacy, memory, according to Nora, casts aside not only context but causation, complexity, contingency, consequences, and even chronology. Events become "sites of memory" as Nora called them: disconnected, hermetic reminiscences that can with extraordinary flexibility be recollected and made meaningful to present moments precisely because they have been stripped of historical context and analysis. Once decontextualized through time and repetition, such memories are open to manipulation; they become part of a collective memory. National identities are nurtured into being through the invented traditions of collective memory.[10] Transforming collective memory into national identity was what Turner was up to when he formulated his frontier thesis. When certain United States pundits and politicians call for casting aside regulatory limitations on oil exploitation, saying simply, "Drill, baby, drill," they are tapping into Americans' collective memories about mining.[11]

The notion that mining is an economic windfall is not only a retrospective view but a prospective one as well. In North America, that prospective view of mining has been critical to two significant processes, both of which are inextricably entangled with the exploitation of natural resources: imperialism and industrialism.[12] As the historian Kent Curtis has argued, the discovery of valuable mineral deposits is rarely happenstance. Societies organized around the exploitation of minerals conceive of the planet as a storehouse of untapped "riches lying in wait." Appointing a moment in time as marking the discovery of a mineral resource is the first act in constructing a narrative of progress, in which the unused waste is put to productive use.[13] Curtis wrote of the nineteenth-century western United States, but his insight applies to other empires in North America as well. Spanish explorers reported the presence of gold in the Americas almost from the first. On Hispaniola, Christopher Columbus reported to Ferdinand of Spain in 1493, the majority of the rivers "contain gold," and the island had "great mines of gold and of other metals."[14]

The voyages of Columbus were part of an imperial outreach by western European states in the late fifteenth century to shorten trade routes and find and exploit new resources to support a rising population.[15] By 1500, Europeans began to spill over their borders in search of productive agricultural lands. In 1492, the Reconquista—the Spanish effort to expel the Moors from the Iberian Peninsula—was completed when the Spanish overran Granada. Spanish conquistadors in the Americas extended the spirit of the Reconquista to the New World. Beginning in the late fifteenth century, the English undertook their own reconquest: in their case, as assertion of their control over Ireland. The English conquest of Ireland was a kind of rehearsal for their colonization of North America, which began in 1497, when Giovanni Caboto, a Genoese-born naturalized Venetian working for England, discovered a "new-found" land.[16]

Mineral wealth was never far from the minds of sixteenth-century European empire builders. Within a decade of their successful conquest of Mexico, Spaniards began to exploit silver at Sultepec and Zumpango near Mexico City; in later decades they expanded northward to Zacatecas (1546), Guanajuato (1550), and San Luis Potosí (1592). Authorities in New Spain drafted thousands of Indians to work the mines; roughly ten thousand men worked in New Spain's mines at the end of the sixteenth century. To supplement native laborers, the Spaniards turned to African slaves; one out of every seven mine workers at the end of the sixteenth century was African.[17] Much of the ore mined at Zacatecas and elsewhere in New Spain contained more lead than silver, so workers heated the ore in smelters to draw out the lead. (Wood for the smelters and for mine shafts consumed enormous amounts of timber in the mining regions; already in 1568 the viceroy of New Spain had to issue a proclamation reserving to the use of the mines the timber in the vicinity of Zacatecas.) After smelting, workers crushed the remaining ore, mixed it with salt and mercury, and spread out the resulting slurry on a patio; teams of bare-legged men with shovels sloshed through the silver-mercury mixture, inhaling the toxic lead and mercury vapors.[18] However costly to the environment and human health, the mines were tremendously productive. From the middle of the sixteenth century until the beginning of the nineteenth, Spanish America was the world's leading producer of silver.[19] The silver mines of the Americas sustained the Spanish empire in the seventeenth and eighteenth centuries.

The English looked on at Spanish colonial silver production with an envy they made no effort to conceal. In 1584, Richard Hakluyt, advising Elizabeth I on the advantages of possessions in the Americas, suggested that England's vagrants and petty thieves might be put to work in America "in mynes of golde, silver, copper, leade, and yron."[20] In 1624, despite having found no evidence of such riches during his time in Virginia between 1607 and 1609, John Smith wrote confidently that the colony possessed mineral wealth. "Copper we may doubt is wanting," he wrote, "but there is good probability that both copper and better minerals are there to be had for their labor."[21] Strictly speaking, Smith was correct. Valuable minerals are widely distributed throughout the world: gold, iron, copper, and other minerals are found in minute amounts in almost all parts of the earth's crust and ocean. Yet mines—and the pollution that results from mining—are forged where these minerals are concentrated. Neither the English nor the French colonies in North America yielded much mineral wealth (though the French established lead mines in the Mississippi Valley in the eighteenth century).

Like the Spanish empire in the New World, the United States empire in the Trans-Mississippi West was grounded in mining. In 1848, Colonel Richard Mason, the military governor of California (newly annexed to the United States from Mexico), reported to President James Polk that in the foothills of the Sierra Nevada, hundreds of men had rapidly accumulated fortunes of thousands of dollars in gold. The prospectors, Mason wrote, "bore testimony that they had found gold in greater or less quantities in the numerous small gullies or ravines that occur in that mountainous region." He concluded: "I have no hesitation now in saying, that there is more gold in the country drained by the Sacramento and San Joaquin Rivers than will pay the cost of the present war with Mexico a hundred times over. No capital is required to obtain this gold, as the labouring man wants nothing but his pick and shovel and tin pan, with which to dig and wash the gravel, and many frequently pick gold out of the crevices of rocks with their knives, in pieces of from one to six ounces."[22] The report—another in the genre that Columbus had started in 1493—incited the migration of 100,000 people to California over the following year. In the 1850s, California produced one-third of the world's gold.

Much as the silver mines had drawn the Spanish deeper into the North American continent in the sixteenth century, the exploitation of minerals was the engine that drove United States imperial expansion in North America in the nineteenth century. Beginning in California in 1848, a boom-and-bust pattern emerged: mineral rushes spurred the in-migration of prospectors, the exploitation of other resources such as timber and pastureland, and eventually clashes between settlers and Natives. That pattern, established in California, was repeated in Nevada and Colorado in the late 1850s and early 1860s, Montana in the 1860s, the Black Hills in the 1870s, Arizona in the 1880s, and the Yukon in the 1890s. To consolidate its hold on the West, the U.S. Army usually followed in the wake of these mineral rushes, prosecuting wars against Native Americans who, fighting to maintain control over resources they needed for their subsistence (such as salmon in California and bison in Colorado and Montana), impeded access to mining lands. The violence against Natives in California was particularly destructive: during the height of the California gold rush between 1848 and the end of the 1850s, the Native population of California fell by 80 percent, from 150,00 to 30,000. Altogether, in the second half of the nineteenth century, lurching from one mineral rush to another, the United States transformed itself from a moderately sized republic to a continental empire.[23]

The exploitation of below-ground mineral resources was critical not only to empire building in North America but to industrialization. The most notable below-ground resources were coal and iron ore (used to manufacture steel); copper (for electric wiring); and petroleum (which provided half of the United States' energy needs by the middle of the twentieth century and was a significant part of the economies of Canada and Mexico). There was no clear dividing line between imperialism and industrialization—indeed, in many respects they were conjoined processes centering on the exploitation of resources. The exploitation of precious minerals spurred industrial production, notably the construction of railroads to transport heavy ore and mining machinery. By the end of the nineteenth century, there were almost 190,000 miles of rail in the United States, and another 25,000 miles in the rest of North America. Altogether, the continent, home to 5 percent of the global population, possessed 45 percent of the world's railroad miles.[24] Railroads, however overcapitalized, mismanaged, inefficient, and liberally subsidized by the federal government, nonetheless facilitated the exploitation of minerals.[25] Rail lines that had once hauled heavy stamp mill machinery to gold and silver mines in Colorado eventually turned to hauling coal. The arrival of the Northern Pacific Railroad in Butte, Montana, in 1881, made possible a shift from the exploitation of the dwindling silver mines to copper. The Southern Pacific never reached Tombstone, Arizona, before the silver mines there were exhausted, but for decades that railroad hauled copper from the mines that opened thereafter in nearby Bisbee on the U.S.-Mexico border. Likewise, when the U.S. firm Phelps-Dodge developed the copper deposits near the former silver mines of Nacozari, Sonora, they built their own rail line to connect Nacozari to U.S. railroads in Arizona.[26]

The steelmaking cities along the banks of the Great Lakes—among them, Chicago, Gary, Detroit, Toledo, Cleveland, and Buffalo in the United States and Hamilton and Oshawa in Canada—developed there because of the ease with which iron ore from the Mesabi Range in northern Minnesota could be shipped via the Great Lakes. The Mesabi Range furnished one-third of all the iron used in the United States in 1900, and 16 percent of the world's supply.[27] In the United States, coal arrived at the steel mills by rail from the mines of Pennsylvania and Illinois; in Canada, the construction of transcontinental railroads made coal resources in the western provinces accessible. In 1850, the United States had produced merely 8.4 million short tons of coal; by 1900 it produced 270 million, almost one-third of the coal in the world. Coal and iron ore mining and steel production have been major sources of pollution since the late nineteenth century: slag from coal and iron ore mines was dumped near mines, where it seeped into the groundwater or into rivers, poisoning water supplies. The smoke emitted into the air during the production of pig iron and steel contained carcinogenic particulates. By the end of the nineteenth century, there were 14,000 smokestacks in and around the greatest steel city in North America, Pittsburgh. By the mid-twentieth century, emissions from the steel mills were so thick in Pittsburgh that on particularly bad days the city kept the streetlights on all day for visibility.[28]

In 1900, coal provided 70 percent of the energy needs in the United States. At that time, petroleum—most of which was produced in Pennsylvania—provided a mere 5 percent.[29] A strike at the Spindletop oil field near Beaumont, Texas, in 1901, transformed the way that North Americans tapped energy sources below the earth's surface. Within a year, 285 wells were clustered around Beaumont, and oil companies were drilling wells elsewhere in Texas, California (for a time in the 1920s, California produced more oil than Texas), and Mexico.[30] By 1945 petroleum was the primary fuel source in the United States, and that nation was the world's largest exporter of oil.

The expansion of oil drilling into Mexico reveals starkly the connections between the exploitation of below-ground resources, industrialization, and a modern imperialism of capital investment.[31] During the Porfiriato—the reign of Mexican president Porfirio Díaz, between 1876 and 1910—the Mexican economy, particularly manufacturing, railroads, mining, and petroleum, grew at a rapid pace as foreign investment flooded into Mexico. In 1901, Mexico produced a mere 10,000 barrels of oil. In 1924, it produced 140 million barrels. By that time, 300 of the 400 oil companies operating in Mexico were controlled by investors from the United States. Most of the rest of the Mexican oil industry was controlled by British or Dutch interests. The foreign control of Mexican petroleum production lasted until a strike by Mexican oil workers in 1937 led the Mexican president, Lázaro Cárdenas, to nationalize 16 foreign oil companies.[32] Oil production in Canada—discoveries near Calgary in 1914 and near Edmonton in 1947 opened up petroleum production in western Canada—was similarly dominated by American investors. Unlike Mexico, however, Canada never threw off American influence: by 1960, nearly 90 percent of the Canadian petroleum industry was owned by non-Canadians, and 80 percent of the foreign ownership was with U.S.-based oil companies.[33] Yet First Nations communities in northern Canadian have proven to be savvy negotiators with oil interests, pressing oil pipeline builders to allow local control, share revenues, and minimize environmental impacts.[34]

In sum, in North America, miners have rushed to exploit underground resources as rapidly as possible, regardless of the ecological costs. They have lurched from silver and gold to mercury, copper, coal, iron ore, petroleum, asbestos, and uranium.[35] Miners have usually had the power of the state behind them. The search for precious minerals was a primary aim of expanding imperial states in North America from the sixteenth to the nineteenth century. Later, governmental authorities in the United States, and to a lesser degree in Canada, eager to see mineral resources funneled into industrial development, opened copper, coal, iron ore, and petroleum to exploitation at nominal costs and with minimal regulatory oversight. The Mexican government, until 1910, invited foreign investors—usually Americans—to exploit its mineral resources, before sharply curtailing such investment in the 1930s. Mexico's abrupt nationalization of foreign oil holdings was neither unprecedented nor unrepeated. The history of mining in North America is not simply a story of exploitation. Groups that have felt the burdens of state-sponsored mining have offered meaningful resistance to it, including Californians and Coloradans downstream of gold mines, Montana ranchers downwind of copper smelters, and Native Americans and First Nations people near iron mines in Minnesota and gold mines in the Northwest Territories.

Nonetheless, mining remains a large part of the economies of the United States, Canada, and Mexico. A 2012 study ranked the mining and minerals sector of the U.S. economy as the eighth-most productive in the world. In that same study, Canada was tenth and Mexico thirteenth.[36] According to the International Energy Agency, in 2014 the United States produced nearly 14 million barrels of petroleum or other liquid fuels (including shale gas) every day, making it the world's leading oil-producing state. Canada, at 4.3 million barrels per day, was fifth; Mexico, at 2.8 million per day, was tenth. Across North America, the exhortation to "drill, baby, drill" continues to resonate.

That exhortation appeals to North Americans' collective memory of mining as an economic godsend and, to no less an extent, an adventure that has stretched from prospectors in gold rush California to wildcatters in the Texas oil boom and, many people imagine, beyond. Americans continue to celebrate that collective memory of adventure and profit at mining simulacra at Knott's Berry Farm in Los Angeles and the Museum of Science and Industry in Chicago. Neither the historical processes such as imperialism and industrialization that buoyed those mineral rushes nor the ecological context and consequences of mineral production have a place in collective memory. In contrast to collective memory, the environmental history of mining in North America reminds us of mining's social costs: not merely the aggregate amount of pollution but the ways in which the consequences of mining are borne unevenly by the inhabitants of North America's diverse societies. In large part, what the environmental history of mining in North America reveals is that mines have not so much *produced* wealth as *rearranged* it. While taking valuable gold or copper or coal from the ground, they have vaulted toxins from below the earth into the atmosphere, soil, rivers, and the bodies of human beings and other animals. This extensive alteration of the environment was premised on the belief that nature was a storehouse of resources that could be extracted as commodities. The environment, however, contradicts that assumption at every turn. Minerals do not cease to be part of nature when they pass from the environment to the human economy as commodities. The flow of minerals as commodities through the market is inseparable from the flow of energy and materials through the environment. Mercury, arsenic, asbestos, and coal, among other things, pass from minerals to commodities and then back to the environment as waste. People are not just economic actors who shepherd those minerals from the environment to the market. As biological entities, we are part of the environment, and thus not only do we feel the effects of a warming climate because of the carbon dioxide produced by the burning of coal and oil, but some of the mercury, arsenic, and asbestos comes to be lodged in our bodies when we unearth and process it. To borrow a phrase from John Muir, "When we try to pick out anything by itself, we find it hitched to everything else in the universe."[37]

NOTES

1. For the environmental costs of coal, see J.R. McNeill, *Something New Under the Sun: An Environmental History of the Twentieth-Century World* (New York: W.W. Norton, 2000), 57–60. For some of the dangers to coal miners, see chapter 5, by Thomas G. Andrews, in this volume. For the dangers to silver miners, see chapter 4, by Robert N. Chester III, in this volume.

2. John Muir, quoted in *Prospectus of the Cataract and Wide West Hydraulic Gravel Mining Co.* (San Francisco: Fluto and Co., 1876), 4–6.

3. For the environmental costs of hydraulic mining in California, see Andrew C. Isenberg, *Mining California: An Ecological History* (New York: Hill and Wang, 2005), 23–51, 161–78. For the environmental costs of twentieth-century gold mining, see chapter 10, by John Sandlos and Arn Keeling, in this volume. For other examples of legal opposition to mining pollution, see chapter 6, by Timothy James LeCain, and chapter 3, by George Vrtis, in this volume.

4. *General Review of Mining in British Columbia* (Victoria, British Columbia: Richard Wolfenden, 1904), 42–43; Peter R. Mulvihill, William R. Morison, and Sherry MacIntyre, "Water, Gold, and Obscurity: British Columbia's Bullion Pit," *Northern Review*, 25–26 (Summer 2005): 197–210.

5. See Janice T. Driesbach, Harvey L. Jones, and Katherine Church, *Art of the Gold Rush* (Berkeley: University of California Press, 1998).

6. Frederick Jackson Turner, "The Significance of the Frontier in American History," American Historical Association, *Annual Report* (1893): 199–227. For Turner's enduring cultural significance, see James R. Grossman, ed., *The Frontier in American Culture: An Exhibition at the Newberry Library, August 26, 1994–January 7, 1995*, essays by Patricia N. Limerick and Richard White (Berkeley: University of California Press, 1994).

7. Doug Owram, *Promise of Eden: The Canadian Expansionist Movement and the Idea of the West, 1858–1900* (Toronto: University of Toronto Press, 1992), 127–32.

8. "Learn How to Pan for Real Gold" (video), Knott's Berry Blog, Knott's Berry Farm, https://www.knotts.com/blog-article/haunt/Learn-How-to-Pan-for-REAL-GOLD (accessed June 24, 2016).

9. Pierre Nora, "Between Memory and History: Les Lieux de Memoire," *Representations*, 26 (Spring 1989): 7–24. See also Kerwin Lee Klein, "On the Emergence of Memory in Historical Discourse," *Representations* 69 (Winter 2000): 127–50.

10. Michael Kammen, *Mystic Chords of Memory: The Transformation of Tradition in American Culture* (New York: Knopf, 1991); Eric Hobsbawm and Terence Ranger, eds., *The Invention of Tradition* (Cambridge: Cambridge University Press, 1983).

11. The phrase "Drill, baby, drill" first appeared as the title of an August 28, 2008, blog written by Erik Rush. "States within the continental U.S. alone contain more reserves than the reserves of Saudi Arabia," Rush wrote. "We should drill like a field full of randy rabbits." See "Drill, Baby, Drill," World Net Daily, August 28, 2008, http://www.wnd.com/2008/08/73559/# (accessed June 26, 2016). Michael Steele, the chair of the Republican National Committee, used the phrase during the Republican National Convention in early September 2008. The Republican vice presidential nominee, Sarah Palin, repeated it during a debate with Vice President Joe Biden on October 2, 2008.

12. For the necessity of environmental historians to engage with thematic subfields, see Isenberg, "Introduction: A New Environmental History," in *The Oxford Handbook of Environmental History*, ed. Andrew C. Isenberg (New York: Oxford University Press, 2014), 1–20.

13. Kent Curtis, "Producing a Gold Rush: National Ambitions and the Northern Rocky Mountains, 1853–1863," *Western Historical Quarterly* 20 (Spring 2009): 275–97.

14. *The Journal of Christopher Columbus*, trans. Cecil Jane (London: Anthony Blond, 1968), 194.

15. While local conditions were variable and thus generalizations difficult, in the middle of the fourteenth century, many parts of western Europe had faced a collective resource shortage: their populations pressed on the limits of medieval agricultural productivity. A reckoning was delayed by the Black Death, but by 1500 or so, the population of western Europe had recovered, and local conditions of scarcity once again appeared. See William Chester Jordan, *The Great Famine: Northern Europe in the Early Fourteenth Century* (Princeton: Princeton University Press, 1996), 7–39; Emmanuel Le Roy Ladurie, *The Peasants of Languedoc*, trans. John Day (Urbana: University of Illinois Press, 1974); William H. McNeill, "The Impact of the Mongol Empire on Shifting Disease Balances, 1200–1500," in *Plagues and Peoples* (1976; reprinted, New York: Anchor, 1998), 161–207.

16. Nicholas P. Canny, "The Ideology of English Colonization: From Ireland to America," *William and Mary Quarterly*, 3rd ser., 30 (October 1973): 575–98.

17. Peter Bakewell, "Mining in Colonial Spanish America," in *The Cambridge History of Latin America, Volume 2: Colonial Latin America*, ed. Leslie Bethell (Cambridge: Cambridge University Press, 1984), 105–51.

18. For other environmental consequences of silver mining in New Spain, see chapter 1, by Daviken Studnicki-Gizbert, and chapter 2, by Antonio Avalos-Lozano and Miguel Aguilar-Robledo, in this volume.

19. Richard L. Gerner, "Long-Term Silver Mining Trends in Spanish America: A Comparative Analysis of Peru and Mexico," *American Historical Review* 93 (October 1988), 898–935. Much of the silver came from the mines of Peru in the seventeenth century, but New Spain was the leading producer in the eighteenth century.

20. See *Documentary History of the State of Maine*, vol. 2 (Cambridge, MA: John Wilson, 1877), 37.

21. *Captain John Smith's America: Selections from His Writings*, ed. John Lankford (New York: Harper, 1967), 17.

22. Colonel Richard Mason, Monterey, to Brigadier General R. Jones, Washington, DC, August 17, 1848, 30th Congress, 2nd Sess., Ex. Doc. 1, 56–64.

23. See Malcolm Rohrbough, *Days of Gold: The California Gold Rush and the American Nation* (Berkeley: University of California Press, 1997); Elliott West, *The Contested Plains: Indians, Goldseekers, and the Rush to Colorado* (Lawrence: University Press of Kansas, 1998); Ari Kelman, *A Misplaced Massacre: Struggling over the Memory of Sand Creek* (Cambridge, MA: Harvard University Press, 2013); Andrew Isenberg, *The Destruction of the Bison: An Environmental History, 1750–1900* (New York: Cambridge University Press, 2000); and Albert L. Hurtado, *Indian Survival on the California Frontier* (New Haven: Yale University Press, 1988).

24. *Statistics of the American and Foreign Iron Trades for 1900* (Philadelphia: American Iron and Steel Association, 1901), 9.

25. For railroads in the U.S. West, see Richard White, *Railroaded: The Transcontinentals and the Making of Modern America* (New York: W. W. Norton, 2011); see also Carroll Van West, *Capitalism on the Frontier: Billings and the Yellowstone Valley in the Nineteenth Century* (Lincoln: University of Nebraska Press, 1993). For the nineteenth-century West as an industrial place, see Isenberg, "Environment and the Nineteenth-Century West, or, Process Encounters Place," and David Igler, "Engineering the Elephant: Industrialism and the Environment in the Greater West," in *A Companion to the American West*, ed. William Deverell (Malden, MA: Wiley-Blackwell, 2004), 77–111; Igler, "The Industrial Far West: Region and Nation in the Late Nineteenth Century," *Pacific Historical Review* 69 (May 2000): 159–92.

26. Mark Wyman, *Hard Rock Epic: Western Miners and the Industrial Revolution* (Berkeley: University of California Press, 1979), 158. See also Samuel Truett, *Fugitive Landscapes: The Forgotten History of the U.S.-Mexico Borderlands* (New Haven: Yale University Press, 2006), 55–56; Timothy LeCain, *Mass Destruction: The Men and Giant Mines That Wired America and Scarred the Planet* (New Brunswick, NJ: Rutgers University Press, 2009); and chapter 6, by LeCain, in this volume.

27. For iron ore in northern Minnesota, see chapter 7, by Jeffrey T. Manuel, and chapter 11, by Nancy Langston, in this volume.

28. See Joel Tarr, "The Metabolism of the Industrial City: The Case of Pittsburgh," in *City, Country, Empire: Landscapes in Environmental History*, ed. Jeffry M. Diefendorf and Kurk Dorsey (Pittsburgh: University of Pittsburgh Press, 2005), 15–37.

29. For Pennsylvania, see Brian Black, *Petrolia: The Landscape of America's First Oil Boom* (Baltimore: Johns Hopkins University Press, 2000).

30. For California, see Paul Sabin, *Crude Politics: The California Oil Market, 1900–1940* (Berkeley: University of California Press, 2005).

31. For the role of international capital in spurring mining development, notably petroleum development, see chapter 12, by Steven M. Hoffman, in this volume.

32. See Jonathan Brown, *Oil and Revolution in Mexico* (Berkeley: University of California Press, 1993).

33. James Laxer, *Oil and Gas: Ottawa, the Provinces, and the Petroleum Industry* (Toronto: James Lorimer, 1983), 6–8.

34. See Paul Sabin, "Voices from the Hydrocarbon Frontier: Canada's Mackenzie Valley Pipeline Inquiry (1974–1977)," *Environmental History Review* (Spring 1995): 17–48. For the ways in which Natives are adversely affected by mining, see Hurtado, *Indian Survival on the California Frontier*; chapters 10 and 11 in this volume.

35. For uranium and asbestos, see chapter 8, by Eric Mogren; chapter 9, by Robynne Mellor; and chapter 13, by Jessica van Horssen, in this volume.

36. International Council on Mining and Metals, "Mining's Contribution to Sustainable Development" (October 2012), https://www.icmm.com/document/4440 (accessed June 25, 2016).

37. Muir, *My First Summer in the Sierra* (Boston: Houghton Mifflin, 1944), 157.

Contributors

MIGUEL AGUILAR-ROBLEDO is Professor of Geography and Dean at the Autonomous University of San Luis Potosí, Mexico. He also golds appointments in the Multidisciplinary Graduate Program in Environmental Science, Environment and Resource Management, as well as the Latin American Graduate Program on Territory, Society and Culture. His research focuses on regional and applied historical geography and environmental history.

THOMAS G. ANDREWS is Professor of History at the University of Colorado at Boulder. He is the author of two books—*Killing for Coal: America's Deadliest Labor War* (Harvard University Press, 2008) and *Coyote Valley: Deep History in the High Rockies* (Harvard University Press, 2015)—and is working on a third, tentatively entitled "An Animals' History of the United States."

ANTONIO AVALOS-LOZANO is Professor of Environmental Science at the Autonomous University of San Luis Potosí, Mexico. His research focuses on mineral production and deforestation in New Spain/Mexico and climate history.

ROBERT N. CHESTER III is an adjunct professor in Sacramento, California. He has published on the topic of food in the journal *Environmental History* and is currently working on a book manuscript entitled "Comstock Creations: The Nature of America's Largest Silver Strike."

STEVEN M. HOFFMAN was Professor of Political Science at St. Thomas University. Sadly, Steven passed away in 2015, while this book was still in production. During his impressive career, he published many books and articles on pressing political and environmental topics, including the coedited volume, with Thibault Martin, *Power Struggles: Hydro Development and First Nations in Manitoba and Quebec* (University of Manitoba Press, 2008).

ANDREW C. ISENBERG is Professor of History at Temple University. He specializes in environmental history and the history of the North American borderlands. He is the author of *The Destruction of the Bison: An Environmental History, 1750–1920* (Cambridge University Press, 2000), *Mining California: An Ecological History* (Hill and Wang, 2005), and *Wyatt Earp: A Vigilante Life* (Hill and Wang, 2013), which was a finalist for the Weber-Clements Prize in Southwestern History. He is the editor of *The Nature of Cities: Culture, Landscape, and Urban Space* (University of Rochester Press, 2006) and *The Oxford Handbook of Environmental History* (Oxford University Press, 2014).

ARN KEELING is Associate Professor of Geography at Memorial University of Newfoundland, Canada. His research focuses on the historical geography of mineral development, pollution, and remediation in the Canadian North. He is the coeditor, with John Sandlos, of *Mining and Communities in Northern Canada: History, Politics, and Memory* (University of Calgary Press, 2015).

NANCY LANGSTON is Professor of Environmental History at Michigan Technological University. She is author of four books, including *Toxic Bodies: Hormone Disruptors and the Legacy of DES* (Yale University Press, 2010) and *Sustaining Lake Superior* (Yale University Press, forthcoming).

TIMOTHY JAMES LECAIN is Associate Professor of History at Montana State University. His first book, *Mass Destruction: The Men and Mines That Wired America and Scarred the Planet* (Rutgers University Press, 2009), won the Best Book Award from the American Society for Environmental History. His latest book, *The Matter of History: How Things Create the Past* (Cambridge University Press, forthcoming), develops a neomaterialist theory and method of history through a comparative environmental history of Japanese and American copper mining.

JEFFREY T. MANUEL is Associate Professor in the Department of Historical Studies at Southern Illinois University at Edwardsville. He is the author of *Taconite Dreams: The Struggle to Sustain Mining on Minnesota's Iron Range, 1915–2000* (University of Minnesota Press, 2015). In addition to his research on the history of extractive industries, he is also active in public and oral history projects.

J. R. MCNEILL is Professor of History and University Professor at Georgetown University. His most recent books are *The Great Acceleration: An Environmental History of the Anthropocene since 1945* (Harvard University Press, 2016), coauthored with Peter Engelke, and *Mosquito Empires: Ecology and War in the Greater Caribbean, 1620–1914* (Cambridge University Press, 2010), which won the Beveridge Prize from the American Historical Association.

ROBYNNE MELLOR is a Ph.D. candidate in history at Georgetown University. Her dissertation, which she is currently writing, draws on approaches from environmental history, diplomatic history, and international comparative history to examine uranium mining and milling in Canada, the United States, and the Soviet Union from 1945 to 1985. Her research has been supported by awards from the Social Science Research Council, the Social Sciences and Humanities Research Council of Canada, and the American Society of Environmental History.

ERIC MOGREN is Associate Professor of History at Northern Illinois University, where he is also a Faculty Associate in the Institute for the Study of the Environment, Sustainability, and Energy. He is the author of *Warm Sands: Uranium Mill Tailings Policy in the American West* (University of New Mexico Press, 2002) and *Native Soil: A History of the DeKalb County Farm Bureau* (Northern Illinois University Press, 2005).

JOHN SANDLOS is Professor of History at Memorial University of Newfoundland, Canada, where he has worked on several projects focusing on the historical and contemporary impacts of abandoned mines in northern Canada. He is the coeditor, with Arn Keeling, of *Mining and Communities in Northern Canada: History, Politics, and Memory* (University of Calgary Press, 2015). His research has been supported by the Social Sciences and Humanities Research Council of Canada and ArcticNet. The chapter in this volume was also generously supported by a writing fellowship from the Rachel Carson Center for Environment and Society in Munich, Germany.

DAVIKEN STUDNICKI-GIZBERT is Associate Professor of History at McGill University, Canada, where he teaches global, Latin American, and environmental history. His research focuses on the long-term political ecology of resource extraction in North America, public interest research on Canadian mining in Latin America, and the use of historical research in advancing indigenous self-determination in western Panama.

JESSICA VAN HORSSEN is Senior Lecturer at Leeds Beckett University in England. She is the author of *A Town Called Asbestos: Environmental Health, Contamination, and Resilience in a Resource Community* (University of British Columbia Press, 2016).

GEORGE VRTIS is Associate Professor of History and Environmental Studies at Carleton College. His research currently focuses on precious-metals mining in Colorado, the environmental relationships that link the Twin Cities with the rest of Minnesota, and the politics of wilderness at Grand Canyon National Park. He is the coeditor, with Christopher W. Wells, of *Nature's Crossroads: The Twin Cities and Greater Minnesota* (University of Pittsburgh Press, forthcoming).

Index

Italic page numbers indicate illustrations.